OPTICAL RADIATION MEASUREMENTS

Volume 1
RADIOMETRY

OPTICAL RADIATION MEASUREMENTS

A Treatise

Edited by FRANC GRUM

OPTICAL RADIATION MEASUREMENTS

Volume 1
RADIOMETRY

FRANC GRUM

*Research Laboratories
Eastman Kodak Company
Rochester, New York*

RICHARD J. BECHERER

*Lincoln Laboratory
Massachusetts Institute of Technology
Lexington, Massachusetts*

ACADEMIC PRESS
New York   San Francisco   London   1979
A Subsidiary of Harcourt Brace Jovanovich, Publishers

ACADEMIC PRESS, INC.
111 Fifth Avenue, New York, New York 10003

United Kingdom Edition published by
ACADEMIC PRESS, INC. (LONDON) LTD.
24/28 Oval Road, London NW1 7DX

Library of Congress Cataloging in Publication Data

Main entry under title:

Optical radiation measurements.

CONTENTS: v. 1. Grum, F., Becherer R. J.
Radiometry.
1. Radiation--Measurement. 2. Optical
measurements. I. Grum, Franc II. Becherer,
Richard J., Date
QC475.067   539'.2'028   78-31412
ISBN 0-12-304901-6 (v. 1)

PRINTED IN THE UNITED STATES OF AMERICA

79 80 81 82   9 8 7 6 5 4 3 2 1

Contents

Foreword

Optical radiant energy has always been the connecting link between the human visual system and the external world, and energy from the sun has always been essential to biological growth and development. Yet it is only within recent years that it has become possible to generate, detect, and control optical radiant energy to meet a wide range of needs. A brief list of current areas of application and research that involve radiant energy in a direct way must include astrophysics, clinical medicine, color science, illumination engineering, laser systems, meteorology, military sensors, photography, radiation heat transfer, solar energy, and visual information display. The spectral regions of interest in these areas span the visible, the infrared, and the ultraviolet.

In each of these areas the pace of development has been rapid. New techniques have been introduced and new measurement methods have arisen to meet pressing needs. It is our plan that these volumes, under the title "Optical Radiation Measurements," will provide information on many aspects of these recent developments. Throughout the treatise a basic viewpoint will be taken in the belief that this will prove most useful. Although recent results will be described, it is not intended that these volumes be simply a compendium of data of the type found in handbooks. Each volume will emphasize principles and generally applicable methods.

FRANC GRUM

Preface

Radiometry embraces a wide range of topics from basic concepts of radiant energy and its tranfer to the calibration of instrumentation. The purpose of this book is to bring together a balanced selection of these topics that will serve as an introduction to the measurement of optical radiant energy and will also be detailed enough to serve as a text for instruction and reference.

The book is divided into three parts: concepts, components, and techniques. Chapters 2–4 describe the basic concepts such as radiation laws, terminology, and the transfer of radiant energy. The emphasis in these early chapters is on fundamentals. Chapters 5–7 survey the major components of radiometric systems. Our intention here has been to introduce increasing amounts of specific performance data, particularly for sources and detectors of radiant energy. The final three chapters, 8–10, describe representative techniques. Radiometry is primarily concerned with measurements and so techniques properly play a central role.

Each chapter is followed by a list of specific references to facilitate further study of a topic of interest. A list of general references is found at the end of Chapter 1. We have included many tables of data and a large number of figures to increase the utility and clarity of the book.

In the matter of units we have adhered to SI (Système Internationale) and SI derived units. The terminology is that now recommended by a number of international standardizing organizations including the CIE (Commision Internationale de l'Éclairage). Throughout the book the symbols used are explicitly defined and units are shown following the related equation.

The reader will realize that in the limited space available in a single book it is not possible, nor is it our intention, to treat all of the applications or techniques of radiometry in detail. As the first volume in the treatise ''Optical Radiation Measurements,'' this book is intended to be introductory in character. Later volumes with chapters by invited authors who are expert in their special areas will complement the present volume in this respect.

Acknowledgments

In preparing this book we have had the benefit of comments and suggestions from a number of our colleagues. We particularly wish to thank Professor F. Billmeyer of Rensselaer Polytechnic Institute; Mr. A. Karoli of Eppley Laboratories; Mr. R. Keyes and Dr. R. Kingston of Lincoln Laboratory, Massachusetts Institute of Technology; and Professor E. Wolf of the University of Rochester for reading and commenting on portions of the manuscript. Their advice has been extremely valuable to us in this undertaking. We are also indebted to Dr. Bruce Steiner of the National Bureau of Standards for a number of helpful discussions over a period of several years regarding the applications of radiometry and radiometric standards. Many of these discussions took place within CORM, the Council for Optical Radiation Measurements.

The cooperation and encouragement of the Eastman Kodak Company and of Lincoln Laboratory, Massachusetts Institute of Technology, are gratefully acknowledged. Early versions and revisions of the manuscript were skillfully typed by Miss Deborah Lionetta, Miss Joyce Craven, and Mrs. Cathy Thayer. Our thanks go to them for their patience. Finally, we wish to express special thanks to our families for their understanding and support in a project that has occupied a considerable amount of personal time over a period of two and one-half years.

1

Introduction

1.1 RADIOMETRY

Radiometry is the science and technology of the measurement of electromagnetic radiant energy. Measurements of radiant energy are conducted throughout the electromagnetic spectrum, as shown in Fig. 1.1, using methods which are suited to the spectral region of interest. Our objective is to describe the concepts, components, and techniques which are used for radiant energy measurement in the optical region of the electromagnetic spectrum.

Typical configurations of radiometric measurement systems are shown in Fig. 1.2. As shown in the figure, a complete radiometric measurement system will generally involve a number of *components* including a source of radiant energy, a transmission medium through which the radiant energy passes, an object which transmits, reflects, or absorbs radiant energy, an optical system, a detector which converts the radiant energy to another form of energy

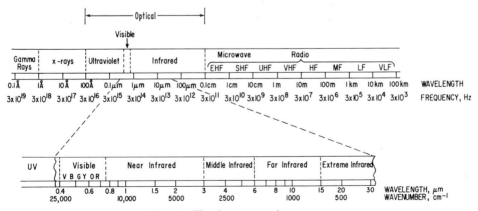

Fig. 1.1   The electromagnetic spectrum.

1

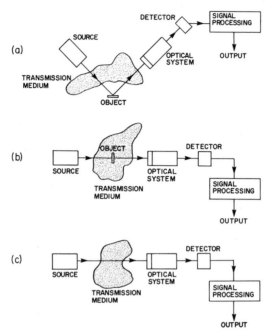

Fig. 1.2 Typical configurations of radiometric measurement systems.

(usually electrical), and a signal processing device. Figure 1.2 shows typical arrangements of these components in three complete radiometric systems. The principal differences among these three configurations is the presence or absence of the object in the path of the radiant energy and the orientation of the optical system to collect the radiant energy which is reflected or transmitted by the object. Although these simplified configurations may not cover all situations, they do represent the arrangements of components used in the majority of all radiometric measurements.

Typical *sources* of radiant energy include the sun, lasers, electrical discharge sources, fluorescent materials, and in general any material body which is heated to a temperature above absolute zero. A theoretically ideal blackbody produces a distribution of radiant power with wavelength which is closely approximated by many practical sources. The characteristics of the radiant energy distribution produced by a blackbody constitute an important starting point for any survey of radiometry.

The *transmission medium* of primary interest is the terrestrial atmosphere. One of the principal characteristics of the terrestrial atmosphere is its variability. Variations in temperature, pressure, water content, and distribution of molecular species cause the optical properties of the atmosphere to vary with time. These variations are in addition to the spatial variations of optical properties which are due to the large-scale gradations which occur in the vertical structure of the atmosphere as a function of altitude above the surface of the earth.

Objects of interest in radiometry are as varied as the applications of radiometry. This variety of objects precludes a comprehensive survey of this component of the radiometric system. However, it is possible to consider the general transmission, reflection, and absorption characteristics of objects which influence the choice among various measurement techniques. This approach is taken in the chapters of this book dealing with techniques.

Optical systems used in radiometric measurement systems will usually consist of lenses, mirrors, apertures, prisms, gratings, filters, interferometers, polarizers, attenuators, diffusers, integrating spheres, fiber optics, or other devices. Much of the effort involved in the development of a radiometric measurement system is directed toward the sophisticated utilization of optical devices.

Detectors play a fundamental role in radiometry. Much of the history of progress in radiometry is associated with progress in the development of detectors. The most important physical detectors of radiant energy are photomultipliers, photographic detectors, pyroelectric detectors, thermo-couples, bolometers, photoconductors, and photodiodes. Another very important detection system which plays a preeminent role in many applications is the human visual system. Actually, the human visual system is probably best regarded as a combination of optical system, detectors, and signal processing device. Although a number of psychophysical measurements have been made which permit the characterization of the simpler properties of the human visual system, it is well known that the adaptive capability of our visual system restricts the general applicability of these measurement results. Nevertheless, a knowledge of these simpler properties has proved highly useful in applications of radiometry, such as colorimetry and photometry, which involve the human observer.

Signal processing is also an essential part of a radiometric measurement. Frequently the capability of the signal processing component will dictate the design of the rest of the radiometric system. Signal processing can involve sophisticated mathematical analyses of measurement data performed with the aid of very large digital data processing facilities. It can also be as simple as a direct reading of the electrical signal coming from the detector. Most of the signal processing devices which are used in performing radiometric measurements are not unique to this field. However, the signal processing functions which are performed are usually uniquely associated with the radiometric components or application of interest. For this reason we discuss signal processing requirements in connection with the discussion of related system components or measurement techniques.

1.2 BASIC CONCEPTS

The basic concepts of radiometry center around the three complementary properties of electromagnetic radiation—ray, wave, and quantum. It is generally true that at microwave and longer wavelengths radiant energy exhibits primarily wave properties while at the opposite end of the spectrum, at x-ray and shorter wavelengths, radiant energy exhibits primarily ray and quantum properties. In the optical region the ray, wave, and quantum properties are all important in varying degrees.

Much of radiometry is based on the geometrical optics of rays. The use of geometrical optics is founded on an assumption that the wave characteristics of radiant energy do not lead to a deviation of the spatial distribution of radiant energy from a path defined by geometrical rays. In particular, geometrical optics ignores the effects of diffraction. *Diffraction* is the wave phenomenon which arises when a beam of radiant energy passes through an aperture. As the radiant energy leaves the aperture it will not exactly follow the path defined by the geometry of the aperture and the incident beam. Instead, it will spread out. The angle through which the beam will spread is given approximately by

$$\theta = \lambda/D, \tag{1.1}$$

where θ is the diffraction angle, λ the wavelength of radiant energy, and D the aperture diameter. Figure 1.3 shows the magnitude of the diffraction

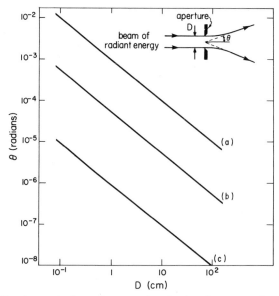

Fig. 1.3 Diffraction angle for various wavelengths λ and aperture diameters D. (a) $\lambda = 10.6$ μm at P20 line of CO_2 IR laser; (b) $\lambda = 555$ nm at peak of luminous efficiency in the visible spectrum; (c) $\lambda = 10$ nm at the boundary of UV and x-ray spectrum.

angle for various wavelengths and aperture diameters. In the optical region of the spectrum and with the apertures typically used in optical systems, this angle is small. For this reason it is frequently possible to use geometrical optics to estimate the path taken by radiant energy and neglect diffraction effects.

A second assumption which is used in much of radiometry is that the radiant energy is incoherent so that interference effects can be ignored. *Interference* is a wave phenomenon which arises when two or more beams of radiant energy which are coherent or partially coherent are added together. Interference can greatly increase or decrease the measured radiant energy in comparison with the sum of the radiant energies contained in the beams separately. Although it is not always possible to separate the spatial and temporal coherence properties of a beam of radiant energy, it is useful to do so to illustrate the nature of the assumption of incoherence. The temporal coherence length l_c of a beam of radiant energy is the coherence distance along the direction of propagation of the beam. It is given approximately by

$$l_c = c/\Delta v = \lambda^2/\Delta \lambda, \tag{1.2}$$

where l_c is the temporal coherence length, c the velocity of light ($= 2.99 \cdot 10^8 \ \mathrm{m\,s^{-1}}$), Δv the frequency interval included in the beam of radiant energy, λ the wavelength, and $\Delta \lambda$ the wavelength interval included in the beam of radiant energy. Figure 1.4 shows a beam of radiant energy incident on a combination of two partially reflecting surfaces. The figure also shows two of the emerging beams, one of which has traveled a total additional path

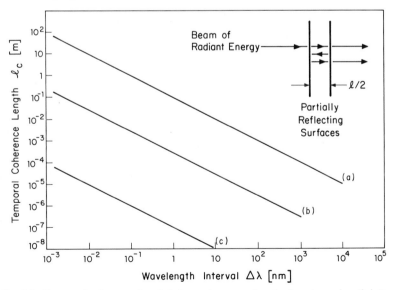

Fig. 1.4 Temporal coherence length l_c for various wavelengths λ and wavelength intervals $\Delta \lambda$. (a) $\lambda = 10.6 \ \mu m$; (b) $\lambda = 555 \ nm$; (c) $\lambda = 10 \ nm$.

length l in comparison with the other beam. If $l \lesssim l_c$, then the beams will interfere. If $l \gg l_c$, then they will not interfere. Figure 1.4 shows the coherence length l_c for various wavelengths as a function of wavelength interval $\Delta\lambda$. In many cases, especially those involving thermal sources, the wavelength interval $\Delta\lambda$ is large enough so that interference effects can be neglected. However, interference effects can occur even with thermal sources if the path length difference l is small enough. Both diffraction and coherence are considered in more detail in Chapter 3.

In addition to these ray and wave properties, radiant energy exhibits quantum properties. Each quantum, or photon, of radiant energy has energy given by

$$Q = hv, \tag{1.3}$$

where Q is the photon energy (J), h is Planck's constant ($= 6.63 \times 10^{-34}$ J s), and v the optical frequency of radiant energy (Hz). At optical frequencies Q is very small. However, in many cases the quantity of energy being measured is also small. The random nature of the arrival of photons at a detector in combination with the small number of photons will then set an ultimate limit to the uncertainty of the measurement. Uncertainty arises in many ways in radiometric measurements, and in many cases is due to effects other than the ultimate limit set by the quantum nature of the energy being measured.

1.3 CHARACTERISTICS OF RADIOMETRY

In comparison with the measurement of other physical quantities, the level of uncertainty involved in the measurement of radiant energy or power is very high. Uncertainties of 1% are considered to be very good and are achieved only in those situations where great care has been taken and where the available components, techniques, and measurement standards are consistent with this level of uncertainty. More typically, uncertainties of 10% or greater are considered adequate and are the result of good measurement technique.

One of the reasons for this high level of uncertainty is that radiant energy is distributed. It is distributed over wavelength, position, direction, time, and polarization. In comparison, electrical measurements which achieve very low levels of uncertainty involve electromagnetic energy which is not distributed or at most is distributed over time. A radiometric measurement system will usually be designed to measure the distribution of power or energy over one parameter while integrating over or holding the other parameters constant. Even when the other parameters have been successfully integrated over or held constant it may be very difficult to employ components, techniques, or standards which will permit adequate measurement of the energy distribution over the one parameter of interest.

Another reason for uncertainty is that all of the atoms and molecules in the components of the radiometric system are constantly scattering, absorbing, and reemitting radiant energy as illustrated in Fig. 1.5. This may involve a relatively simple problem of scattered light inside a grating monochromator. It may also involve a more complex problem of multiple scattering in the atmospheric transmission medium with the amount of scattering depending on source geometry, wavelength, and polarization.

In addition to high levels of measurement uncertainty, the field of radiometry has been characterized by confusion of terminology, symbols, definitions, and units. Although this is not a fundamental physical problem it has proven to be a significant hindrance. Specialists in various subfields and application areas. have in the past developed their own practices. The principal subfields of radiometry in which systems of nomenclature have been established are (1) illumination engineering, (2) military infrared applications, (3) atmospheric physics and meteorology, and (4) radiant heat transfer. There is now some evidence that the first of these systems will be

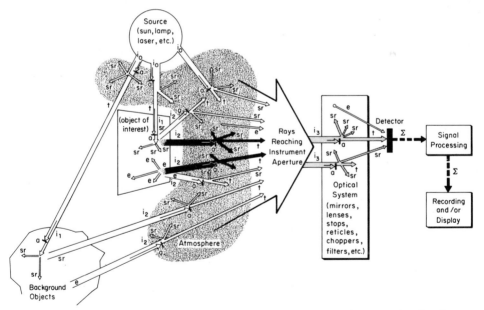

Fig. 1.5 Generalized radiometric configuration. The components shown consist of: i, the incident radiant energy (UV, visible, and/or IR), a, the absorbed radiant energy, sr, the scattered or reflected radiant energy, t, the transmitted radiant energy, e, the emitted radiant energy, and Σ, the signal (usually electrical). The solid arrows indicate source ("desired") rays and the dot-filled arrows indicate source rays mixed with some background ("noise") rays. [From Nicodemus (1970).]

more generally adopted. Various national and international organizations have issued recommendations and statements of intention in favor of adopting this system. This is discussed in more detail in Chapter 2.

1.4   APPLICATIONS

Radiometry is of fundamental importance in an extremely wide range of applications. To some extent this is due to our living in an environment of radiant energy. Because we are surrounded by natural sources of electromagnetic energy, many requirements arise for its measurement and control.

Certainly many of these applications involve visible radiant energy and systems which reproduce the information which would be detected by the human visual system. Examples of this type of application are photometry, photography, television, visual information displays, applications of color science, and vision research.

Scientific areas also include widespread applications of radiometry. Some of these applications are in the areas of planetary astronomy, astrophysics, meteorology and atmospheric physics, materials science, photobiology, and photochemistry. Characteristics of the molecular and atomic structure of the source or transmission medium play a central role in determining which spectral regions are involved. For example, if the surface of a planet is expected to be at temperatures of about 100–300 K, then the measurements will most likely be made in the infrared region, where the surface radiance is greatest.

Radiometry has also grown in importance for applications in military defense areas. The primary emphasis here has been in extending the range of human vision over longer distances or to include environmental conditions where unaided human vision is limited. These conditions include night and weather effects. Night vision devices have been developed using either active or passive sensing. Much of the passive sensing device work has been concentrated in the infrared region of the spectrum, where the radiant energy from objects at ambient temperatures reaches its peak.

Recently the invention and development of lasers in all spectral regions from the ultraviolet (UV) through the visible and infrared (IR) has led to a closer examination of the fundamentals of radiometry. Previously, the available sources of radiant energy were essentially incoherent. The introduction of the laser with its high degree of spatial and temporal coherence has led to the development of new types of radiometric instrumentation as well as the expansion of related theoretical concepts.

A brief summary of some of the application areas and spectral regions of primary interest is shown in Table 1.1.

TABLE 1.1 Some Applications of Radiometry
and the Corresponding Spectral Regions
of Primary Interest

	UV	VIS	IR
Astrophysics	–	–	–
Clinical medicine	–	–	–
Colorimetry		–	
Earth resources satellites		–	–
Illumination engineering		–	
Laser measurements	–	–	–
Materials science	–	–	–
Meteorology and atmospheric physics		–	–
Military electrooptical sensors		–	–
Photobiology and photochemistry	–	–	–
Photographic systems	–	–	–
Photometry		–	
Radiation heat transfer			–
Solar energy		–	
Television systems		–	
Visual information display		–	
Vision research		–	

GENERAL REFERENCES

The following is a list of general references related to radiometry and radiometric measurements.

Bauer, G. (1965). "Measurement of Optical Radiations." Focal Press, London and New York.

Born, M., and Wolf, E. (1964). "Principles of Optics," 2nd rev. ed. Pergamon, New York.

Bramson, M. A. (1968). "Infrared Radiation—A Handbook for Applications." Plenum Press, New York.

CIE (1970). "International Lighting Vocabulary". CIE Publ. CIE No. 17(E-1.1). International Commission on Illumination (CIE), Paris, France.

Committee on Colorimetry of OSA (1953). "The Science of Color." Crowell, New York.

Driscoll, W. G., and Vaughan, W. (eds.) (1978). "Handbook of Optics." McGraw-Hill, New York.

Elenbaas, W. (1972). "Light Sources," Philips Technical Library. Macmillan, New York.

Forsythe, W. E. (1937). "Measurement of Radiant Energy." McGraw-Hill, New York.

Heard, H. G. (ed.) (1968). "Laser Parameter Measurements Handbook." Wiley, New York.

Holter, M. R. *et al.* (1962). "Fundamentals of Infrared Technology." Macmillan, New York.

Hudson, R. D., Jr., and Hudson, J. W. (eds.) (1975). "Infrared Detectors." Halsted Press and Wiley, New York.

Jamieson, J. A. *et al.* (1963). "Infrared Physics and Engineering." McGraw-Hill, New York.

Jenkins, F. A., and White, H. E. (1957). "Fundamentals of Optics," 3rd ed. McGraw-Hill, New York.

Keyes, R. J. (ed.) (1977). "Optical and Infrared Detectors." Springer-Verlag. Berlin and New York.

Keitz, H. A. E. (1971). "Light Calculations and Measurements," Philips Technical Library, 2nd rev. ed. St. Martin's Press, New York.

Kingslake, R. (ed.) (1969). "Applied Optics and Optical Engineering," Vols. I–V. Academic Press, New York.

Kingston, R. (1978). "Detection of Optical and Infrared Radiation." Springer-Verlag, Berlin and New York.

Koller, L. R. (1965). "Ultraviolet Radiation," 2nd ed. Wiley, New York.

Kortüm, G. (1969). "Reflectance Spectroscopy." Springer-Verlag, Berlin and New York.

Kruse, P. W., McGlauchlin, L. D., and McQuistan, R. B. (1962). "Infrared Technology: Generation, Transmission, and Detection." Wiley, New York.

Mavrodineanu, R., Schultz, J. I., and Menis, O. (eds.) (1973). Accuracy in Spectrophotometry and Luminescence Measurements. National Bureau of Standards Spec. Publ. 378.

Nicodemus, F. E. (ed.) (1970). "Radiometry: Optical Resource Letter." American Institute of Physics, New York.

Nicodemus, F. E. (ed.) (1976). Self Study Manual on Optical Radiation Measurements. National Bureau of Standards Tech. Note 910-1, March.

Nimeroff, I. (ed.) (1972). Precision Measurement and Calibration: Colorimetry. National Bureau of Standards Special Publ. 300, Vol. 9, June.

Planck, M. (1959). "Theory of Heat Radiation." Dover, New York.

Siegel, R., and Howell, J. R. (1972). "Thermal Radiation Heat Transfer." McGraw-Hill, New York.

Smith, R. A., Jones, F. E., and Chasmar, R. P. (1968). "The Detection and Measurement of Infrared Radiation," 2nd ed. Oxford Univ. Press, London and New York.

Sparrow, E. M., and Cess, R. D. (1978). "Radiation Heat Transfer." Augmented ed., McGraw-Hill, New York.

Stone, J. M. (1963). "Radiation and Optics." McGraw-Hill, New York.

Swindell, W. (ed.) (1975). "Polarized Light." Wiley, New York.

Vanasse, G. A. (ed.) (1977). "Spectrometric Techniques." Academic Press, New York.

Walsh, J. W. (1956). "Photometry." Dover, New York.

Willardson, R. K., and Beer, A. C. (eds.) (1970). "Semiconductors and Semimetals," Vol. 5 Infrared Detectors. Academic Press, New York.

Wolfe, W. L. (ed.) (1965). "Handbook of Military Infrared Technology." Office of Naval Research, Washington, D.C.

Wyszecki, G. (ed.) (1970). Colorimetry: Optical Resource Letter. American Institute of Physics, New York.

Wyszecki, G., and Stiles, W. S. (1967). "Color Science." Wiley, New York.

2

Terminology and Units

2.1 INTRODUCTION

Confusion often arises due to improper usage of terms and symbols pertaining to optical radiation quantities. This is especially true when new units are introduced.

In this chapter we shall discuss units and terms most commonly used in optical radiation measurements. The basic concepts underlying the quantities discussed in this chapter are illustrated in Chapter 3.

The reader is reminded of the distinction between radiant and luminous (photometric) quantities. Pure physical quantities for which radiant energy is evaluated in energy units are defined here. Each one of these corresponds to another quantity for which radiant energy is evaluated by means of a standard photometric observer. These two quantities are represented by the same principal symbol and are distinguished only by the subscript. The subscript "e" or no subscript is used in the case of physical quantities, and the subscript "v" in the case of photometric quantities.

When certain quantities are considered for monochromatic radiant energy, they are functions of wavelength; hence, their designation must be preceded by the adjective "spectral." The symbol itself, for such quantities, is followed by the symbol for wavelength (λ), e.g., spectral irradiance has the symbol $E(\lambda)$ or $E_e(\lambda)$.

If spectral concentration of a quantity X is considered, it may also be preceded by "spectral" but the symbol is now subscripted, i.e., $X_\lambda = dX/d\lambda$.

The use of standardized symbols and definitions for various quantities is important for interdisciplinary communication and for international understanding. The importance of a single, practical, worldwide system of nomenclature has been recognized and addressed by a number of national and international technical organizations.

We shall adhere to the CIE system of units (CIE, 1970) since it is, for the most part, in agreement with the SI system of units (Page and Vigoureux, 1974). The standard terms used by the CIE have been widely accepted by many scientific organizations (ANSI, 1967; Jones, 1963; Bell, 1959; Thomas, 1973).

In this chapter we shall present and/or discuss radiometric and photometric terms and units. The most frequently used terms are given in tabular form; others are given descriptively.

We cannot claim that all the terms pertaining to optical radiation measurement are given here. It would be literally impossible to do so for it would take a whole volume to do just that. For additional definitions of terms and units and particularly for colorimetric terms, not given in this chapter, the reader is referred to CIE (1970) and CIE (1977) and other texts given in the references.

2.2 RADIOMETRIC QUANTITIES

Table 2.1 defines the most frequently used fundamental quantities concerned with the measurement of radiant energy.

TABLE 2.1   Fundamental Radiometric Quantities[a]

Quantity	Symbol	Defining equation	Units
Radiant energy	Q, Q_e		J (joule)
Radiant energy density	w, w_e	$w = dQ/dV$	$\mathrm{J\,m^{-3}}$
Radiant power or flux	Φ, Φ_e	$\Phi = dQ/dt$	W (watt)
Radiant exitance	M, M_e	$M = d\Phi/dA$	$\mathrm{W\,m^{-2}}$
Irradiance	E, E_e	$E = d\Phi/dA$	$\mathrm{W\,m^{-2}}$
Radiant intensity	I, I_e	$I = d\Phi/d\omega$	$\mathrm{W\,sr^{-1}}$
Radiance	L, L_e	$L = d^2\Phi/d\omega\,(dA\cos\theta)$ $= dI/(dA\cos\theta)$	$\mathrm{W\,m^{-2}\,sr^{-1}}$
Emissivity	ε	$\varepsilon = M/M_{bb}$	—

[a] $d\omega$ is the element of solid angle through which flux from the point source is radiated, $d\theta$ the element of angle between line of sight and normal to the surface considered, and M and M_{bb} are, respectively, the radiant exitance of the surface and that of a blackbody at the same temperature.

2.3 PHOTOMETRIC QUANTITIES

Table 2.2 defines some of the fundamental quantities concerned with the measurement of light. In this table the abbreviations are: lm is lumen, sr is steradian, and cd is candela.

TABLE 2.2 Fundamental Photometric Quantities

Quantity	Symbol	Defining equation	Units
Luminous energy	Q_v	$Q_v = K_m \int_0^\infty V(\lambda) Q_\lambda \, d\lambda$	lm s
Luminous energy density	w_v	$w_v = dQ_v/dV$	lm s m^{-3}
Luminous flux	Φ_v	$\Phi_v = dQ_v/dt$	lm
Luminous exitance	M_v	$M_v = d\Phi_v/dA$	lm m^{-2}
Illuminance	E_v	$E_v = d\Phi_v/dA$	lm m^{-2}
Luminous intensity	I_v	$I_v = d\Phi_v/d\omega$	cd = lm sr^{-1}
Luminance	L_v	$L_v = d^2\Phi_v/d\omega(dA \cos\theta)$ $= dI_v/dA \cos\theta$	cd m^{-2}
Luminous efficacy	K	$K = \Phi_v/\Phi$	lm W^{-1}
Photopic luminous efficiency	$V(\lambda)$	$V(\lambda) = K(\lambda)/K_m$	—
Maximum spectral luminous efficacy	K_m	$K_m \cong 680$ at $\lambda = 555$ nm	lm W^{-1}

2.4 PHOTON QUANTITIES

2.4.1 Photon Number

Number of photons emitted by a source of radiant energy: symbol, N_p; unit, 1.

Note. The number of photons N_p is derived from the spectral concentration of the radiant energy Q_λ or Q_v by the following equations:

$$dN_p = Q_v \, dv \, (1/hv) = Q_\lambda \, d\lambda \, (\lambda/hc),$$

$$N_p = \int dN_p = (1/h) \int (Q_v/v) \, dv = (1/hc) \int (Q_\lambda \lambda) \, d\lambda,$$

where v is the frequency, h Planck's constant, λ the wavelength, and c the velocity of light.

2.4.2 Photon Flux

Number of photons emitted, transferred, or received in an element of time: symbol, Φ_p,

$$\Phi_p = dN_p/dt;$$

unit, 1 per second (s^{-1}).

2.4.3 Photon Intensity (of a Source in a Given Direction)

Quotient of the photon flux leaving the source, propagated in an element of solid angle containing the given direction, by the element of solid angle:

symbol, I_p,

$$I_p = d\Phi_p/d\omega;$$

unit, 1 per second per steradian ($s^{-1}\,sr^{-1}$).

2.4.4   Photon Radiance (in a Given Direction, at a Point on the Surface of a Source or a Receptor, or at a Point in the Path of a Beam)

Quotient of the photon flux leaving, arriving at, or passing through an element of surface at this point and propagated in directions defined by an elementary cone containing the given direction, by the product of the solid angle of the cone and the area of the orthogonal projection of the element of surface on a plane perpendicular to the given direction: symbol, L_p,

$$L_p = d^2\Phi_p/d\omega\,dA\cos\theta;$$

unit, 1 per second per steradian per square meter ($s^{-1}\,sr^{-1}\,m^{-2}$).

2.4.5   Photon Irradiance (at a Point of a Surface)

Quotient of the photon flux incident on an element of the surface containing the point, by the area of that element: symbol, E_p,

$$E_p = d\Phi_p/dA;$$

unit, 1 per second per square meter ($s^{-1}\,m^{-2}$).

2.4.6   Photon Exposure (at a Point of a Surface)

Surface density of the number of photons received: symbol, H_p,

$$H_p = dN_p/dA = \int E_p\,dt;$$

unit, 1 per square meter (m^{-2}).

Note. An equivalent definition is the product of a photon irradiance and its duration.

2.4.7   Photon Exitance

Quotient of the photon flux leaving an element of the surface containing the point, by the area of that element: symbol, M_p,

$$M_p = d\Phi_p/dA = \int L_p\cos\theta\,d\omega;$$

unit, 1 per second per square meter ($s^{-1}\,m^{-2}$).

2.5   SPACE QUANTITIES

2.5.1   Space Irradiance

Integral of the incident radiance taken over the solid angle: symbols, E_0, E_{e0},

$$E_0 = \int L \, d\omega;$$

unit, watts per square meter ($\mathrm{W\,m^{-2}}$).

Note. If L is the radiance incident from the solid angle 4π sr on a small sphere, E_0 is the quotient of the total incident radiant flux by the cross section of this sphere.

2.5.2   Space Exposure

Integral of the space irradiance taken over time: symbols, H_0, H_{e0},

$$H_0 = \int E_0 \, dt = \iint L \, d\omega \, dt;$$

unit, joules per square meter ($\mathrm{J\,m^{-2}}$).

Note. Equivalent definition: product of a space irradiance and its duration.

2.6   TERMS CONNECTED WITH DETECTORS

2.6.1   Responsivity

Quotient of the output quantity Y by the input quantity X, $R = Y/X$.

X may stand for	Y may stand for
irradiance E	current i
illuminance E_v	voltage v
radiant flux Φ	conductance G
luminous flux Φ_v	

Note 1.   If X is used to stand for radiometric quantities, the term is radiant responsivity R; if X stands for photometric quantities, the term is luminous responsivity R_v. Using the terms spectral responsivity $R(\lambda)$ and relative spectral energy distribution $S(\lambda)$ one obtains

$$\text{radiant responsivity}\quad R = \frac{\int X_\lambda R(\lambda)\,d\lambda}{\int X_\lambda \, d\lambda} = \frac{\int S(\lambda) R(\lambda)\,d\lambda}{\int S(\lambda)\,d\lambda}$$

$$\text{luminous responsivity}\quad R_v = \frac{\int X_\lambda R(\lambda)\,d\lambda}{K_m \int X_\lambda V(\lambda)\,d\lambda} = \frac{\int S(\lambda) R(\lambda)\,d\lambda}{K_m \int S(\lambda) V(\lambda)\,d\lambda}$$

Note 2. The responsivity depends on the relative spectral energy distribution; moreover, it may depend on the polarization and the direction of the incident radiant energy, on the spatial uniformity of irradiation and the temperature of the detector as well as on the electrical circuit. If the output quantity and the input quantity are not proportional, the responsivity also depends on the value of the input quantity.

2.6.2 Spectral Responsivity $R(\lambda)$

At the wavelength λ in the infinitesimal wavelength range $d\lambda$ about λ the quotient of the output quantity $dY(\lambda) = Y_\lambda(\lambda)\,d\lambda$ by the radiometric input quantity $dX(\lambda) = X_\lambda(\lambda)\,d\lambda$:

$$R(\lambda) = dY(\lambda)/dX(\lambda).$$

2.6.3 Relative Spectral Responsivity $r(\lambda)$

Ratio of the spectral responsivity $R(\lambda)$ at wavelength λ and the spectral responsivity at reference wavelength λ_0:

$$r(\lambda) = R(\lambda)/R(\lambda_0).$$

Note 1. Sometimes it may be useful to replace $R(\lambda_0)$ by a mean value R_m given by

$$R_\mathrm{m} = \int_{\lambda_1}^{\lambda_2} R(\lambda)\,d\lambda \Big/ (\lambda_2 - \lambda_1) \approx \sum_{i=1}^{i=n} R(\lambda_i)\,\Delta\lambda_i \Big/ (\lambda_n - \lambda_1).$$

Note 2. If the output quantity is not proportional to the input quantity, it is possible to define a relative spectral responsivity independent of the value of the output quantity, if one refers to *equal* values of the output quantity. With finite values ΔX and ΔY one obtains

$$r(\lambda) = R(\lambda)/R(\lambda_0) = \Delta X(\lambda_0)/\Delta X(\lambda) \qquad \text{for} \quad \Delta Y(\lambda) = \Delta Y(\lambda_0).$$

2.6.4 Actinity (Relative Responsivity) $a(Z)$

Of a radiant energy distribution Z, ratio of the responsivity $R(Z)$ when the photosensitive device is irradiated with the distribution Z and the responsivity $R(N)$ when it is irradiated with a reference distribution N:

$$a(Z) = R(Z)/R(N).$$

Note 1. If the spectral responsivity $R(\lambda)$ and the spectral input quantities $(X_\lambda)_N = (dX/d\lambda)_N$, $(X_\lambda)_Z = (dX/d\lambda)_Z$ are known, the actinity may be calculated. It is sufficient to have the relative values $r(\lambda)$ and the relative spectral energy distributions $S(\lambda)_N$, $S(\lambda)_Z$.

Hence

$$a(Z) = \frac{\int S(\lambda)_Z r(\lambda)\, d\lambda \int S(\lambda)_N\, d\lambda}{\int S(\lambda)_Z\, d\lambda \int S(\lambda)_N r(\lambda)\, d\lambda},$$

$$a_v(Z) = \frac{\int S(\lambda)_Z r(\lambda)\, d\lambda \int S(\lambda)_N V(\lambda)\, d\lambda}{\int S(\lambda)_Z V(\lambda)\, d\lambda \int S(\lambda)_N r(\lambda)\, d\lambda}.$$

Note 2.   If the ouput quantity is not proportional to the input quantity, it is possible to define an actinity independent of the value of the output quantity if one refers to *equal* values of the output quantity. When also the equations of Note 1 hold, one obtains

$$a(Z) = R(Z)/R(N) = X(N)/X(Z) \qquad \text{for} \quad Y(N) = Y(Z).$$

2.6.5   Response Time: Rise (Fall) Time

The time required for the output quantity Y to rise (fall) from a stated low (high) percentage to a stated higher (lower) percentage of the maximum value when a steady level of radiant energy is instantaneously applied (removed). The percentages used must be stated.

Note.   If the signal output rises (falls) exponentially with time, the time required for it to change from its initial value by the fraction $(1 - 1/e)$ of the final change is called the *time constant*.

2.6.6   Noise Equivalent Power NEP

The value of input power which, when modulated in a stated manner, produces an output quantity equal to the output quantity caused by noise, both in a stated bandwidth.

2.6.7   Detectivity *D*

Reciprocal of the noise equivalent power NEP,

$$D = 1/\text{NEP}.$$

2.6.8   Normalized Detectivity *D**

A normalized value of the detectivity, given by the product of the detectivity D, the square root of the detector area A, and the square root of the frequency bandwidth B,

$$D^* = D(AB)^{1/2}.$$

2.7   REFLECTION, TRANSMISSION, AND ABSORPTION

2.7.1   Reflection

Regular Reflection; Specular Reflection. Reflection without diffusion in accordance with the laws of optical reflection as in a mirror.

Diffuse Reflection. Diffusion by reflection in which, on the macroscopic scale, there is no regular reflection.

Isotropic Diffuse Reflection. Diffuse reflection in which the spatial distribution of the reflected radiant energy is such that the radiance (luminance) is the same in all directions of the hemisphere of incidence.

Mixed Reflection. Partly regular and partly diffuse reflection.

Retroreflection; Reflex Reflection. Reflection in which radiant energy is returned in directions close to the direction from which it came, this property being maintained over wide variations of the direction of the incident radiant energy.

Reflectance. Ratio of the reflected radiant (luminous) flux to the incident flux: symbol, ρ,

$$\text{spectral reflectance} \quad \rho(\lambda) = \Phi_{\lambda\rho}/\Phi_\lambda,$$

where $\Phi_{\lambda\rho}$ is reflected spectral radiant flux and Φ_λ is incident spectral radiant flux.

Note. Where mixed reflection occurs, the (total) reflectance may be divided into two parts, regular reflectance ρ_r and diffuse reflectance ρ_d, corresponding, respectively, to the two modes of reflection referred to in definitions of *regular reflection* and *diffuse reflection*:

$$\rho = \rho_r + \rho_d.$$

Regular Reflectance. Ratio of the radiant (luminous) flux that has undergone regular reflection to the incident flux: symbol, ρ_r.

Diffuse Reflectance. Ratio of the radiant (luminous) flux that has undergone diffuse reflection to the incident flux: symbol, ρ_d.

Reflectivity. Reflectance of a layer of material of such a thickness that there is no change of reflectance with increase in thickness: symbol, ρ_∞.

Reflection (Optical) Density. Logarithm to base ten of the reciprocal of the reflectance: symbol, D,

$$\text{spectral reflection (optical) density} \quad D(\lambda) = \log[1/\rho(\lambda)].$$

Spectral Fresnel Reflectance. If a transparent, absorbing body (refractive index n, absorption index κ) has smooth surfaces and no disturbing surface layers, the Fresnel reflectances $\bar{\rho}_P(\lambda)$ and $\bar{\rho}_S(\lambda)$ may be calculated for the components of the flux which are polarized parallel or normal to the plane of incidence, respectively.

For a nonabsorbing material ($\kappa = 0$) with an angle of incidence θ and an angle of refraction θ' the two components may be calculated as follows (in these equations the symbol λ, indicating wavelength dependence, is omitted):

$$\bar{\rho}_S = \left[\frac{\cos\theta - (n^2 - \sin^2\theta)^{1/2}}{\cos\theta + (n^2 - \sin^2\theta)^{1/2}}\right]^2 = \frac{\sin^2(\theta - \theta')}{\sin^2(\theta + \theta')},$$

$$\bar{\rho}_P = \left[\frac{n^2\cos\theta - (n^2 - \sin^2\theta)^{1/2}}{n^2\cos\theta + (n^2 - \sin^2\theta)^{1/2}}\right]^2 = \frac{\tan^2(\theta - \theta')}{\tan^2(\theta + \theta')}.$$

For unpolarized radiant energy the total Fresnel reflectance is given by

$$\bar{\rho}_t = \tfrac{1}{2}(\bar{\rho}_S + \bar{\rho}_P).$$

With incident angle $\theta = 0$ (index "0") the above equations are identical:

$$\bar{\rho}_0 = \bar{\rho}_{t,0} = \bar{\rho}_{s,0} = \bar{\rho}_{p,0} = [(n-1)/(n+1)]^2.$$

Radiance (Luminance) Factor (at a representative element on the surface of a nonself-radiating medium, in a given direction, under specified conditions of irradiation [illumination]). Ratio of the radiance (luminance) of the medium to that of a perfect reflecting or transmitting diffuser identically irradiated (illuminated): symbol, β,

$$\text{spectral radiance factor}\quad \beta(\lambda) = L_\lambda / L_{\lambda w},$$

where L_λ is the spectral radiance of the medium and $L_{\lambda w}$ is the spectral radiance of a perfect reflecting diffuser.

Note.   In case of fluorescent media the radiance (luminance) factor is the sum of two portions, the reflected radiance (luminance) factor β_S and the fluorescent radiance (luminance) factor β_L (CIE, 1977; Grum and Costa, 1977):

$$\beta_T = \beta_S + \beta_L.$$

Reflected Radiance (Luminance) Factor (at a representative element of the surface of a medium, in a given direction, under specified conditions of irradiation [illumination]). Ratio of the radiance (luminance) due to reflection of the medium to that of a perfect reflecting diffuser identically irradiated (illuminated): symbol, β_S.

Fluorescent Radiance (*Luminance*) *Factor* (at a representative element of the surface of a medium, in a given direction, under specified conditions of irradiation [illumination]). Ratio of the radiance (luminance) due to fluorescence of the medium to the radiance (luminance) of a perfect reflecting diffuser identically irradiated (illuminated).

Note. The spectral fluorescent radiance (luminance) factor is a material constant that depends also on the spectral distribution of the irradiating energy.

Reflectance Factor (at a representative element of a surface, for the part of the reflected radiant energy contained in a given cone with apex at the representative element of the surface, and for incident radiant energy of a given spectral composition and geometrical distribution). Ratio of the radiant (luminous) flux reflected in the directions delimited by the cone to that reflected in the same directions by a perfect reflecting diffuser identically irradiated (illuminated): symbol, R.

Note 1.   For specularly reflecting surfaces that are irradiated (illuminated) by a source of small solid angle, the reflectance factor may be much larger than one if the cone includes the mirror image of the source. In this case the value of reflectance factor obtained with an instrument will depend on the measuring geometry (specifically sample size, source–sample distance, and sample–receptor distance) unless the sample size is sufficiently small to permit all of the specularly reflected flux to be incident on the receiver.

Note 2.   If the solid angle of the cone approaches 2π sr, the reflectance factor approaches the reflectance. In this case a regular component (if present) must be included. This can be excluded if the regular component is sufficiently defined and a suitable trap is used; then the reflectance factor approaches the diffuse reflectance.

Note 3.   If the solid angle of the cone approaches zero sr, the reflectance factor approaches the radiance (luminance) factor, provided there is no regular reflection in the direction of observation.

Note 4.   The term "directional reflectance" is used currently in the U.S. in this sense.

Reflectometer Value. Value measured by means of a particular reflectometer: symbol, R'.

Note.   The reflectometer employed should be specified. The measured reflectometer value depends on the geometric characteristics of the reflectometer, on the illuminant, on the spectral responsivity of the receptor (even when equipped with filters), and on the reference standard used. If necessary, these conditions should be specified.

2.7.2 Transmission

Transparent Medium. Medium in which the transmission is mainly regular and which has a high regular transmittance. Objects are seen distinctly through such a medium if its geometrical form is suitable.

Translucent Medium. Medium which transmits light entirely, or almost entirely, by diffuse transmission. In general, objects are not seen distinctly through such a medium.

Opaque Medium. Medium which transmits no radiant energy or practically none.

Regular Transmission; Direct Transmission. Transmission without diffusion.

Diffuse Transmission. Transmission in which diffusion occurs independently, on the macroscopic scale, of the laws of refraction.

Isotropic Diffuse Transmission. Diffuse transmission in which the spatial distribution of the transmitted radiant energy is such that the radiance (luminance) is the same in all directions of the hemisphere in which the radiant energy is transmitted.

Mixed Transmission. Partly regular and partly diffuse transmission.

Transmittance. Ratio of the transmitted radiant or luminous flux to the incident flux: symbol, τ,

$$\text{spectral transmittance} \quad \tau(\lambda) = \Phi_{\lambda\tau}/\Phi_{\lambda},$$

where $\Phi_{\lambda\tau}$ is the transmitted spectral radiant flux and Φ_{λ} the incident spectral radiant flux.

Note. Where mixed transmission occurs, the (total) transmittance may be divided into two parts, regular transmittance τ_r and diffuse transmittance τ_d: symbol, $\tau = \tau_r + \tau_d$.

Regular Transmittance. Ratio of the radiant (luminous) flux, that after passing through the medium has undergone regular transmission, to the incident radiant (luminous) flux: symbol, τ_r.

Diffuse Transmittance. Ratio of the radiant (luminous) flux that has undergone diffuse transmission to the incident radiant (luminous) flux: symbol, τ_d.

Spectral Internal Transmittance (of a homogeneous nondiffusing layer). Ratio of the spectral radiant flux $(\Phi_{\lambda})_{ex}$, reaching the exit surface of the layer to the spectral flux $(\Phi_{\lambda})_{in}$, leaving the entry surface: symbol, $\tau_i(\lambda)$,

$$\tau_i(\lambda) = (\Phi_{\lambda})_{ex}/(\Phi_{\lambda})_{in}.$$

Note. For a given layer the spectral internal transmittance depends on the path length of the radiant energy in the layer and hence, in particular, on the angle of incidence.

Spectral Transmissivity (of an absorbing material). Spectral internal transmittance of a layer of the material such that the path of the radiant energy is of unit length and under conditions in which the boundary of the material has no influence: symbol, $\tau_{i0}(\lambda)$.

Note. The path length used should be specified.

2.7.3 Absorption

Absorption is the transformation of radiant energy to a different form of energy by interaction with matter.

Absorptance. Ratio of the absorbed spectral radiant (luminous) flux to the incident flux: symbol, α,

$$\text{spectral absorptance} \quad \alpha(\lambda) = \Phi_{\lambda\alpha}/\Phi_\lambda,$$

where $\Phi_{\lambda\alpha}$ denotes the absorbed spectral radiant flux and Φ_λ the incident spectral radiant flux.

Transmission (Optical) Density. Logarithm to base ten of the reciprocal of the transmittance: symbol, D,

$$\text{spectral transmission (optical) density} \quad D(\lambda) = \log[1/\tau(\lambda)].$$

Spectral Internal Transmission Density (spectral internal absorbance). Logarithm to the base ten of the reciprocal of the internal spectral transmittance: symbol, $A(\lambda)$,

$$A(\lambda) = \log[1/\tau_i(\lambda)].$$

Note 1. For a given layer, the spectral internal transmission density depends on the path length of the radiant energy in the layer and hence, in particular, on the angle of incidence.

Note 2. Occasionally the natural logarithm is used instead of the common logarithm; the corresponding quantity is then called the Napierian internal transmission density:

$$A_n(\lambda) = \ln[1/\tau_i(\lambda)].$$

Spectral Internal Absorptance (of a homogeneous nondiffusing layer). Ratio of spectral radiant flux absorbed between the entry and the exit surfaces of the layer to the spectral radiant flux which leaves the entry surface:

symbol, $\alpha_i(\lambda)$,

$$\alpha_i(\lambda) = \frac{(\Phi_\lambda)_{in} - (\Phi_\lambda)_{ex}}{(\Phi_\lambda)_{in}},$$

where $(\Phi_\lambda)_{in}$ denotes the flux leaving the entrance surface and $(\Phi_\lambda)_{ex}$ the flux entering the exit surface.

Note. For a given layer, the spectral internal absorptance depends on the path length of the radiant energy in the layer and hence, in particular, on the angle of incidence.

Spectral Absorptivity (of an absorbing material). Spectral internal absorptance of a layer of the material such that the path of the radiant energy is of unit length and under conditions in which the boundary of the material has no influence: symbol, $\alpha_{i0}(\lambda)$.

Note. The unit path length used should be specified.

Linear Spectral Absorption Coefficient (of an absorbing medium). Quotient of the Napierian spectral internal transmission density by the path length l traversed by the radiant energy: symbol, $a(\lambda)$,

$$a(\lambda) = A_n(\lambda)/l$$
$$\cong (2.30/l)A(\lambda);$$

unit, m^{-1}.

Note. The linear spectral absorption coefficient is also the part of the linear spectral attenuation coefficient that is due to absorption.

Spectral Absorption Index. The spectral absorption index is used for heavily absorbing materials instead of the linear spectral absorption coefficient: symbol, $\kappa(\lambda)$,

$$\kappa(\lambda) = (\lambda/4\pi)a(\lambda).$$

Absorption Coefficients. The spectral absorption coefficients of some optically clear materials (for example, isotropic crystals, homogeneous liquids, and gases at standard temperature and pressure) are characteristics of the materials. For solutions of absorbing materials in nonabsorbing solvents, the relative spectral absorption coefficients, defined below, are obtained by dividing by the concentration. These coefficients are characteristics of the dissolved substances, according to Beer's law, provided that their states of aggregation and solvation are independent of the nature of the solvent and of the concentration.

Molar Spectral Absorption Coefficient. Quotient of the spectral absorption coefficient $a(\lambda)$ and the molar concentration c of a solution of an absorbing material in a nonabsorbing solvent: symbol, $a_c(\lambda)$,

$$
\begin{aligned}
a_c(\lambda) &= a(\lambda)/c \\
&= (1/c)A_n(\lambda)/l \\
&\cong (2.30/cl)A(\lambda);
\end{aligned}
$$

unit, liter $mol^{-1} cm^{-1}$.

Spectral Mass Absorption Coefficient. Quotient of the spectral absorption coefficient $a(\lambda)$ and the mass density of the medium.

Spectral Volume Absorption Coefficient. For material in nonabsorbing media, the quotient of the spectral absorption coefficient $a(\lambda)$ either by the volume concentration c_v of the material or by its percent volume concentration (PVC $= 100c_v$).

Linear Spectral Attenuation Coefficient (of an absorbing and diffusing medium, for a collimated beam of energy). Quotient of the relative decrease in spectral concentration of the radiant flux of a collimated beam of energy during traversal with normal incidence of an infinitesimal layer of the medium, by the thickness dl of that layer: symbol, μ,

$$\mu(\lambda) = (1/\Phi_\lambda)(d\Phi_\lambda/dl);$$

unit, m^{-1}.

Note. $\mu(\lambda)/\rho$, where ρ is the density of the medium, is called the spectral mass attenuation coefficient.

2.7.4   4π Transmittance

4π transmittance is the ratio of all fluxes (backward and forward) leaving the sample to the incident flux; it is achieved when the sample is positioned into a spherical integrator (integrating sphere) (Costa *et al.*, 1976): symbol, $\tau_{4\pi}$,

$$\tau_{4\pi} = \Phi_{\tau,4\pi}/\Phi_0.$$

2.7.4.1   4π ABSORPTANCE

The complement to unity of 4π transmittance: symbol, $\alpha_{4\pi}$,

$$\alpha_{4\pi} = 1 - \tau_{4\pi}.$$

2.8 SUPPLEMENTARY TERMS

2.8.1 Radiometric Terms

Isotropic Point Source. A point source which emits with equal radiant intensity in all directions.

Radiant Exposure. Surface density of the radiant energy received: symbols, H, H_e,

$$H = dQ/dA;$$

unit, joule per square meter ($J\,m^{-2}$).

Note 1. The radiant exposure is also equal to the time integral of irradiance $H = \int E\,dt$.

Note 2. Equivalent definition: product of an irradiance and its duration.

Note 3. In photobiology this quantity is called *dose* (International Photobiology Committee, 1954).

Light Exposure. Surface density of the luminous energy received: symbol, H_v,

$$H_v = dQ_v/dA;$$

unit, lumen seconds per square meter ($lm\,s\,m^{-2}$).

Note 1. The light exposure is also equal to the time integral of illuminance: $H_v = \int E_v\,dt$.

Note 2. Equivalent definition: product of an illuminance and its duration.

Thermal Radiator; Thermal Source. Radiator (source) emitting by thermal radiation.

Blackbody; Full Radiator, Planckian Radiator. Thermal radiator that absorbs completely all incident energy whatever the wavelength, the direction of incidence, or the polarization. This radiator has, for any wave length and in any direction, the maximum spectral concentration of radiance for a thermal radiator in thermal equilibrium at a given temperature.

Kirchhoff's Law. At a point on the surface of a thermal radiator at each temperature and for each wavelength, the spectral (directional) emissivity in any given direction is equal to the spectral (directional) absorptance for radiant energy incident from the same direction. This holds for any polarized component and, hence, for unpolarized radiant energy as well.

Nonselective Radiator. Thermal radiator whose spectral emissivity is constant with respect to wavelength over the range considered and less than 1.

Note. A nonselective thermal radiator may also be called a graybody.

Quantity of Light, Luminous Energy. Time integral of luminous flux: symbol, Q_v,

$$Q_v = \int \Phi_v \, dt;$$

unit, lumen seconds (lm s).

Note.   Equivalent definition: product of a luminous flux and its duration.

Spectral Distribution (of a radiometric quantity). The spectral concentration of the radiometric quantity as a function of wavelength.

Relative Spectral Distribution (of a radiometric quantity). Ratio of the spectral concentration of a radiometric quantity $X_\lambda(\lambda)$ at wavelength λ and the spectral concentration of this quantity at reference wavelength λ_0: symbol, $S(\lambda)$,

$$S(\lambda) = X_\lambda(\lambda)/X_\lambda(\lambda_0).$$

Note 1.   The dimensionless function $S(\lambda)$ describes the spectral character of radiant energy in a general manner, independent of a special radiometric quantity.

Note 2.   Sometimes it may be useful to replace $X_\lambda(\lambda_0)$ by a mean value $(X_\lambda)_m$ given by

$$(X_\lambda)_m = \int_{\lambda_1}^{\lambda_2} X_\lambda(\lambda) \, d\lambda \bigg/ (\lambda_2 - \lambda_1) \approx \sum_{i=1}^{i=n} X_\lambda(\lambda_i) \, \Delta\lambda_i \bigg/ (\lambda_n - \lambda_1).$$

2.8.2   Instrumental Terms

Spectroradiometer. Instrument for measuring the spectral concentration of radiometric quantities.

Spectrophotometer. Instrument for measuring the ratio of two spectral concentrations of a radiometric quantity at the same wavelength.

Dispersion. 1. Phenomenon of the change in velocity of propagation of radiant energy in an optical medium as a function of its frequency, which causes a separation of the monochromatic components of a complex distribution. 2. Property of an optical device securing the separation of the monochromatic components of a complex distribution.

Solid Angle. The solid angle of a cone is defined as the ratio of the area cut out on a spherical surface (with its center at the apex of that cone) to the square of the radius of the sphere: symbols, ω, Ω; unit, steradian (sr). 1 sr is the solid angle which, having its vertex in the center of a sphere, cuts off an area of the surface of the sphere equal to that of a square with sides of length equal to the radius of the sphere.

Slit Function. Relative spectral transmittance of a monochromator for a given spectral adjustment and a given setting of its slit widths.

Note 1.   The pattern of the slit function is usually determined by using a spectral source yielding thin lines sufficiently separated in order that a single one only falls within the spectral range being transmitted. The spectral adjustment of the monochromator is then varied and the relative responses are recorded. However, it should be noted that, in such a way, the recorded response is the symmetrical image of the slit function with respect to the wavelength of the line.

Note 2.   The shape of the slit function is triangular when the monochromatic image of the entrance slit exactly fills the exit slit and trapezoidal in other cases. The width at midheight of the profile always corresponds to the bandwidth of the monochromator.

Dispersion Prism. Prism made of a transparent material of high dispersion and intended to secure the separation of the spectral components of a complex radiant energy distribution by means of refraction. The effect of the change of refractive index with wavelength caused by dispersion is used.

Diffraction Grating. A transparent or reflecting plane or curved plate with a periodic structure whose pitch is of the order of magnitude of the wavelength of the energy to be analyzed. By interference among different coherent beams of diffracted radiant energy there occur deviations from the direction of regular transmittance or regular reflectance depending on the wavelength. Therefore an incident complex radiant energy distribution may be separated into its spectral components.

Monochromator. Optical instrument intended to isolate a narrow part of the spectrum of a source (thus it provides nearly *monochromatic* radiant energy). There are two essentially different possibilities:

Prism—or Grating—Monochromator. It comprises a dispersion prism or a diffraction grating able to be turned around an axis, two slits (an entrance and an exit slit), and a system of lenses and mirrors securing a series of monochromatic images of the entrance slit in the plane of the exit slit.

Filter Monochromator. It consists usually of a series of narrow-band filters. Also a special interference filter having a continuous change of the transmitted wavelengths along a certain direction is used, the desired wavelength range being set by displacing the filter in front of a slit.

Double Monochromator. Optical instrument consisting of two successive monochromators for the purpose of reducing stray light. In such a system the exit slit of the first monochromator is the entrance slit for the second (it is usually called the intermediate slit).

Slit. Small aperture whose width is usually much smaller than its height.

Note.   Such slits are normally placed at the entrance and the exit of a monochromator and their widths and heights are normally adjustable.

Spectrum. 1. Spatial display of a complex radiant energy distribution produced by separation of its monochromatic components. 2. Composition of a complex radiant energy distribution.

Note 1.   Examples of sense 2: continuous spectrum, line spectrum.

Note 2.   The radiant energy X_K of a *continuum* is usually described by the spectral concentration $X_\lambda(\lambda)_K$ of the quantity concerned:

$$X_K = \int X_\lambda(\lambda)_K \, d\lambda.$$

The radiant energy of a *line* with a mean wavelength λ_i is usually described by the total radiant energy $X(\lambda_i)_L$ of the line, its bandwidth $\Delta\lambda = \lambda_2 - \lambda_1$ being very small:

$$X(\lambda_i)_L = \int_{\lambda_1}^{\lambda_2} X_\lambda(\lambda)_L \, d\lambda, \qquad \lambda_1 < \lambda_i < \lambda_2.$$

For a complex spectrum consisting of a part X_K of continuous radiant energy distribution and a part X_K contributed by n lines, one obtains, for the total quantity X,

$$X = X_K + X_L = \int X_\lambda(\lambda)_K \, d\lambda + \sum_{i=1}^{i=n} X(\lambda_i)_L.$$

Note 3.   Using a monochromator set at wavelength λ_i with a bandwidth $\Delta\lambda$, the value of $X_\lambda(\lambda)_K$ may be assumed to be constant $X_\lambda(\lambda_i)_K$, and if only radiant energy of one line with wavelength λ_i leaves the exit slit the total radiant energy $X(\lambda_i)$ passing the monochromator is

$$X(\lambda_i) = X_\lambda(\lambda_i)_K \, \Delta\lambda + X(\lambda_i)_L.$$

Note 4.   If some lines with very small differences of wavelengths cover a wavelength region $\overline{\Delta\lambda}$, it is recommended to adjust the bandwidth of the monochromator to $\Delta\lambda > \overline{\Delta\lambda}$. Then the radiant energy of all these lines passes the monochromator.

REFERENCES

ANSI (1967). "American Standard Nomenclature and Definitions for Illuminating Engineering," ANSI Publ. Z7.1.
Bell, E. E. (1959). *Proc. IRE* **47**, 1432.
CIE (1970). "International Lighting Vocabulary." CIE Publ. No. 17 (E-1.1). International Commission on Illumination (CIE), Paris, France.
CIE (1977). CIE Tech. Rep. Publ. 38, TC-2,3.

Costa, L. F., Grum, F., and Wightman, T. (1976). *J. Color Res. Appl.* **1**, 193.
Grum, F., and Costa, L. F. (1977). *TAPPI Monogr. Ser.* **60**, 119.
Jones, R. C. (1963). *J. Opt. Soc. Am.* **53**, 1314.
Page, C. H., and Vigoureux, P. (1974). "The International System of Units (SI)," National Bureau of Standards Special Publ. 330.
Thomas, W. (1973). "SPSE Handbook of Photographic Science and Engineering," pp. 144–149. Wiley, New York.

3

Transfer of Radiant Energy

3.1 INTRODUCTION

Radiometry is based on a number of optical concepts which can be used to describe the transfer of radiant energy from a source to a detector and to explain the relationship between this energy and the system components. The radiometric system *components* which directly involve optical concepts are the

(1) Source,
(2) Transmission medium,
(3) Reflecting or transmitting objects,
(4) Optical system, and
(5) Detector.

Obviously, this list includes all the components of a radiometric system with the exception of signal processing. In a sense, the signal processing can also be said to involve optical concepts in an indirect way since the design of signal processing functions will often depend on a knowledge of the optical effects associated with the other components of the radiometric system. The important *optical properties* of these components are

(1) Spatial (position and direction),
(2) Spectral (optical wavelength),
(3) Temporal (modulation),
(4) Polarization, and
(5) Coherence.

Our purpose in this chapter is to introduce or review, as the case may be, those optical concepts which are useful in a wide range of applications and are generally applicable to more than one component. Clearly, there are some optical concepts which are important but are uniquely associated with a single component. These are introduced as required in Chapters 4–7, which describe individual components.

The approach taken in the following sections is as follows. We first establish the connection between Maxwell's equations of electromagnetic theory and the ray concept of geometrical optics. The reason for establishing the connection is to clarify the concept of a ray and also to indicate the conditions under which the concept breaks down requiring a return to the more involved methods of wave optics described in Section 3.3. Next we establish the fundamentally important facts that within the limits of geometrical optics the radiance and basic radiance are conserved along a ray and so play a central role in radiometric calculations. These concepts of rays and radiance are then used to formulate the general problem of radiant energy transfer from an incoherent radiating surface to a receiving surface. Since many applications involve the use of imaging components between source and detector, we also consider the geometrical optics of these systems and define the spatial properties which are relevant to radiometry.

Having explored the use of geometrical opticals we turn to the more involved but more accurate methods of wave optics. Interference and diffraction effects are described and the spatial distribution of radiant power due to diffraction at rectangular and circular apertures is calculated. Coherence effects and calculations are considered. The radiometry of partially coherent sources is an area of current research and some of the concepts are still evolving. However, we describe the basic concepts and generally useful relationships. These discussions of diffraction and coherence then lead to a brief description of the spatial distribution of radiant energy in imaging systems based on wave optics.

We next define the changes in radiant energy which occur due to transmission and reflection at an interface. Absorption and scattering, and emission, are treated from a simple phenomenological point of view which corresponds to current measurement practice. The polarization of polychromatic sources is presented in terms of the measurable Stokes parameters. Finally, we present an elementary derivation and description of the characteristics of photon and thermal noise which dictate the fundamental limit of accuracy in all radiometric measurements.

3.2 SPATIAL DISTRIBUTION OF RADIANT ENERGY: GEOMETRICAL OPTICS

3.2.1 Foundations of Geometrical Optics

Many of the simplest and most useful methods of radiometry are based on the ray concept of geometrical optics. Geometrical optics provides in many cases an adequate description of the propagation or transfer of radiant energy from a source, through the other components of a radiometric system, and to a detector. This section is a brief summary of the relationship between geometrical optics and the more fundamental description of electromagnetic fields provided by Maxwell's equations.

It is well known (Born and Wolf, 1964) that Maxwell's equations lead to the wave equation

$$\nabla^2 U = (1/v^2)(\delta^2 U/\delta t^2), \tag{3.1}$$

where $U = U(\bar{r}, t)$ represents the magnitude of one component of the electric or magnetic field in an appropriate coordinate system, $v = c/n$, c is the velocity of light, and n the index of refraction. The solution of (3.1) is

$$U(\bar{r}, t) = U(\bar{r}) \exp\{ik[S(\bar{r}) - vt]\}, \tag{3.2}$$

where $S(\bar{r})$ is real. With the definition $U(\bar{r}) = \exp b(\bar{r})$ we can substitute (3.2) in (3.1) and equate real parts to obtain

$$(\nabla S)^2 - n^2 = (\lambda^2/4\pi^2)[\nabla^2 b + (\nabla b)^2], \tag{3.3}$$

where ∇^2 is the Laplacian, ∇ the gradient, and λ the wavelength. Equation (3.3) is useful in that it shows the nature of the basic approximation which leads to the ray concept of geometrical optics. When the wavelength becomes small so that $\lambda \to 0$, then (3.3) becomes

$$(\nabla S)^2 = n^2, \tag{3.4}$$

where ∇S is the gradient of S. The function $S(\bar{r})$ is known as the *eikonal* and Eq. (3.4) is the *eikonal equation*. It is the basic equation of geometrical optics. Since (3.4) is based on the approximation that the right side of (3.3) vanishes it cannot be expected to be valid when spatial changes in the amplitude distribution $b(\bar{r})$ are appreciable with respect to $1/\lambda$ as they are, for example, near a well-defined focus of a beam of radiant energy or near the sharp geometrical edge of a shadow.

The surfaces

$$S(\bar{r}) = \text{const.} \tag{3.5}$$

are the geometrical wave fronts or surfaces of constant optical phase. Figure 3.1 illustrates the concept of the geometrical wave front and the gradient ∇S which is normal to the wave front. It can be shown (Born and Wolf, 1964) that the time-averaged Poynting vector which defines the direction and magnitude of energy flow in an electromagnetic field is also in the direction of the normal to this geometrical wave front and that its magnitude is equal to the product of the average energy density and the velocity c/n. This indicates that within the accuracy of geometrical optics the average energy density is propagated with the velocity c/n.

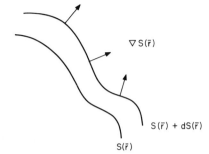

Fig. 3.1 Geometrical wave front and gradient ^1S normal to wave front. The direction of ^1S is the direction of the ray.

The *ray* is then defined simply as the normal to the geometrical wave front. On this basis we can say that the ray concept of geometrical optics defines the direction of energy flow. The magnitude of the radiant energy flow is defined in the next section in terms of the radiometric quantity *radiance*.

3.2.2 Radiance and Basic Radiance

3.2.2.1 RADIANCE

The radiometric quantity *radiance* was defined in Chapter 2 along with other radiometric quantities. In this section we will show (Nicodemus, 1963) that radiance occupies a central position in radiometry because within the limitations of geometrical optics the radiance is found to be conserved along a ray within a homogeneous isotropic medium. More generally the *basic radiance* is conserved even across the boundary between media. It then becomes possible to follow the process of radiant energy transfer from a source to a detector by following the geometrical rays.

Consider an elementary beam of incoherent radiant energy in a homogeneous isotropic medium as shown in Fig. 3.2. By definition, the elementary beam is composed of a single central ray of interest and a small bundle of rays which consists of all rays which pass through both area elements dA_1 and dA_2 around the central ray. The solid angle subtended at the opposite end by each area element is $d\omega_1 = \cos\theta_1 \, dA_1/D^2$ and $d\omega_2 = \cos\theta_2 \, dA_2/D^2$,

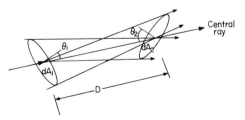

Fig. 3.2 Elementary beam of radiant energy.

where θ_1 and θ_2 are the angles between the central ray and the surface normals and D the separation. The products of solid angle and projected area defined as $dG_1 = d\omega_2 \cos\theta_1 dA_1$ and $dG_2 = d\omega_1 \cos\theta_2 dA_2$ are then found to be equal. This quantity

$$dG = d\omega_1 \cos\theta_2 dA_2 = d\omega_2 \cos\theta_1 dA_1 \quad (\text{m}^2 \text{sr}) \tag{3.6}$$

is the *geometrical extent* of the elementary beam of rays. As shown by (3.6) it is the product of a projected area element and the solid angle subtended at this element by a second area element. It is a useful invariant characteristic of the elementary beam of rays since the two surfaces chosen are arbitrary.

In Chapter 2 the radiometric quantity radiance is defined as

$$L = d^2\Phi/dA \cos\theta \, d\omega \quad (\text{W m}^{-2} \text{sr}^{-1}) \tag{3.7}$$

where $dA \cos\theta \, d\omega$ is interpreted in the same way as in (3.6). Now consider Fig. 3.2 again. Since the same rays pass through dA_1 and dA_2, geometrical optics implies that the power $d\Phi$ which is associated with the rays which pass through both areas is equal. Then $d^2\Phi_1 = d^2\Phi_2$. From this consideration and from (3.6) and (3.7) we obtain the fundamentally important result

$$L_1 = L_2 = L \quad (\text{W m}^{-2} \text{sr}^{-1}), \tag{3.8}$$

where L_1 and L_2 are the radiances at dA_1 and dA_2 along the direction of the central ray. The important interpretation of (3.8) is that the *radiance* defined by (3.7) as the radiant power per unit solid angle (in the direction of the ray) per unit projected area (perpendicular to the ray) has the same value at any point along a ray within a homogeneous isotropic lossless medium. Because the radiance is conserved in a lossless medium it is convenient to characterize losses due to absorption or scattering within a medium in terms of reduction of radiance. This is done in Section 3.5.

3.2.2.2 BASIC RADIANCE

The concept of radiance conservation represented by (3.8) applies to a single medium. Many situations involve a smooth boundary between two homogeneous isotropic media. This is represented in Fig. 3.3. Consider an elementary beam of rays to be incident from the left on a small area element dA at the boundary. The rays in the beam are refracted at the boundary according to Snell's law ($n \sin\theta = n' \sin\theta'$), where the primes denote the second medium. In the absence of reflection losses at the boundary the power in the beam is the same on both sides of the boundary. From (3.7) this power is $d^2\Phi = L \cos\theta \, dA \, d\omega = L' \cos\theta' \, dA \, d\omega'$ where again the primes denote the second medium. By differentiating Snell's law and using it together with the solid angle definitions $d\omega = \sin\theta \, d\theta \, d\phi$ and $d\omega' = \sin\theta' \, d\theta' \, d\phi$ in this expression for $d^2\Phi$ we find that

$$L/n^2 = L'/n'^2 \quad (\text{W m}^{-2} \text{sr}^{-1}). \tag{3.9}$$

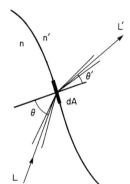

Fig. 3.3 Smooth boundary between two homogeneous isotropic media.

That is, the quantity L/n^2 in the direction of a ray is invariant along that ray even across a smooth boundary between different media in the absence of reflection at the boundary. The quantity L/n^2 is the *basic radiance*. Again, because the basic radiance is conserved across the boundary, it is convenient to characterize losses due to reflection at the boundary in terms of the reduction of basic radiance. These losses are considered in Section 3.4.

The physical interpretation of Eq. (3.9) is of course that on crossing the boundary the elementary beam retains the same power $d^2\Phi$ and the same area dA, but the solid angle $d\omega$ and the projected area $\cos\theta\, dA$ of the beam change due to refraction. The consequent increase or decrease in the beam radiance is just given by the factor n'^2/n^2.

The dependence on wavelength is not explicitly shown in (3.7), (3.8), or (3.9). However, these relations do apply to each wavelength component separately since the radiance L and the index n may depend on wavelength.

3.2.2.3 Measurement of Radiance

In principle it is possible to measure the radiance of an incoherent source or an incoherent beam of radiant energy with the simple radiometer shown in Fig. 3.4. The radiometer consists of a detector and an aperture in an otherwise opaque tube. The length of the tube is l. We assume that the ratio

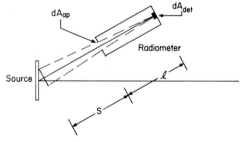

Fig. 3.4 Simple radiometer for measuring the radiance of an incoherent source or an incoherent beam of radiant energy.

of l to the linear dimension of the detector is large enough so that the aperture subtends the same solid angle at all points of the detector.

We first consider the measurement of the radiance of a source. The projected area of the aperture on the source is $dA_{ap}(l + S)^2/l^2$. This defines the area of the source from which radiant energy reaches the detector. The solid angle subtended by the detector when viewed from the source is $dA_{det}/(l+S)^2$. From the definition of radiance in Eq. (3.7) we see that the product of projected area and solid angle is

$$dA \cos \theta \, d\omega = dA_{ap} \, dA_{det}/l^2. \tag{3.10}$$

If the output of the detector is calibrated in watts $d^2\Phi$, then the radiance of the source is

$$L = d^2\Phi/dA \cos \theta \, d\omega = d^2\Phi/(dA_{ap} \, dA_{det}/l^2) \quad (\text{W}\,\text{m}^{-2}\,\text{sr}^{-1}). \tag{3.11}$$

If we had assumed that the radiance in the aperture was the same as the radiance at the source and calculated the product of projected area and solid angle at the aperture, we would have obtained the same result. This is the principle used in measuring the radiance of the beam of radiant energy rather than the radiance of a source.

3.2.3 Transfer of Radiant Energy

It is now possible to use the concepts of rays and radiance conservation to describe the transfer of radiant energy from a radiating source surface to a receiving surface. This is done here under the assumption that the medium which fills the space between the two surfaces is homogeneous and isotropic and, in particular, that there are no imaging or focusing elements between the two. The description of radiant energy transfer through imaging systems is reserved for Section 3.2.4. We also assume here that the medium is lossless. Radiant energy transfer in the presence of absorption or scattering within the medium are considered in Section 3.5.

3.2.3.1 GENERAL METHOD

The most general, but frequently required, formulation for determining the amount of radiant power transferred from a source surface to another surface within a lossless homogeneous isotropic medium is as follows. Consider Fig. 3.5. The radiant power transferred to dA_2 from a source area element dA_1 is found from (3.7) and (3.8) to be

$$d^2\Phi_{12} = L_1 \, dA_1 \, dA_2 \, (\cos \theta_1 \cos \theta_2/r_{12}^2) \quad (\text{W}), \tag{3.12}$$

where r_{12} is the distance between dA_1 and dA_2, θ_1 and θ_2 the angles between the ray along r_{12} and the surface normals, and $\cos \theta_1 \, dA_1/r_{12}^2$ the solid angle

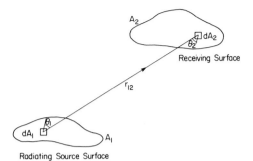

Fig. 3.5 Geometry of radiant energy transfer from area element dA_1 to area element dA_2.

$d\omega_1$. The total radiant power received at surface 2 from surface 1 is then

$$\Phi_{12} = \int_{A_1} \int_{A_2} L_1(\cos\theta_1 \cos\theta_2/r_{12}^2)\, dA_1\, dA_2 \quad \text{(W)}, \qquad (3.13)$$

where changes in L_1, θ_1, θ_2, and r_{12} must be considered when integrating. Numerical methods can be used to evaluate the integrals in (3.13) and are often required when no further assumptions can be made about the source radiance or the geometry.

The linear addition of power which is represented by Eq. (3.13) involves an assumption that the contributions from the various source elements are incoherent, i.e., that there are no interference effects present. The transfer of radiant power from coherent or partially coherent sources is considered in Section 3.3.4.

3.2.3.2 CONFIGURATION FACTOR

When the source surface is *Lambertian*, then L_1 is constant by definition and, as shown in Chapter 4,

$$L_1 = M_1/\pi \quad \text{(W m}^{-2}\text{ sr}^{-1}), \qquad (3.14)$$

where M_1 is the radiant exitance (W m^{-2}). With (3.14), (3.13) becomes

$$\Phi_{12} = (M_1/\pi) \int_{A_1} \int_{A_2} (\cos\theta_1 \cos\theta_2/r_{12}^2)\, dA_1\, dA_2. \qquad (3.15)$$

The total power leaving surface 1 is $\Phi_1 = M_1 A_1$, where A_1 is the total area of surface 1. The fraction of the total power leaving surface 1 which reaches surface 2 is then

$$F_{12} = (1/\pi A_1) \iint (\cos\theta_1 \cos\theta_2/r_{12}^2)\, dA_1\, dA_2$$

$$= \frac{\text{total power reaching surface 2}}{\text{total power leaving surface 1}}, \qquad (3.16)$$

where F_{12} is the dimensionless *configuration factor* from A_1 to A_2. In terms of this factor the total power transferred from A_1 to A_2 is seen from (3.15) to be

$$\Phi_{12} = M_1 A_1 F_{12} = \pi L_1 A_1 F_{12}$$

$$= \text{total power transferred from } A_1 \text{ to } A_2. \quad (3.17)$$

Notice that when surface 2 is a hemisphere, then $F_{12} = 1$. In general we can see that $F_{12} \leq 1$. Configuration factors have been calculated for many geometries and are often available as formulas or tabulated data (Siegel and Howell, 1972; Sparrow and Cess, 1978). These can save considerable time in radiative transfer calculations. Since the geometrical factors in (3.16) are independent of wavelength, F_{12} is also independent of wavelength. It can be shown that

$$A_1 F_{12} = A_2 F_{21}, \quad (3.18)$$

where F_{21} is the configuration factor from A_2 to A_1. This is fairly obvious from the symmetry of the double integral in (3.16).

3.2.3.3 GEOMETRICAL EXTENT AND OPTICAL EXTENT

The *extent* of a beam of radiant energy (Jacquinot, 1965) is another geometrical optics concept which is widely used to describe the amount of radiant power transferred from one surface to another. It is more restricted than the configuration factor in its range of validity but is easier to evaluate. Because of its ease of evaluation it is widely used in approximate descriptions of radiant energy transfer.

Figure 3.5 shows two surfaces. The general equation describing the transfer of radiant power between these two surfaces is Eq. (3.13). When the radiance L_1 is uniform and when the geometry of the radiating and receiving surfaces is such that the integrals over A_1 and A_2 are separable, then (3.13) becomes

$$\Phi_{12} = LA\Omega \quad \text{(W)}, \quad (3.19)$$

where

$$A = \int_{\text{surf } 1} dA_1 \quad (\text{m}^2) \quad (3.20)$$

is the area of surface 1 and

$$\Omega = \int_{\text{surf } 2} (\cos \theta_1 \cos \theta_2 / r_{12}^2)\, dA_2 = \int_{\text{surf } 2} \cos \theta_1 \, d\omega_2 \quad (\text{sr}) \quad (3.21)$$

the *projected solid angle* subtended by surface 2 at surface 1. In Eq. (3.19) the product $G = A\Omega$ is the *geometrical extent* of the finite beam of radiant power defined by A_1 and A_2. Note that Eq. (3.19) can apply to transfer in either direction.

When the two surfaces are not in the same medium it is possible to define an invariant quantity known as the *optical extent*

$$G_n = n^2 A\Omega \quad \text{(m}^2\,\text{sr)}, \tag{3.22}$$

where A, Ω, and the index n are in the medium of interest. Radiant power Φ_{12} is then the product of basic radiance (3.9) and optical extent (3.22).

3.2.3.4 MULTIPLE REFLECTIONS

When the receiving surface is radiometrically *black* it absorbs, by definition, all power incident on it. Then (3.13), (3.17), or (3.19) represent not only the power incident on, but also the power absorbed by, a black surface. For nonblack surfaces reflections from surface 2 must be considered and, if surface 1 is also nonblack, this may lead to multiple reflections. Procedures for taking account of multiple reflections are available but can be fairly complex (Siegel and Howell, 1972).

3.2.3.5 THE CIRCULAR SOURCE OF UNIFORM RADIANCE

As an important example of the transfer of radiant energy between a radiating source and a receiving surface we now evaluate the configuration factor from a large circular source to a small area element on the axis of the source. From the configuration factor we then find the radiant power transferred and the irradiance. The arrangement is shown in Fig. 3.6. The source radius is a and the receiving area element dA_2 is at distance b.

The configuration factor is defined by Eq. (3.16). From the geometry, $\theta_1 = \theta_2 = \theta$. The configuration factor is then

$$F_{12} = (1/\pi A_1)\, dA_2 \int (\cos^2\theta/r_{12}^2)\, dA_1. \tag{3.23}$$

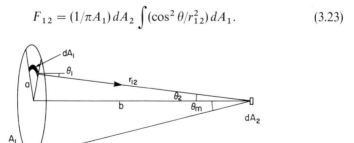

Fig. 3.6 Circular source of uniform radiance with receiving area element dA_2 on the axis.

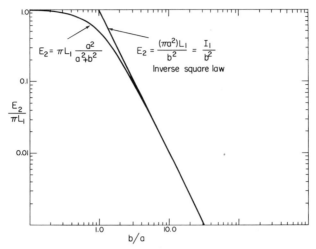

Fig. 3.7 Irradiance on the axis from a circular source of uniform radiance L_1. The parameter b is the distance from source to the receiver and a the source radius.

Using $r_{12}^2 = b^2/\cos^2\theta$ and $dA_1 = 2\pi b^2 \sin\theta \, d\theta/\cos^3\theta$ we find that

$$F_{12} = (2/A_1) \, dA_2 \int_0^{\theta_m} \sin\theta \cos\theta \, d\theta \tag{3.24}$$

$$F_{12} = (dA_2/A_1)\sin^2\theta_m. \tag{3.25}$$

From (3.17) the radiant power transferred from A_1 to dA_2 is

$$d\Phi_{12} = \pi L_1 \sin^2\theta_m \, dA_2, \tag{3.26}$$

so that the irradiance at dA_2 is

$$E_2 = d\Phi_{12}/dA_2 = \pi L_1 \sin^2\theta_m. \tag{3.27}$$

In terms of the distances a and b this is

$$E_2 = \pi L_1(a^2/(a^2 + b^2)). \tag{3.28}$$

Equation (3.28) is shown graphically in Fig. 3.7 as a function of the parameter b/a.

At large distances from the source ($b/a \to \infty$) Eq. (3.28) becomes

$$E_2 = (\pi a^2)L_1/b^2 = I_1/b^2, \tag{3.29}$$

where I_1 is the radiant intensity of the source. Equation (3.29) is the *inverse square law* and indicates that at large distances the source acts as a point source. Figure 3.8 shows the fractional difference between (3.28) and (3.29) as a function of b/a. When $b/a \geq 10$ it is found that the fractional error involved in treating the extended circular source as a point source following the inverse square law (3.29) is less than 1%.

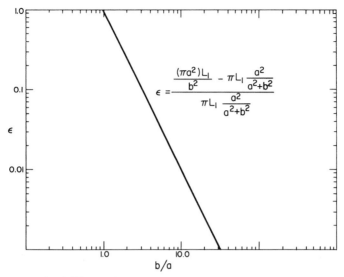

Fig. 3.8 Fractional difference between Eqs. (3.28) and (3.29) showing the range of validity of the inverse square law approximation for a finite circular source.

3.2.4 Imaging Systems

In Section 3.2.3 we assumed that the medium between the source surface and the receiving surface was homogeneous and isotropic. If the receiving surface of interest is a detector, then the presence of imaging or focusing elements between source and detector means that this assumption does not apply. As a result it is necessary to consider separately the transfer of radiant power through imaging systems. In this section we first define the relevant geometric properties of rays in imaging systems. We then use the concept of conservation of basic radiance as a ray passes through media of different indices of refraction to follow the transfer of radiant power from a source, through the imaging elements, and to a detector. We neglect losses due to absorption, scattering, or surface reflections. These are considered separately in Section 3.4.

3.2.4.1 PARAXIAL OPTICS

Figure 3.9 shows a schematic representation of a coaxial imaging system. The focusing elements, which may be refracting or reflecting, are represented here only by the first and last optical surfaces encountered between the source and its image. A number of highly developed techniques exist for tracing rays through the elements of a coaxial optical system on the basis of geometrical optics (Kingslake, 1978). Examination of these methods shows that in general they involve the angles which the ray makes with

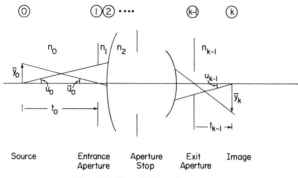

Fig. 3.9 Coaxial imaging system.

the optical axis and with the normals to the surfaces of the focusing elements. When we consider only those rays which make small angles with the optical axis we are in the realm of *paraxial optics.* Paraxial optics is characterized by the approximation $\sin \theta \cong \theta$, where θ is the ray angle with the axis. In this approximation it is found that the equations of the ray paths take a simple linear form. The use of sign conventions required for the definition and use of these equations is as follows.

With reference to Fig. 3.9 the *sign conventions* are

(1)   Rays travel from left to right.
(2)   Surfaces are numbered in the order in which the rays strike the surface.
(3)   Thickness t is the distance from one surface to another. Negative t indicates that the surface is to the left of the reference surface. The subscript for t is the same as the reference surface from which it is measured.
(4)   If the center of curvature of a surface is to the right of a surface, then the radius is positive.
(5)   Index of refraction is denoted by a subscript corresponding to the surface just to its left.

The height of the ray at the nth surface is denoted by y_n. The angle of the ray which leaves the nth surface is u_n with respect to the optical axis. The two equations which describe the ray paths are then the *paraxial refraction equation*

$$n_{-1}u_{-1} - nu = (n - n_{-1})y/r \qquad (3.30)$$

and the *paraxial translation equation*

$$y_{+1} = y + (t/n)(nu) \qquad (3.31)$$

where the subscripts denote any one surface and the immediately preceding (-1) or following $(+1)$ surface. Reflective focusing surfaces are accommodated simply by adopting the convention that at a reflecting surface $n_{+1} = -n$. With this convention Eq. (3.30) can be used for reflection. Then Eq. (3.30) shows the change in the direction of the ray at each surface and Eq. (3.31) shows the change in ray height between surfaces. Numerical procedures based on either matrix algebra (Brouwer, 1964) or the *ynu* tabular method (Kingslake, 1978) are available and are based on (3.30) and (3.31).

3.2.4.2   CARDINAL POINTS

Paraxial optics is used to define a number of geometrical characteristics of imaging systems which are important in radiometry. These include the *cardinal points* and the *imaging equations*. Figure 3.10 shows a ray from an axial source point which is located so far to the left of the optical system that the incoming ray is essentially parallel to the optical axis. This ray crosses the optical axis at the *second focal point* F_2. Continuing this ray backward or forward until it crosses the incoming ray defines the *second principal point* P_2. The distance from P_2 to F_2 is by definition the *second focal length* f'. A similar procedure for a ray coming from the right and parallel to the optical axis defines the *first focal point* F_1, the *first principal point* P_1, and the *first focal length* f. It is also possible to define a third type of point, the *nodal point*, as

$$N_2 F_2 \doteq (P_2 F_2) n_0 / n_{k-1}, \tag{3.32}$$

where N_2 is the *second nodal point*, $N_2 F_2$ the distance from N_2 to F_2, $P_2 F_2$ the distance from P_2 to F_2, and n_0 and n_{k-1} the indices of refraction at the source and at the image. A first nodal point can also be defined. The nodal point has the interesting property that a ray incident on the first nodal point N_1 at a given angle will emerge from the optical system in the same direction from N_2. Notice that when the source and image are in the same medium so that $n_0 = n_{k-1}$, the nodal points coincide with the principal points. The six points F_1, F_2, P_1, P_2, N_1, N_2 are the *cardinal points* of the optical system.

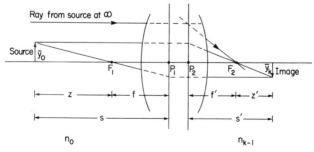

Fig. 3.10   Cardinal points of an imaging system based on paraxial optics.

3.2.4.3 IMAGING EQUATIONS

On the basis of these definitions of the cardinal points it can be shown by using the geometry of the rays which pass through F_1 and F_2 as shown in Fig. 3.10 that

$$zz' = ff', \tag{3.33}$$

where z and z' define the source and image locations with respect to the focal points F_1 and F_2. Equation (3.33) is the *Newtonian equation*. It is possible to measure positions from the principal points P_1 and P_2 by defining $s = z + f$ and $s' = z' + f'$ so that (3.33) becomes

$$(f/s) + (f'/s') = 1. \tag{3.34}$$

This is the *Gaussian equation* for the source and image positions. The lateral magnification of the system can be shown from the geometry of Fig. 3.10 to be

$$m = \bar{y}_k/\bar{y}_0 = -z'/f' = -fs'/f's. \tag{3.35}$$

When the image and source are in the same medium so that $n_{k-1} = n_0$, then $f = f'$ and Eqs. (3.33)–(3.35) simplify.

3.2.4.4 ENTRANCE AND EXIT APERTURES

Figure 3.9 shows two rays leaving the source. The ray from the axial source point leaving at angle u_0 is the *axial ray*. As the angle u_0 is increased, some physical obstruction within the optical system will limit the maximum angle at which rays from the source will pass through the system. This obstruction is the *aperture stop*. The paraxial image of the aperture stop which is formed by all optical surfaces and elements preceding it is the *entrance aperture*. Similarly, the image of the aperture stop which is formed by all surfaces and elements following it is the *exit aperture*. The ray shown in Fig. 3.9 which leaves the axial point of the source at $y_0 = 0$ and passes through the edge of the entrance aperture is the *paraxial marginal ray*. The second ray, for which the overbar is used on the height and angle coordinates, leaving the extreme edge of the source at height $\bar{y}_0$ and passing through the axial point of the entrance aperture is the *paraxial chief ray*. It can be seen from the figure that these rays define the angle subtended by the entrance aperture at the source and by the source at the entrance aperture.

It should be noted that the location of the aperture stop and therefore the entrance and exit apertures can change with a change in the position of the source along the optical axis. Also, for the axial source point the cone of rays which is limited by the entrance aperture will be circular as long as the aperture stop is circular. This is not the case for off-axis source points. The cone of rays for an off-axis source point is noncircular and the total solid angle is usually smaller than for an axial source point. This phenomenon is known as *vignetting* and leads to a reduction of the total power transferred through the optical system. The analysis of vignetting involves projecting all the lens apertures or other limiting apertures into the plane of the entrance aperture by using Eqs. (3.30) and (3.31). The clear area of the entrance aperture which remains unobstructed then defines the solid angle of the cone of the rays which will pass through the system. A corresponding unobstructed area can be found in the exit aperture since the exit aperture is the image of the entrance aperture.

3.2.4.5 RADIANT ENERGY TRANSFER TO AN IMAGE

The definitions of entrance and exit apertures just given can be used to advantage in describing the transfer of radiant energy from a source, through an optical system, and to an image. These apertures have the advantage that (1) they define the area of the beam at that point, or alternatively they define the solid angle of the beam when viewed from the source or the image, and (2) the medium between the source and entrance aperture or the image and exit aperture is usually homogeneous and isotropic. Based on these properties of the entrance and exit apertures it is possible to evaluate Eq. (3.13) as in a nonimaging situation if the radiance is known. The radiance is found by tracing rays from the source through the optical system. This can be based on general ray tracing methods or on the simpler approximate methods of paraxial optics described in previous sections, as required.

The radiant power in an image is usually dependent on position in the image plane. The radiant power in a small area dA_2 as shown in Fig. 3.11 is found from Eq. (3.13) to be

$$d\Phi_{12} = dA_2 \int_{\text{exit aper.}} L_x(\cos\theta_2 \cos\theta_2'/r_{12}^2)\, dA_2', \qquad (3.36)$$

where L_x is the radiance in the exit aperture. When the source and image are in the same medium then $L_x = L_1$, the source radiance along each ray.

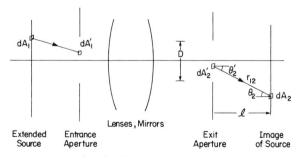

Fig. 3.11 Transfer of radiant power from a source to an image.

In the absence of losses the irradiance in the image is then

$$E_2 = d\Phi_{12}/dA_2 = \int_{\text{exit aper.}} L_1(\cos\theta_2 \cos\theta_2'/r_{12}^2)\,dA_2'. \qquad (3.37)$$

The concept of calculating the irradiance in an image by assuming that the exit aperture acts as a source of the same radiance as the actual source along each ray is a useful one. Any transmission losses along each ray path through the optical system should also be taken into account.

3.2.4.6   The $\cos^4$ Law

As an example of the calculation of the irradiance in an image we consider an on-axis source which is large and of uniform radiance. Then the radiance in the exit aperture is also uniform. If the exit aperture is small enough so that the angle θ_2' as shown in Fig. 3.11 is the same for all points in the aperture, then $\theta_2 = \theta_2'$. The distance r_{12} in Eq. (3.37) is then $l/\cos\theta_2$, where l is the distance from the exit aperture to the image plane. Then Eq. (3.37) becomes

$$E_2 = L_1(\pi D^2/4l^2)\cos^4\theta_2 \quad (\text{W m}^{-2}), \qquad (3.38)$$

where D is the exit aperture diameter.

When the source is at a large distance from the entrance aperture so that l/D is approximately equal to f/no, the f-number, then

$$E_2 = (\pi L_1/4)(1/(f/\text{no})^2)\cos^4\theta_2 \quad (\text{W m}^{-2}), \qquad (3.39)$$

where $(f/\text{no}) = f/D$ is the f-number of the imaging system. Figure 3.12 shows $E_2(\theta_2)/E_2(0)$ as a function of θ_2. For very small angles the angle dependence is completely negligible but at larger angles it leads to a significant decrease in irradiance in the image.

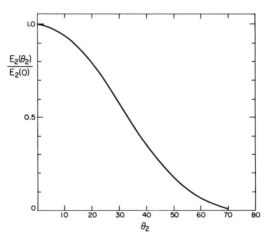

Fig. 3.12   The $\cos^4$ law for the irradiance in the image of an extended uniform source.

If the source radiance is nonuniform or the exit aperture is so large that θ_2 varies significantly over the pupil, then (3.37) can be used as a basis for numerical calculation of the image irradiance.

3.2.4.7 GEOMETRICAL EXTENT OF AN IMAGING SYSTEM

The product of area and *projected solid angle* which was defined in Section 3.2.3.3 also plays a central role in describing in an approximate way the radiometry of imaging systems. Under the same conditions of (1) uniform radiance, and (2) independence of the area and solid angle factors, the radiant power in the image is found from the general equation (3.13) and definitions (3.20) and (3.21) to be

$$\Phi = LA\Omega = LG \tag{3.40}$$

We are assuming that the source and image are in the same medium. The area and solid angle can be defined in a number of ways as shown in Fig. 3.13. The possible definitions are:

(1) Source

$$G = A_s\Omega_s; \tag{3.41}$$

(2) Optical system (source and instrument)

$$G = A_e\Omega_e' = A_x\Omega_x' = A_s\Omega_s' = A_i\Omega_i'; \tag{3.42}$$

(3) Instrument

$$G = A_e\Omega_e = A_x\Omega_x; \tag{3.43}$$

(4) Detector

$$G = A_D\Omega_D, \tag{3.44}$$

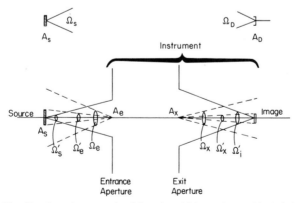

Fig. 3.13   Identification of areas and solid angles which can be used in defining the geometrical extent of a source, optical system, instrument, or detector.

where A_s, A_e, A_x, A_i, and A_D are the areas of the source, entrance aperture, exit aperture, source image, and detector. The angles Ω_s and Ω_D are, respectively, the solid angle over which the source radiates energy and the solid angle over which the detector can receive energy, while the solid angles Ω_s', Ω_e', Ω_x', and Ω_i' are the actual solid angles subtended as shown in Fig. 3.13 in a specific combination of source and instrument.

When maximum utilization of available radiant energy is a consideration, the instrument should be designed so that $A_e\Omega_e$ is at least as great as $A_s\Omega_s$. Since this is usually not possible, the geometrical extent of an instrument is a measure of its radiant energy collecting capability.

When transmission losses due to reflection, absorption, or scattering are taken into account, the radiant power (3.40) becomes

$$\Phi = LG\tau \quad (\text{W}), \tag{3.45}$$

where τ is the radiant transmittance defined in Chapter 2.

In the geometrical optics of a perfect imaging system it is found (Born and Wolf, 1964) that the quantity $ny\tan\alpha$ is invariant where, for example, v is the source height as shown in Fig. 3.9 and α the half-angle subtended at the source by the entrance aperture. For systems free from coma the corresponding quantity is $ny\sin\alpha$. For small angles (paraxial approximation) these become

$$g = ny\alpha \tag{3.46}$$

which for circular source and aperture is related to the *optical extent* defined in (3.22) by

$$G_n = \pi^2 g^2. \tag{3.47}$$

3.2.4.8 THE IMAGE OF A POINT SOURCE

When an optical system forms an image of a point source the size of the image is principally influenced by

(1) diffraction,
(2) aberrations of the optical instrument, and
(3) atmospheric turbulence.

Diffraction limitations are fundamental and always present. The required analysis involves the methods of wave optics as presented in Section 3.3. Aberrations of the instrument can frequently be reduced by proper design and construction to the level at which diffraction limits the image size. In large aperture systems covering a wide field of view this is not possible and in this case aberrations limit the image size. The aberrations of optical instruments are discussed in a number of excellent textbooks (Born and Wolf, 1964; Palmer, 1971). The importance of the third factor, atmospheric turbulence, depends on (a) the strength of turbulence, (b) total optical path length through the turbulent medium, (c) wavelength, and (d) the size of the instrument entrance aperture. In most laboratory applications of radiometry the strength of the turbulence and the path length are small enough so that atmospheric effects on image size are negligible.

Independent of the mechanism responsible for determining the image size, it is possible to obtain a simple approximate relationship between the energy radiated by a point source and the irradiance in its image. Strictly speaking, the arguments used in this section are not all based on the concepts of rays and radiance conservation as defined and used in previous sections of this chapter. They are more closely described as based on dimensional considerations and heuristic arguments. More accurate analysis, however, substantiates their approximate validity.

Consider the general imaging arrangement shown in Fig. 3.14. A point source is imaged by an optical instrument, possibly through atmospheric turbulence. The instrument entrance and exit aperture diameters are D_e and D_x. The distances from source to entrance aperture and from exit aperture to image are r and l. When diffraction constitutes the fundamental limitation to image size, the angle subtended by the image at the exit aperture is approximately

$$\theta_x = \lambda/D_x \qquad (3.48)$$

as shown in Section 3.3.2. We can also define a diffraction angle $\theta_e = \lambda/D_e$ which is the corresponding diffraction angle on the source side of the instrument. In using the term *point source* what we then mean is that the geometrical angle subtended by the source at the entrance aperture of the instrument is much less than θ_e.

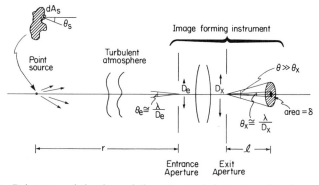

Fig. 3.14 Point source being imaged through a turbulent atmosphere by an optical instrument.

When aberrations or atmospheric turbulence limit the image size, the angle θ subtended by the image at the exit aperture will be found to be much greater than θ_x. In any case, it is convenient, for simplicity, to assume that in all three situations the energy in the image of the point source is uniformly distributed over an area δ. More detailed calculations of the spatial energy distribution in the image would of course show that the energy is nonuniformly distributed. The quantity δ which we use in the following could then be defined as the effective area of the image.

A point source is characterized by its radiant intensity

$$I_s = d\Phi_s/d\omega = \int_{\text{source area}} L_s \cos \theta_s \, dA_s \quad (\text{W sr}^{-1}), \qquad (3.49)$$

where L_s is the source radiance, θ_s the angle between the normal to each source area element and the direction of $d\omega$, and dA_s a source area element. The total radiant power at the instrument entrance aperture is then

$$\Phi_e = \tau_A \int_{\text{ent. aper.}} I_s/r^2 \, dA_e, \qquad (3.50)$$

where $dA_e/r^2 = d\omega$, τ_A the transmittance of the atmosphere, dA_e the area element in the entrance aperture, and it is assumed that the source is on the optical axis. For constant radiant intensity and distance r over the entrance aperture (3.50) becomes

$$\Phi_e = (I_s/r^2)\tau_A A_e \quad (\text{W}). \qquad (3.51)$$

With the assumption that all energy collected by the entrance aperture and transmitted from the entrance aperture to the image is uniformly distributed over the image area δ, the image irradiance is

$$E_i = \Phi_e \tau_0/\delta \quad (\text{W m}^{-2}), \qquad (3.52)$$

where τ_0 is the transmittance of the instrument. The simple approximate relationship (3.52) shall now be examined for each of the three factors: diffraction, aberrations, and atmospheric turbulence. Equation (3.52) can also be written in the simple form

$$E_i = \tau_0(E_e A_e/\delta) \quad (\text{W m}^{-2}) \qquad (3.53)$$

where $E_e = \tau_A I_s/r^2$ is the total irradiance at the entrance aperture and $A_e = \pi D_e^2/4$ the entrance aperture area.

(1) *Diffraction.* When diffraction determines image size, the quantities δ and D_x are related. The approximate relation is found from (3.48) to be

$$\delta = (\pi/4)(l\theta_x)^2 = (\pi/4)l^2\lambda^2/D_x^2. \qquad (3.54)$$

Using (3.54) and (3.51) in (3.52) with $A_e = (\pi/4)D_e^2$ leads to the general expression for image irradiance given by

$$E_i = \tau_0\tau_A I_s D_e^2 D_x^2/r^2 l^2 \lambda^2. \tag{3.55}$$

In most applications the point source is at a large distance from the entrance pupil so that $l/D_x = f/no$, the f-number. Using this we find that for a distant point source (3.55) is

$$E_i = \tau_0\tau_A I_s D_e^2/r^2\lambda^2(f/no)^2 \quad (\text{W m}^{-2}). \tag{3.56}$$

When diffraction determines image size Eq. (3.56) shows the significant irradiance advantage in using a large entrance aperture and small f/no. However, a complete radiometric evaluation would include detector and background energy considerations. Background effects can be analyzed with the aid of Eq. (3.39). The relation between detector size and performance is discussed in Chapter 6.

(2) *Aberrations of the Optical Instrument.* When instrument design or construction produce an instrument which exhibits aberrations, the relationship between image size and instrument parameters can be complex. In special situations where one aberration such as third-order spherical aberration, coma, astigmatism, field curvature, or distortion predominates, it may be possible to estimate (Smith, 1966) the image area δ for use in Eq. (3.52) or (3.53). Also, it may be possible to estimate δ from numerical geometrical ray tracing and spot diagrams (Smith, 1966). In general this approach will show that the power is not uniformly distributed over an area δ. Regarding the area δ_G as the effective area of the image as determined by geometrical aberrations we can write (3.53) as

$$E_i = I_s\tau_A\tau_0(\pi D_e^2/4r^2\delta_G) \quad (\text{W m}^{-2}). \tag{3.57}$$

Although (3.57) is generally valid, its use in design calculations must be done with the realization that optical design and construction limitations will usually result in an increase in δ_G as D_e is increased.

(3) *Atmospheric Turbulence.* Atmospheric turbulence can limit the minimum size of the image which is formed with an optical instrument of large aperture. For a distant point source $\delta = (\pi/4)l^2\theta^2 = (\pi/4)f^2\theta_T^2$ where f is the focal length and θ_T is the turbulence blur angle. Then

$$E_i = \tau_0\tau_A I_s D_e^2/r^2 f^2\theta_T^2 \quad (\text{W m}^{-2}), \tag{3.58}$$

where θ_T must be determined from theory or measurement.

When θ_T is fixed, (3.58) shows that the image irradiance is directly proportional to $\pi D_e^2/4$, the area of the entrance aperture. Thus there is a significant radiometric advantage in using an instrument of large aperture even when the image size and corresponding spatial resolution are limited by atmospheric turbulence.

3.3 SPATIAL DISTRIBUTION OF RADIANT ENERGY: WAVE OPTICS

3.3.1 Diffraction

The methods of geometrical optics presented in Section 3.2 are based on the concept of a ray. This concept arises from the eikonal equation (3.4) which depends on the assumption that $\lambda \to 0$, or at least that the wavelength is negligibly small in comparison with the distances over which changes occur in the spatial energy distribution. This assumption is found to break down (a) in the region of a well-defined focus, and (b) near the edge of a geometrical shadow. Geometrical optics is also found to be inadequate (c) at distances greater than or approximately equal to D^2/λ, where D is the transverse dimension of a beam-defining aperture.

The spatial distribution of radiant energy is found to be accurately described in situations (a)–(c) as well as in other situations by wave optics (Born and Wolf, 1964). Wave optics takes into account the finite size of the optical wavelength λ. Wave optics provides a comprehensive description of the spatial distribution of radiant energy and includes geometrical optics as a special case in the limit of $\lambda \to 0$.

When a well-defined beam of radiant energy passes through an aperture, the spatial distribution of the radiant energy which has passed through the aperture is found to deviate from the straight line paths indicated by geometrical optics. This is due to *diffraction*. Diffraction leads to a redistribution of the radiant energy both within and outside the geometrical boundary of the beam. Because diffraction theory is based on wave concepts it describes the propagation or transfer of radiant energy in terms of a scalar complex wave amplitude $U(x, y, z)$, where (x, y, z) are spatial coordinates. This wave amplitude is usually taken to represent one cartesian component of the electric field vector. For a plane wave the amplitude $U(x, y, z)$ is then related to the energy density at (x, y, z) in a dielectric medium by

$$w(x, y, z) = (\varepsilon/2)\left|U(x, y, z)\right|^2 \quad (\text{W m}^{-3}), \qquad (3.59)$$

where ε is the permittivity. Since the irradiance at x, y, z is related to the energy density by $E(x, y, z) = (c/n)w(x, y, z)$, the irradiance is

$$E(x, y, z) = (\varepsilon c/2n)\left|U(x, y, z)\right|^2 \quad (\text{W m}^{-2}), \qquad (3.60)$$

where n is the index of refraction and c the velocity of light in vacuum. In (3.59) and (3.60) we assume that the energy density and irradiance represent

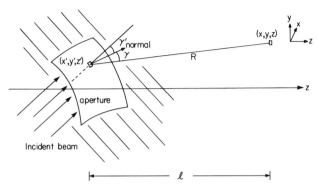

Fig. 3.15 Geometry used in analyzing diffraction effects.

averages over times which are much longer than the optical period $1/v = \lambda/c$. At optical frequencies $v = 10^{12} - 10^{17}$ Hz, detectors are not available which can follow the instantaneous values of the electromagnetic field.

Figure 3.15 defines the geometry we shall use in analyzing diffraction effects. A beam of radiant energy is incident on the aperture from the left. The radiant energy which passes through the aperture is to be observed at the point (x, y, z). From scalar diffraction theory (Born and Wolf, 1964) the complex wave amplitude $U(x, y, z)$ at the observation point (x, y, z) is given by the *Fresnel–Kirchhoff formula*

$$U(x, y, z) = (-i/\lambda) \iint_{\text{aper.}} U(x', y', z')\tfrac{1}{2}(\cos \gamma + \cos \gamma')(e^{ikR}/R)\,dx'\,dy', \quad (3.61)$$

where (x', y', z') are the spatial coordinates of points in the aperture, $U(x', y', z')$ in the aperture is taken to be the complex wave amplitude which would be produced at (x', y', z') by the incident beam in the absence of the aperture, and outside the aperture is taken to be zero, γ the angle between the aperture surface normal at (x', y', z') and the direction to the observation point, γ' the angle between the aperture surface normal at (x', y', z') and the direction of the incident beam, R the distance from (x', y', z') to (x, y, z) and $k = 2\pi/\lambda$.

In Eq. (3.61) the angle γ is always well defined. However, the angle γ' is not well defined unless the beam incident at each aperture point (x', y', z') comes from only one direction. The question of coherence between the portions of the incident beam coming from different directions is examined in Section 3.3.4. All of the radiometric formulas previously derived in Section 3.2 on the basis of ray optics assumed incoherence.

Equations (3.61) and (3.60) can be used directly for numerical calculation of the spatial distribution of radiant energy. However, there are several approximations which can be made to simplify the form of (3.61) as well as to afford considerable insight into the effects of diffraction.

3.3.2   Fraunhofer Diffraction

Fraunhofer diffraction occurs in two situations. These are:

(a)   in a nonfocusing geometry when the observation point (x, y, z) is at a distance from the diffracting aperture which is somewhat greater than D^2/λ where D is the transverse dimension of the diffracting aperture, and

(b)   In a focusing geometry when the observation point is in the focal plane.

The special form of (3.61) for Fraunhofer diffraction is found as follows. When the angles γ and γ' are small,

$$(\cos \gamma + \cos \gamma')/2 \cong 1. \tag{3.62}$$

When γ is small, it is also found that the distance R in (3.61) can be expanded in the coordinates (x, y, z) and (x', y', z') as

$$R = l + \frac{x^2 + y^2}{2l} + \frac{x'^2 + y'^2}{2l} - \frac{xx' + yy'}{l} + \cdots, \tag{3.63}$$

where $l = z - z'$ and the higher-order terms can be ignored. In situation (a) when $l \gg D^2/\lambda$, the third term of (3.63) is negligible when used in e^{ikR} of (3.61). The condition

$$l \gg D^2/\lambda \tag{3.64}$$

is the *far-field condition* and is critical in the definition of Fraunhofer diffraction for situation (a). In situation (b) the same third term of (3.63) is negligible but for a different reason (Born and Wolf, 1964) which is associated with the choice of a spherical reference surface in the diffracting aperture with its center of curvature at the focal point.

In either situation (a) or (b), Eq. (3.62) and modified (3.63) can be used to write (3.61) as

$$U(x, y, z) = (C/\lambda l) \iint\limits_{\text{aper.}} U(x', y', z') \exp[(-ik/l)(xx' + yy')]\, dx'\, dy', \tag{3.65}$$

where $C = -ie^{ikl} \exp[ik(x^2 + y^2)/2l]$ is a phase factor which is unimportant when (3.65) is used in (3.60) to calculate irradiance.

Equation (3.65) describes Fraunhofer diffraction. The form of the relation between the incident and diffracted fields in (3.65) is significant. A change in either the distance l, the aperture dimensions (x', y'), or the wavelength λ results in only a linear scaling of the dimensions of $U(x, y, z)$ along the x and y axis with the functional form remaining unchanged. Also the integral in (3.65) is in the form of a two-dimensional Fourier transform of $U(x', y', z')$. This means that previously calculated results and general relationships from Fourier analysis can be applied to the solution of Fraunhofer diffraction problems (Bracewell, 1965; O'Neill, 1963). Close examination of the definition of a Fourier transform shows that (3.65) is not exactly in the standard form. When a change of variables is made to put (3.65) into the standard form, it is found that the two variables which are reciprocally related by the Fourier transform are $(x'/\lambda, y'/\lambda)$ and $(x/l, y/l)$. The first pair represents the dimensions in the diffracting aperture in units of wavelength and the second represents the angles of the observation point at the diffracting aperture. This identification of the reciprocally related variables in (3.65) is found to be generally useful. The application of (3.65) presented later in this section illustrates this reciprocal relationship.

The complex wave amplitude in the diffracting aperture can be written as

$$U(x', y', z') = U_m(x', y', z') \exp[ik W(x', y', z')], \tag{3.66}$$

where U_m is the magnitude and kW the phase. In the focusing geometry of an imaging system, $W(x', y', z')$ is the *aberration function* and is specified on the reference sphere in the exit aperture. When $W = 0$ for all points in the exit aperture the system is *diffraction limited*.

3.3.2.1   FRAUNHOFER DIFFRACTION AT A CIRCULAR APERTURE

We now apply (3.65) to determine the irradiance in the Fraunhofer diffraction pattern of a circular aperture. Figure 3.16 shows the geometry. With proper interpretation the results are applicable to either focusing or nonfocusing geometries. With uniform wave amplitude U_m and phase $W = 0$ in (3.66) we have $U(x', y', z') = U_m$. Using this in (3.65) and changing to cylindrical coordinates,

$$U(\rho, \theta, z) = \frac{CU_m}{\lambda l} \int_{\rho'=0}^{D/2} \int_{\theta'=0}^{2\pi} \exp\left[\frac{-ik}{l} \rho\rho'(\cos\theta\cos\theta' + \sin\theta\sin\theta')\right]\rho'\, d\rho'\, d\theta'. \tag{3.67}$$

Using a trigonometric identity and a change of angle variables in (3.67) leads to

$$U(\rho, \theta, z) = (CU_m/\lambda l)(\pi D^2/4)[2J_1(\pi\rho D/\lambda l)/(\pi\rho D/\lambda l)], \tag{3.68}$$

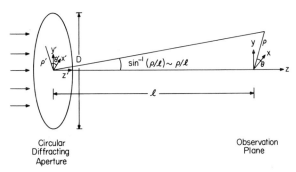

Fig. 3.16   Circular aperture geometry for diffraction.

where J_1 is the first-order Bessel function. The irradiance associated with (3.68) is found from (3.60) to be

$$E(\rho, z) = (\varepsilon c/2n)(U_m{}^2 A^2/\lambda^2 l^2)|2J_1(\pi\rho D/\lambda l)/(\pi\rho D/\lambda l)|^2 \quad (\text{W m}^{-2}), \quad (3.69)$$

where $A = \pi D^2/4$. The last factors in (3.68) and (3.69) are shown in Fig. 3.17 and in Table 3.1. The first zero occurs at $\pi\rho_0 D/\lambda l = 3.83$ or

$$\rho_0 = 1.22(\lambda l/D). \quad (3.70)$$

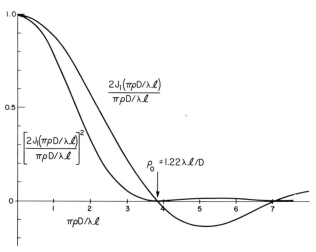

Fig. 3.17   Fraunhofer diffraction pattern for a circular aperture.

TABLE 3.1   Some Values of $\dfrac{2J_1(\pi\rho D/\lambda l)}{\pi\rho D/\lambda l}$ and $\left[\dfrac{2J_1(\pi\rho D/\lambda l)}{\pi\rho D/\lambda l}\right]^2$

$\pi\rho D/\lambda l$	$\dfrac{2J_1(\pi\rho D/\lambda l)}{\pi\rho D/\lambda l}$	$\left\|\dfrac{2J_1(\pi\rho D/\lambda l)}{\pi\rho D/\lambda l}\right\|^2$
0	1 (max)	1 (max)
0.50	0.969	0.939
1.00	0.880	0.775
1.50	0.744	0.553
2.00	0.577	0.333
2.50	0.398	0.158
3.00	0.226	0.051
3.50	0.079	0.006
3.83	0	0
5.14	−0.132 (min)	0.018 (max)
7.02	0	0
8.42	0.065 (max)	0.004 (max)
10.17	0	0

In an imaging system where D is the diameter of the exit aperture and the source is far enough from the entrance aperture so that l is the focal length, $\rho_0 = 1.22\lambda(f/no)$. The value of ρ_0 in (3.70) defines the *Airy radius*. In angular coordinates the corresponding angle is $\rho_0/l = 1.22\lambda/D$.

Using (3.60), (3.69) can be written as

$$E(\rho, z) = (E_a A^2/\lambda^2 l^2)|2J_1(\pi\rho D/\lambda l)/(\pi\rho D/\lambda l)|^2 \quad (\text{W m}^{-2}), \qquad (3.71)$$

where $E_a = (\varepsilon c/2n)U_m{}^2$ is the uniform irradiance in the aperture. For the imaging situation the diffracting aperture is the exit aperture and

$$E_a = E_x = E_e \tau_0/m^2, \qquad (3.72)$$

where E_x and E_e are the irradiances in the exit and entrance apertures, m the lateral optical magnification of the imaging system, and τ_0 the transmittance from entrance to exit aperture.

When a small circular detector of diameter D_d is located in the image plane to collect the radiant power in the image of the point source, the radiant power at the detector is found (Born and Wolf, 1964) from (3.71) and (3.72) to be

$$\Phi_d = \int_{\theta=0}^{2\pi} \int_{\rho=0}^{D_d/2} E(\rho, z)\rho\, d\rho\, d\theta \quad (\text{W}) \qquad (3.73)$$

$$\Phi_d = E_x A_x [1 - J_0{}^2(\pi D_d D_x/2\lambda l) - J_1{}^2(\pi D_d D_x/2\lambda l)], \qquad (3.74)$$

where D_x is the diameter of the exit aperture and J_0 and J_1 are the zero and first-order Bessel functions. The collected fraction of the available power is $\Phi_d/E_x A_x$. Figure 3.18 and Table 3.2 show this fraction as a function of the

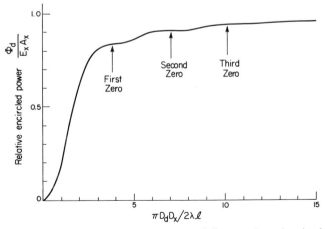

Fig. 3.18   Encircled power for a circular detector of diameter D_d and a circular aperture diameter D_x.

TABLE 3.2   Encircled Power for Circular Exit
Aperture of Diameter D_x and Circular
Detector of Diameter D_d

$\pi D_d D_x/2\lambda l$	$1 - J_0^2(\pi D_d D_x/2\lambda l) - J_1^2(\pi D_d D_x/2\lambda l)$
0	0
1.0	0.221
2.0	0.617
3.0	0.817
4.0	0.838
5.0	0.861
6.0	0.901
7.0	0.910
8.0	0.915
9.0	0.932
10.0	0.938
11.0	0.939
12.0	0.948
13.0	0.952
14.0	0.953

dimensionless parameter $\pi D_d D_x/2\lambda l$. Figure 3.18 is the *encircled power* diagram. The positions of the zeros of (3.68) and (3.71) are indicated in the figure. Notice that the square of the dimensionless parameter which determines the collected power Φ_d in (3.74) is in the form

$$(\pi^2 D_d^2/4)(D_x^2/l^2)(1/\lambda^2) = (\text{area} \times \text{solid angle})/\lambda^2. \tag{3.75}$$

3.3.2.2 FRAUNHOFER DIFFRACTION AT A RECTANGULAR APERTURE

As a second application we now determine the irradiance in the Fraunhofer diffraction pattern of a rectangular aperture. The result is used to define the spectral resolution of a monochromator in Chapter 7. Consider the geometry shown in Fig. 3.19 which defines the parameters involved. From

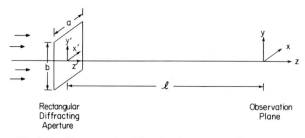

Fig. 3.19   Geometry for diffraction by a rectangular aperture.

(3.65) the complex wave amplitude at the observation point (x, y, z) is

$$U(x, y, z) = \frac{CU_m}{\lambda l} \int_{-a/2}^{a/2} \exp\left(\frac{-ikxx'}{l}\right) dx' \int_{-b/2}^{b/2} \exp\left(\frac{-ikyy'}{l}\right) dy', \quad (3.76)$$

where U_m is the uniform amplitude over the aperture. On performing the integrals, (3.76) becomes

$$U(x, y, z) = \frac{CU_m A}{\lambda l} \frac{\sin(\pi ax/\lambda l)}{\pi ax/\lambda l} \frac{\sin(\pi by/\lambda l)}{\pi by/\lambda l}, \quad (3.77)$$

where $A = ab$ is the area of the aperture. The irradiance is found from (3.77) and (3.60) to be

$$E(x, y, z) = \frac{E_a A^2}{\lambda^2 l^2} \left|\frac{\sin(\pi ax/\lambda l)}{\pi ax/\lambda l}\right|^2 \left|\frac{\sin(\pi by/\lambda l)}{\pi by/\lambda l}\right|^2 \quad (\text{W m}^{-2}), \quad (3.78)$$

where $E_a = (\varepsilon c/2n)U_m^2$ is the irradiance in the aperture. Figure 3.20 and Table 3.3 show the spatial variation in the x direction of (3.77) and (3.78). The first zero occurs at $\pi ax_0/\lambda l = \pi$ or

$$x_0 = \lambda l/a. \quad (3.79)$$

In the imaging geometry of the output of a monochromator the distance x_0 determines the spectral resolution capability of the monochromator. This is discussed in Chapter 7.

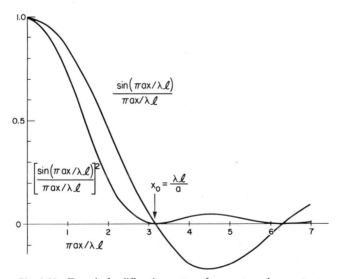

Fig. 3.20   Fraunhofer diffraction pattern for a rectangular aperture.

TABLE 3.3 Values of $\dfrac{\sin(\pi ax/\lambda l)}{\pi ax/\lambda l}$ and $\left|\dfrac{\sin(\pi ax/\lambda l)}{\pi ax/\lambda l}\right|^2$

$\pi ax/\lambda l$	$\dfrac{\sin(\pi ax/\lambda l)}{\pi ax/\lambda l}$	$\left[\dfrac{\sin(\pi ax/\lambda l)}{\pi ax/\lambda l}\right]^2$
0	1	1
0.5	0.959	0.919
1.0	0.841	0.708
1.5	0.665	0.442
2.0	0.455	0.207
2.5	0.239	0.057
3.0	0.047	0.002
3.5	-0.100	0.010
4.0	-0.189	0.036
4.5	-0.217	0.047
5.0	-0.192	0.037
5.5	-0.128	0.017
6.0	-0.047	0.002
6.5	0.033	0.001
7.0	0.093	0.008

3.3.2.3 APERTURE GAIN

The peak irradiance in either the rectangular or the circular aperture diffraction pattern occurs at the origin $x = 0$, $y = 0$, or at $\rho = 0$. This is shown by (3.71) and (3.78) and the corresponding figures. In both cases the peak irradiance is

$$E_p = E_a A^2/\lambda^2 l^2 \quad (\text{W m}^{-2}) \tag{3.80}$$

$$E_p = \Phi_a(4\pi A/\lambda^2)(1/4\pi l^2), \tag{3.81}$$

where $\Phi_a = E_a A$ is the total power in the aperture. The quantity $4\pi A/\lambda^2 = G_a$ is the *gain* of the aperture. It is a measure of the ability of the aperture to concentrate energy in a given direction. In comparing the form of (3.81) with the form of (3.29) it can be seen that the peak irradiance E_p is the same as would be produced by a point source of radiant intensity

$$I = \Phi_a G_a/4\pi \tag{3.82}$$

located at a distance l from the observation point. The definition of radiant intensity as power/solid angle then shows the interpretation of G_a. If $G_a = 1$, the radiant intensity (3.82) would correspond to a point source radiating isotropically into 4π sr. The gain is then the ratio of the effective radiant intensity I to the radiant intensity which would be produced by an isotropically radiating point source. Also from (3.81) and (3.82)

$$G_a = \frac{\text{peak irradiance from aperture}}{\text{irradiance from isotropic point source}} \tag{3.83}$$

when both the aperture and point source have the same total power. The peak irradiance equation (3.81) is then

$$E_p = \Phi_a G_a / 4\pi l^2. \tag{3.84}$$

Gain parameters can be used in (3.84) to find peak irradiance values and the peak irradiance can then be used in (3.71) or (3.78) to find the spatial distribution of radiant energy. The gain is a dimensionless quantity.

3.3.3 Fresnel Diffraction

The second type of diffraction effects which we consider are classified as *Fresnel diffraction*. These include the diffraction effects which do not take place in the far field ($l \gg D^2/\lambda$) or in the focal plane. Actually, Fresnel diffraction theory is more general than Fraunhofer diffraction theory and also includes the Fraunhofer case when we apply it to the far field or focal plane.

Because it is more general, Fresnel diffraction is more complex. It is found that the form of a Fresnel diffraction pattern does not simply scale linearly with changes in wavelength, aperture dimensions, or observation distances as the Fraunhofer pattern does. Also no simple Fourier transform relation exists between the wave amplitude in the aperture and the wave amplitude in the diffraction pattern. For these reasons most Fresnel diffraction problems are solved by using numerical or graphical methods. To facilitate the use of numerical tables or graphs we now put the Fresnel diffraction formula in a standardized form.

The basic diffraction equation is (3.61). At small angles we can make assumption (3.62) and use the expansion

$$R = l + \frac{(x'-x)^2}{2l} + \frac{(y'-y)^2}{2l} + \cdots, \tag{3.85}$$

where $l = z - z'$. When the amplitude in the aperture is uniform, (3.61) then is

$$U(x, y, z) = (-iU_m/\lambda l) \exp(ikl) \iint_{\text{aper.}} \exp\{(ik/2l)[(x'-x)^2 + (y'-y)^2]\} \, dx' \, dy',$$
$$\tag{3.86}$$

where U_m is the uniform amplitude in the aperture. Equation (3.86) is generally applicable to all small-angle Fresnel diffraction situations.

3.3.3.1 FRESNEL DIFFRACTION AT A RECTANGULAR APERTURE

The two aperture configurations most commonly encountered are the rectangular and circular. Consider the rectangular first. In a rectangular geometry it is convenient to make a change of variables and express (3.86) as

$$U(x, y, z) = (-iU_m/2) \exp(ikl) \int_{\alpha_1}^{\alpha_2} \exp(\tfrac{1}{2} i\pi\alpha^2) \, d\alpha \int_{\beta_1}^{\beta_2} \exp(\tfrac{1}{2} i\pi\beta^2) \, d\beta, \tag{3.87}$$

where the dimensionless variables are

$$\alpha = (2/\lambda l)^{1/2}(x' - x), \qquad \beta = (2/\lambda l)^{1/2}(y' - y),$$
$$\alpha_1 = (2/\lambda l)^{1/2}(x + \tfrac{1}{2}a), \qquad \beta_1 = (2/\lambda l)^{1/2}(y + \tfrac{1}{2}b), \qquad (3.88)$$
$$\alpha_2 = (2/\lambda l)^{1/2}(x - \tfrac{1}{2}a), \qquad \beta_2 = (2/\lambda l)^{1/2}(y - \tfrac{1}{2}b),$$

and a and b are the dimensions of the aperture along the x and y axes. The integrals

$$C(\alpha_0) = \int_0^{\alpha_0} \cos \tfrac{1}{2}\pi\alpha^2 \, d\alpha, \qquad S(\alpha_0) = \int_0^{\alpha_0} \sin \tfrac{1}{2}\pi\alpha^2 \, d\alpha \qquad (3.89)$$

are the *Fresnel integrals* and are available in tabular form as in Table 3.4.

TABLE 3.4   Fresnel Integrals $C(\alpha)$ and $S(\alpha)$

α	$C(\alpha)$	$S(\alpha)$	α	$C(\alpha)$	$S(\alpha)$	α	$C(\alpha)$	$S(\alpha)$
0.00	0.0000	0.0000	3.00	0.6058	0.4963	5.55	0.4456	0.5181
0.10	0.1000	0.0005	3.10	0.5616	0.5818	5.60	0.4517	0.4700
0.20	0.1999	0.0042	3.20	0.4664	0.5933	5.65	0.4926	0.4441
0.30	0.2994	0.0141	3.30	0.4058	0.5192	5.70	0.5385	0.4595
0.40	0.3975	0.0334	3.40	0.4385	0.4296	5.75	0.5551	0.5049
0.50	0.4923	0.0647	3.50	0.5326	0.4152	5.80	0.5298	0.5461
0.60	0.5811	0.1105	3.60	0.5880	0.4923	5.85	0.4819	0.5531
0.70	0.6597	0.1721	3.70	0.5420	0.5750	5.90	0.4486	0.5163
0.80	0.7230	0.2493	3.80	0.4481	0.5656	5.95	0.4566	0.4688
0.90	0.7648	0.3398	3.90	0.4223	0.4752	6.00	0.4995	0.4470
1.00	0.7799	0.4383	4.00	0.4984	0.4204	6.05	0.5424	0.4689
1.10	0.7638	0.5365	4.10	0.5738	0.4758	6.10	0.5495	0.5165
1.20	0.7154	0.6234	4.20	0.5418	0.5633	6.15	0.5146	0.5496
1.30	0.6386	0.6863	4.30	0.4494	0.5540	6.20	0.4676	0.5398
1.40	0.5431	0.7135	4.40	0.4384	0.4622	6.25	0.4493	0.4954
1.50	0.4453	0.6975	4.50	0.5261	0.4342	6.30	0.4760	0.4555
1.60	0.3655	0.6389	4.60	0.5673	0.5162	6.35	0.5240	0.4560
1.70	0.3283	0.5492	4.70	0.4914	0.5672	6.40	0.5496	0.4965
1.80	0.3336	0.4508	4.80	0.4338	0.4968	6.45	0.5292	0.5398
1.90	0.3944	0.3734	4.90	0.5002	0.4350	6.50	0.4816	0.5454
2.00	0.4882	0.3434	5.00	0.5673	0.4992	6.55	0.4520	0.5078
2.10	0.5815	0.3734	5.05	0.5450	0.5442	6.60	0.4690	0.4631
2.20	0.6363	0.4557	5.10	0.4998	0.5624	6.65	0.5161	0.4549
2.30	0.6266	0.5531	5.15	0.4553	0.5427	6.70	0.5467	0.4915
2.40	0.5550	0.6197	5.20	0.4389	0.4969	6.75	0.5302	0.5362
2.50	0.4574	0.6192	5.25	0.4610	0.4536	6.80	0.4831	0.5436
2.60	0.3890	0.5500	5.30	0.5078	0.4405	6.85	0.4539	0.5060
2.70	0.3925	0.4529	5.35	0.5490	0.4662	6.90	0.4732	0.4624
2.80	0.4675	0.3915	5.40	0.5537	0.5140	6.95	0.5207	0.4591
2.90	0.5624	0.4101	5.45	0.5269	0.5519			
			5.50	0.4784	0.5537			

In terms of (3.89) the form of (3.87) is

$$U(x, y, z) = (-iU_{m}/2)\exp(ikl)\{[C(\alpha_2) - C(\alpha_1)] + i[S(\alpha_2) - S(\alpha_1)]\}$$
$$\times \{[C(\beta_2) - C(\beta_1)] + i[S(\beta_2) - S(\beta_1)]\}. \tag{3.90}$$

Notice that along the x axis the complex wave amplitude is determined by the difference of two complex numbers $C(\alpha_2) + iS(\alpha_2)$ and $C(\alpha_1) + iS(\alpha_1)$, and similarly along the y axis. This suggests that the evaluation of (3.87) or (3.90) can also be done by representing the points $C + iS$ on an Argand diagram in which C and S are the real and imaginary axes. This diagram is shown in Figure 3.21. The appropriate parameter $\alpha_1, \alpha_2, \beta_1,$ or β_2 as defined in (3.88) is the parameter along the spiral. Figure 3.21 is the *Cornu spiral*. The difference of the two complex numbers is then

$$[C(\alpha_2) + iS(\alpha_2)] - [C(\alpha_1) + iS(\alpha_1)] = \Delta_x e^{i\gamma_x} \tag{3.91}$$

and

$$[C(\beta_2) + iS(\beta_2)] - [C(\beta_1) + iS(\beta_1)] = \Delta_y e^{i\gamma_y} \tag{3.92}$$

where Δ_x and Δ_y are the magnitudes of the vectors joining α_1 and α_2 and β_1 and β_2, respectively.

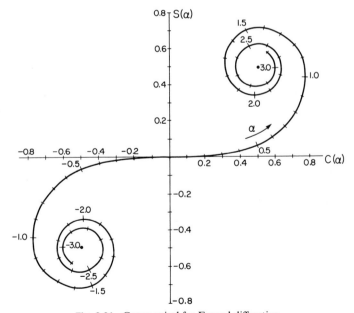

Fig. 3.21 Cornu spiral for Fresnel diffraction.

All of this analysis, which confirms our original statement that Fresnel diffraction effects are more complex than Fraunhofer diffraction effects, then finally enables us to write (3.90) as

$$U(x, y, z) = (-iU_m/2)\exp(ikl)\,\Delta_x\,\Delta_y\exp[i(\gamma_x - \gamma_y)]. \qquad (3.93)$$

Based on (3.60) and (3.93) the irradiance is

$$E(x, y, z) = (\varepsilon c/2n)(U_m^2/4)\,\Delta_x^2\,\Delta_y^2 \qquad (3.94)$$

$$E(x, y, z) = (E_a/4)\,\Delta_x^2\,\Delta_y^2. \qquad (3.95)$$

The irradiance relation (3.95) can be used with the associated Fig. 3.21 or Table 3.4 to determine the spatial distribution of radiant energy due to Fresnel diffraction from a rectangular aperture. Due to the complexity of Fresnel diffraction effects few general rules can be extracted. The effects must usually be determined in each situation with the numerical or graphical techniques presented (Stone, 1963; Born and Wolf, 1964)

3.3.3.2   FRESNEL DIFFRACTION AT A CIRCULAR APERTURE

Figure 3.22 shows the geometry we use to analyze Fresnel diffraction effects for the circular aperture. It is convenient to use polar coordinates in the aperture and at the observation point. It can be shown that on the axis (3.86) reduces to

$$U(0, 0, z) = U(0, z) = -i\pi U_m \int_0^{q_m} \exp(i\pi q)\,dq \qquad (3.96)$$

$$U(0, 0, z) = U_m[1 - \exp(i\pi q_m)], \qquad (3.97)$$

where q is a parameter which represents the number of half wavelengths of optical path between the point (x', y', z') in the aperture and the observation point $(0, 0, z)$. The value q_m is the value of q associated with the edge of the circular aperture. The irradiance is found with the aid of (3.60) to be

$$E(0, z) = 2E_a(1 - \cos\pi q_m) = 4E_a\sin^2(\pi q_m/2) \qquad (3.98)$$

where $E_a = (\varepsilon c/2n)U_m^2$ is the irradiance in the aperture. Equation (3.98) indicates that the irradiance at $(0, z)$ can vary between zero and four times the irradiance of the aperture E_a.

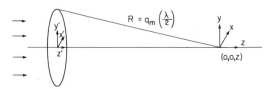

Fig. 3.22   Geometry for Fresnel diffraction by a circular aperture.

Equation (3.98) can be used to evaluate the Fresnel diffraction effects due to a circular aperture by specifying the optical path length q_m. If the irradiance is due to an extended incoherent source, then the total irradiance at the observation point $(0, z)$ is the sum of the irradiances due to each source point. A typical radiometric measurement involving Fresnel diffraction considerations has been evaluated by Blevin (1970) and by Steel *et al.* (1972) and the differences between irradiance levels as determined by geometrical and wave optics have been determined.

3.3.4 Coherence

The radiant energy level at a detector is determined by source elements which are, in general, distributed over a range of positions or directions in space. One of the important consequences of the wave nature of the electromagnetic fields from these source elements is that they can interfere. This interference can greatly increase or decrease the detected radiant energy in comparison with what would be expected from a simple summation of the energy contribution from each source element.

Analysis shows (Born and Wolf, 1964; Beran and Parrent, 1964) that the degree of interference at the detector can be specified in terms of the *degree of coherence* of the fields at the source elements. The coherence properties of the source must be specified in both space and time. The time properties are important since fields which leave two source elements at different times may cover different optical paths and reach the detector simultaneously.

It is generally recognized that the electromagnetic fields produced by a laser have a high degree of coherence in both space and time. All but the simplest radiometric measurements involving this source require consideration of possible interference effects. It is also true of course that incoherent sources give rise to electromagnetic fields which become increasingly spatially coherent as the distance from the source increases. This property is the basis for some of the more advanced techniques of radiometry in astrophysics (Hanbury Brown, 1974), involving both the optical and the microwave portions of the spectrum. Coherence effects can also arise in the laboratory in connection with optical instruments of high spatial resolution or, for example, in a monochromator which uses narrow slits to define the spectral bandwidth. However, aside from these applications, the radiometric measurements which produce the most prominent coherence effects are those which involve a laser.

In addition to the limiting cases of completely coherent or completely incoherent electromagnetic fields there are intermediate cases of partial coherence (Beran and Parrent, 1964). The radiometry of coherent or partially coherent fields is a subject of current research. Although the more quantitative radiometric aspects are still incompletely defined, the essential physical phenomena and some of the quantitative aspects are known.

3.3.4.1 DEGREE OF COHERENCE

The degree of coherence between the electromagnetic fields at two points in space is a measure of their ability to interfere. To show this we consider the experiment shown in Fig. 3.23 which shows a source of electromagnetic radiant energy and two pinholes in an otherwise opaque screen. The purpose of the pinholes is to isolate the fields at the points P_1 and P_2 which arise from the source. By allowing the fields from the two points P_1 and P_2 to propagate to an observation plane and observing the interference effects there we can measure the degree of coherence between the fields at P_1 and P_2. This is shown as follows.

The field at an observation point P_0 is the sum of the fields from the two pinholes. Thus, representing this field as U_0,

$$U_0 = k_1 U_1(t - (r_1/c)) + k_2 U_2(t - (r_2/c)), \tag{3.99}$$

where k_1 and k_2 are complex constants and r_1 and r_2 are the path lengths from P_1 to P_0 and from P_2 to P_0. The time-averaged irradiance at P_0 is then, from Eq. (3.60),

$$E_0 = (\varepsilon c/2n)|U_0|^2 = E_1 + E_2 + (\varepsilon c/2n)2R_e k_1 k_2 \langle U_1(t + \tau)U_2(t) \rangle$$
$$= \text{time-averaged irradiance at } P_0, \tag{3.100}$$

where the angle brackets represent a time average, $\tau = (r_1 - r_2)/c$ is the time difference along the two paths, and we have assumed that the time average

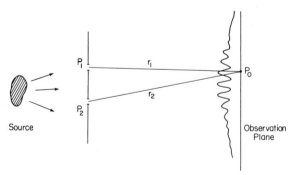

Fig. 3.23 Experiment to measure the degree of coherence between the fields at two points P_1 and P_2.

is a function only of this time difference. The quantity in the brackets is the *mutual coherence function* defined as

$$\Gamma_{12}(\tau) = \langle U_1(t + \tau)U_2^*(t)\rangle \qquad (3.101)$$

and plays a central role in the description of coherence phenomena. It is the cross-correlation function of the fields at the two points P_1 and P_2 for the time difference τ.

We can rewrite Eq. (3.100) as

$$E_0 = E_1 + E_2 + 2(E_1E_2)^{1/2}R_e\gamma_{12}(\tau), \qquad (3.102)$$

where $\gamma_{12}(\tau)$ is the *complex degree of coherence* defined by

$$\gamma_{12}(\tau) = \Gamma_{12}(\tau)/[\Gamma_{11}(0)\Gamma_{22}(0)]^{1/2} \qquad (3.103)$$

which can be seen to have the property $0 \le |\gamma_{12}(\tau)| \le 1$. When the irradiance due to each pinhole is equal, then $E_1 = E_2 = E$ and

$$E_0 = 2E[1 + |\gamma_{12}(\tau)|\cos\phi_{12}(\tau)], \qquad (3.104)$$

where $\phi_{12}(\tau)$ is the phase of $\gamma_{12}(\tau)$.

The interference fringes which are formed at the observation plane can be characterized by their "visibility" which is defined as

$$v = (E_{0\,max} - E_{0\,min})/(E_{0\,max} + E_{0\,min}).$$

Using this definition with (3.104) we find that

$$\text{fringe visibility} = |\gamma_{12}(\tau)| \qquad (3.105)$$

a result which shows the interpretation of the degree of coherence of the wave fields at P_1 and P_2 in terms of the experimentally measurable fringe visibility in the interference pattern.

If the source shown in Fig. 3.23 is a thermal (nonlaser) source centered on the optical axis and having a small spectral bandwidth so that $\Delta\lambda/\lambda$ is small, the following measurements can be made. By observing the fringe visibility near the optical axis where the time difference $\tau \cong 0$ we measure $\gamma_{12}(0)$, the spatial coherence. On the other hand, by making the source very small so that the fields are spatially the same at P_1 and P_2 and measuring the fringes off-axis we essentially measure $\gamma_{11}(\tau)$, the temporal coherence.

Although this experiment shows conceptually the meaning of the degree of coherence $\gamma_{12}(\tau)$, it is not necessarily the best practical method of measuring spatial or temporal coherence. A number of methods have been developed and are described in the literature (Françon and Mallick, 1967).

3.3.4.2   SPATIAL AND TEMPORAL COHERENCE

The spatial and temporal coherence properties of fields are not always easily separated. However, when a beam of radiant energy is *quasi-monochromatic* they are separable as follows. We can define the spectrum $\hat{\gamma}_{12}(v)$ by the Fourier transform relation

$$\gamma_{12}(\tau) = \int_0^\infty \hat{\gamma}_{12}(v)e^{2\pi i v \tau}\, dv. \tag{3.106}$$

When the beam is quasi-monochromatic, then $v - \bar{v} = \Delta v$ is much less than $\bar{v}$, the average frequency. Then (3.106) can be written as

$$\gamma_{12}(\tau) = e^{2\pi i \bar{v}\tau} \int_0^\infty \hat{\gamma}_{12}(v)e^{2\pi i \tau(\Delta v)}\, dv. \tag{3.107}$$

When the time delay τ is much less than the reciprocal of the bandwidth Δv, then the exponential involving $\tau\,\Delta v$ is approximately one and

$$\gamma_{12}(\tau) \cong \gamma_{12}(0)e^{2\pi i \bar{v}\tau}, \tag{3.108}$$

where

$$\tau \ll 1/\Delta v. \tag{3.109}$$

Temporal coherence effects can be associated with path length differences along the direction of the beam of radiant energy. The length defined by

$$l_c = c/\Delta v = \text{coherence length} \tag{3.110}$$

can be regarded as the *coherence length* of the beam of radiant energy. It approximately represents the maximum separation along the path of the beam for which interference effects can be observed. It can be shown by applying diffraction theory that the maximum transverse separation of two pinholes across the beam for which interference effects can be observed at some distance from an incoherent source is approximately

$$D_c = \lambda l/D_s = \text{coherence diameter}, \tag{3.111}$$

where l is the distance to the source, D_s the source diameter, and the source is itself incoherent. The diameter D_c in (3.111) is the separation of points P_1 and P_2 for which the visibility of the interference fringes represented by $\gamma_{12}(0)$ in (3.108) has been reduced to approximately zero.

On the basis of definitions (3.110) and (3.111) we can define a coherence volume of the same dimensions. This is shown in Fig. 3.24. The fields at any two points within this volume are capable of producing interference fringes if allowed to interfere and so have some finite degree of coherence.

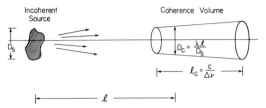

Fig. 3.24   Coherence volume due to incoherent source of diameter D_s and spectral width Δv.

Temporal coherence effects are approximately represented by Eq. (3.110) and more exactly represented by (3.106) which expressed the relation between the degree of coherence as a function of the time difference τ and the spectral distribution $\hat{\gamma}_{12}(v)$. When points 1 and 2 coincide,

$$\gamma_{11}(\tau) = \int_0^\infty \hat{\gamma}_{11}(v)e^{2\pi i v \tau}\,dv \tag{3.112}$$

and $\hat{\gamma}_{11}(v)$ is proportional to the spectral power distribution.

Spatial coherence effects are approximately represented by (3.111) when the source is incoherent. This result, as well as the cases of coherent and partially coherent sources, is examined now in more detail.

3.3.4.3   INCOHERENT SOURCES:
THE VAN CITTERT–ZERNIKE THEOREM

We will now examine the basis for the approximate relation (3.111). Consider the arrangement of incoherent source and geometry shown in Fig. 3.25. The source can be an aperture or the actual radiating area of a material source. The wave amplitude at (x, y, z) is given by the Fresnel–Kirchhoff formula of Eq. (3.61). We assume also that the source is quasi-monochromatic in the sense of (3.109).

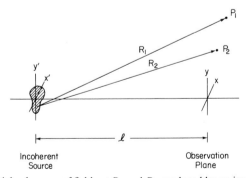

Fig. 3.25   Spatial coherence of fields at P_1 and P_2 produced by an incoherent source.

The assumption of incoherence at the source means that the mutual coherence function $\Gamma_{12}(\tau)$ if calculated at the source is zero except when the two points coincide. When the two points coincide, then (3.101) and (3.60) show that at the source

$$\Gamma_{11}(0) = \langle U_1(t)U_1{}^*(t)\rangle = (2n/\varepsilon c)E_s, \qquad (3.113)$$

where E_s is the field irradiance or, in the case of a source, the radiant exitance M. Using the notation $\Gamma_{12}(0) = \Gamma(P_1, P_2)$ we can then use Eq. (3.61) to calculate the mutual coherence at the observation points P_1, P_2 at some distance from the source. The mutual coherence is $\Gamma_{12}(\tau) = \Gamma(P_1, P_2)e^{2\pi i\bar{\nu}\tau}$, where

$$\Gamma(P_1, P_2) = \left(\frac{1}{\lambda^2}\right)\left(\frac{2n}{\varepsilon c}\right)\iint\limits_{\text{source}} M(x', y')\,\frac{\exp[ik(R_1 - R_2)]}{R_1 R_2}\,dx'\,dy', \quad (3.114)$$

where R_1 is the distance from (x', y') to P_1, R_2 the distance from (x', y') to P_2, and $M(x', y')$ the radiant exitance at the source point (x', y'). Equation (3.105) shows the very interesting and useful result that the mutual coherence at a distance from an incoherent source depends on the *spatial power distribution of the source*. If we specify the location of the observation points to be in a plane at a distance l from the source, then (3.114) is approximately equal to

$$\Gamma(x_1 - x_2, y_1 - y_2)$$

$$= \left(\frac{1}{\lambda^2 l^2}\right)\left(\frac{2n}{\varepsilon c}\right)\iint\limits_{\text{source}} M(x', y')\exp\left\{\frac{ik}{l}[x_1 - x_2)x' + (y_1 - y_2)y']\right\}dx'\,dy'.$$

$$(3.115)$$

Equation (3.115) is the *van Cittert–Zernike theorem*. It shows that the spatial coherence in the observation plane is a function of the separation of the points P_1 and P_2 and that the integral is a Fourier transform of the source radiant exitance. This means that the results of Fraunhofer diffraction calculations can be applied to these coherence problems. In particular, the circular source calculations which led to Eq. (3.68) apply here. Examination of Eqs. (3.68) and (3.70) show that for a circular source of diameter D_s the spatial coherence should extend over a distance $D_c \cong \lambda l/D_s$. This is the result used in (3.111) to define the coherence diameter. More exact results may be obtained by calculating $\Gamma(x_1 - x_2, y_1 - y_2)$ using the same procedure as was employed in Section 3.3.2.1 for Fraunhofer diffraction from a circular aperture. Similar considerations apply for apertures of other shapes such as the rectangular aperture evaluated in Section 3.3.2.2.

3.3.4.4   COHERENT SOURCES

The fields arising from a completely coherent source will remain completely coherent as the wave fronts propagate through a homogeneous isotropic medium. For this reason it is not necessary to use the mathematical techniques of coherence theory in this situation. It is sufficient to follow the course of the complex wave amplitude $U(x, y, z)$ for each spectral component using (3.61). The irradiance at a detector or receiving surface is then given by (3.60).

However, the presence of inhomogeneities, especially small-scale, random inhomogeneities, can reduce the spatial coherence of electromagnetic fields. Some prominent examples are the inhomogeneities associated with atmospheric turbulence (Lawrence and Strohbehn, 1970; Fante, 1975; Strohbehn, 1978) and the inhomogeneities associated with reflection from a rough surface (Beckmann, 1963), In these and other cases the time scale of changes in the inhomogeneities is also important. In addition to changing the coherence, the spatial distribution of power can be changed at some distance from the inhomogeneities. In some cases the inhomogeneities are so prominent that the coherent field becomes completely incoherent. The methods of the previous section then apply. In other cases the fields in the region of the inhomogeneities become partially coherent. This situation is described in the following section.

The laser is a prominent example of a coherent source. The coherence lengths l_c defined by Eq. (3.110) can vary from a few centimeters to many kilometers, depending on the spectral width Δv. The beam of radiant energy from a laser is usually completely spatially coherent so that the prominent spatial coherence questions are those associated with the effects of random media.

3.3.4.5   PARTIALLY COHERENT SOURCES

In general, the coherence properties of a partially coherent source are represented by the mutual coherence function $\Gamma_{12}(\tau)$ defined in (3.101) or its normalized form the complex degree of coherence $\gamma_{12}(\tau)$ defined by (3.103). The radiometry of partially coherent sources and of the fields produced by them is a subject of current research and precise relationships have not yet been completely established. In this section we present a brief description of the relevant theory and an indication of the physical relationships which are emerging. For more detailed information the reader is referred to the literature (Walther, 1968, 1973; Marchand and Wolf, 1974; Wolf, 1978).

Consider the arrangement of a partially coherent source and observation points shown in Figure 3.26. There are two dimensions of interest in characterizing the properties at the source: the transverse dimension D_{sp} over

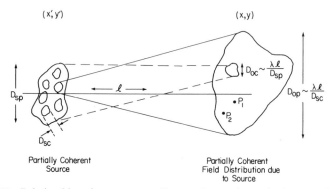

Fig. 3.26 Relationships of source power diameter D_{sp} to observed coherence diameter D_{oc} and source coherence diameter D_{sc} to observed power diameter D_{op} for a partially coherent source.

which the source power distribution extends and the dimension D_{sc} over which the source is spatially coherent. Similarly, in the region at some distance from the source the fields can be approximately characterized by two approximate dimensions: the dimension D_{oc} over which the observed fields are coherent and D_{op} over which the observed power distribution extends. It will always be true that $D_{sc} \leq D_{sp}$ and $D_{oc} \leq D_{op}$. That is, the coherence dimension must be less than or equal to the power distribution dimension.

Now, rather than use coordinates (x_1', y_1') and (x_2', y_2') to describe the position of points in the source, we change to average position and difference coordinates defined as

$$x_a' = \tfrac{1}{2}(x_1' + x_2'), \qquad\qquad (3.116)$$

$$x_d' = x_1' - x_2', \qquad\qquad (3.117)$$

and similarly for the y coordinates. The mutual coherence function at the source can then be specified in terms of these new coordinates as $\Gamma_s(x_a', y_a'; x_d', y_d')$ where we assume that the source is quasi-monochromatic. Similarly, the mutual coherence function of the fields at some distance from the source can be expressed in average and difference coordinates as $\Gamma(x_a, y_a; x_d, y_d)$. The interpretation of the average and difference coordinates is that the dependence on the average coordinates essentially represents the power distribution while the dependence on the difference coordinates essentially represents the spatial coherence properties in the same region. It can then be shown that at large distances a Fourier transform relation exists between the source power distribution and the observed spatial coherence, and similarly between the source spatial coherence and the distant field power distribution. That is,

$$\text{source power distribution} \xleftrightarrow{\substack{\text{Fourier} \\ \text{transform}}} \text{field spatial coherence}$$

$$\text{source spatial coherence} \xleftrightarrow{\substack{\text{Fourier} \\ \text{transform}}} \text{field power distribution}$$

The reciprocal relationship between the associated power distribution and coherence dimensions is shown in Fig. 3.26. We see that

$$D_{oc} = \text{observed spatial coherence diameter} \cong \lambda l / D_{sp}, \qquad (3.118)$$

where D_{sp} is the source power distribution dimension, and

$$D_{op} = \text{observed power dimension} \cong \lambda l / D_{sc}, \qquad (3.119)$$

where D_{sc} is the source spatial coherence diameter.

Recent publications show the approximate relationships described here in more exact mathematical form (Marchand and Wolf, 1974; Walther, 1968, 1973; Wolf, 1978). However, the simple dimensional relations of (3.118) and (3.119) are useful in estimating the approximate scale of the quantities involved.

3.3.5 Image Formation

3.3.5.1 TYPES OF IMAGING SYSTEMS

Many radiometers employ image-forming optical systems to define the spatial position on the source from which radiant power is received at the detector. From another viewpoint, the imaging system defines the angle or direction from which radiant power is received. Figure 3.27 shows two types

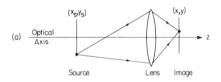

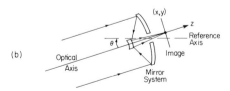

Fig. 3.27 Imaging optical systems: (a) nonscanning refractive; (b) scanning reflective.

of imaging systems. System (a) is nonscanning. The image may be recorded by using a distributed detecting system in the image plane such as photographic film or an array of discrete detectors. System (b) is a scanning system which rotates about an axis perpendicular to the plane of the page so that its observation direction is defined by the angle θ from a reference direction. This second system can receive radiant power from a range of directions with only a single detector by using its angular agility.

In system (a) the complex wave amplitude at the image plane can be expressed in terms of the complex wave amplitude at the source for each wavelength component by

$$U(x, y, z) = \iint_{\text{source}} U(x_s, y_s, z_s) K(x, y, z; x_s, y_s, z_s) \, dx_s \, dy_s \, dz_s, \quad (3.120)$$

where (x, y, z) are the image point coordinates, (x_s, y_s, z_s) the source point coordinates, and K describes the propagation from a single source point to a single image point. Thus K defines the imaging system properties. A relation similar to (3.120) can be written for the scanning system (b) in terms of the angles involved.

Relation (3.120) involves an integral because the amplitude at the image point (x, y, z) is in general a sum of the amplitudes from many source points surrounding the source point which is geometrically conjugate to the image point (x, y, z). Diffraction, aberrations, atmospheric turbulence, scattering in the optical elements, and other factors all contribute to the form of K.

3.3.5.2   INCOHERENT SOURCES

In general, the spatial distribution of radiant power in the image will depend on the degree of coherence in the source. Methods for characterizing these general problems of image formation with partially coherent sources are described in the literature (Beran and Parrent, 1964). For now we confine our attention to incoherent sources. When the source is incoherent the irradiance in the image can be obtained from (3.120) and (3.60). It is

$$E(x, y, z) = (\varepsilon c/2n)|U(x, y, z)|^2 \quad (\text{W m}^{-2})$$

$$= \iint_{\text{source}} M(x_s, y_s, z_s)|K(x, y, z; x_s, y_s, z_s)|^2 \, dx_s \, dy_s \, dz_s, \quad (3.121)$$

where M is the radiant exitance of the source at (x_s, y_s, z_s). The quantity $|K|^2$ is known as the *point spread function*. It represents the spatial distribution of radiant power in the image of a point source located at (x_s, y_s, z_s). In general, the form of the point spread function will vary with the position of this point source in the source plane. This may be due to the nature of the aberrations present in the image-forming optical elements, to vignetting, or to a combination of factors. In any case, any imaging system of the type shown in Fig. 3.27(a) will show this property. Systems of the type shown in Fig. 3.27(b) will not. Because this system rotates its optical axis to scan the source, the form of $|K|^2$ will not change as the system points at various parts of the source. For systems of type (a) the form of $|K|^2$ will remain approximately fixed over small regions of the source or image plane. These regions are known as *isoplanatic* regions. Within them, or for all situations of type (b), the point spread function depends only on coordinate differences. Also, in many applications of interest the source is planar and so no integration over z is involved. Under these conditions (3.121) becomes

$$E(x, y) = \iint_{\text{source}} M(x_s, y_s)|K(x - x_s, y - y_s)|^2 \, dx_s \, dy_s, \qquad (3.122)$$

where it is understood that z and z_s are fixed. Equation (3.122) expresses the image irradiance as a *convolution* of the source radiant exitance with the point spread function.

3.3.5.3 THE TRANSFER FUNCTION

From Fourier analysis (O'Neill, 1963; Goodman, 1968) (3.122) can be transformed into

$$e(f_x, f_y) = m(f_x, f_y)t(f_x, f_y), \qquad (3.123)$$

where e, m, and t are the spatial Fourier transforms at E, M, and $|K|^2$, and (f_x, f_y) are spatial frequency variables defined by

$$t(f_x, f_y) = \iint |K(x, y)|^2 \exp[-i2\pi(f_x x + f_y y)] \, dx \, dy$$
$$= \text{Fourier transform of } |K(x, y)|^2, \qquad (3.124)$$

and similarly for E and M.

The Fourier transform of $|K|^2$ is $t(f_x, f_y)$ and is known as the *transfer function*. It plays a useful role in the evaluation of the spatial distribution of radiant power in an image-forming system because of the simple multiplicative relation shown in (3.123) and also because a number of techniques now exist for its measurement (Murata, 1966). The transfer function can be used to characterize the imaging system whether the image quality is determined by diffraction, aberration, atmosphere, or other causes.

In a scanning system the spatial frequency variables (f_x, f_y) can be related to the temporal frequencies associated with the scan speed s. If the scan is in the x direction so that the spatial frequency variable f_x is involved, then

$$\omega_t = 2\pi f_x s \quad (\text{s}^{-1}), \tag{3.125}$$

where ω_t is the temporal radian frequency. The combination of the spatial transfer function $t(f_x, f_y)$ with the temporal transfer function of the detector and signal processing system is then straightforward.

3.3.5.4   RELATION OF TRANSFER FUNCTION TO EXIT APERTURE

All imaging radiometers are ultimately limited in spatial or angular resolution by diffraction effects. The exit aperture as defined in Section 3.2.4.4 constitutes the limiting aperture on the image side of the system. Diffraction at this aperture is then the ultimate determinant of image resolution. Similarly, diffraction at the entrance aperture limits the spatial resolution on the source side of the optics.

Figure 3.28 shows the geometry of the exit aperture and image plane. The coordinates in the exit aperture are (x', y') and those in the image are (x, y). In the imaging situation we have Fraunhofer diffraction at a focus as discussed in Section 3.3.2. From Eq. (3.65) the wave amplitude $U(x, y)$ in the image is then proportional to the Fourier transform of the wave amplitude $U(x', y')$ in the exit aperture where

$$U(x', y') = U_m(x', y')\exp[ikW(x', y')] = \text{wave amplitude in exit aperture.}$$

$$\tag{3.126}$$

When the exit aperture is circular and $W = 0$ (diffraction limited), then the analysis of Section 3.3.2.1 applies and the point spread function $|K|^2$ is proportional to (3.71). Similarly, when the exit aperture is rectangular and diffraction limited, the point spread function is proportional to (3.78). In the previous section we defined the transfer function $t(f_x, f_y)$ as the Fourier transform of the point spread functions. Figure 3.29 shows these transfer functions (O'Neill, 1963).

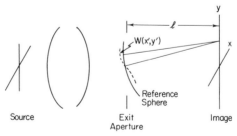

Fig. 3.28   Exit aperture and image plane of an imaging radiometer. $W(x', y')$ is the optical path difference between the actual wave front in the exit aperture and the reference sphere which is centered on the reference image point.

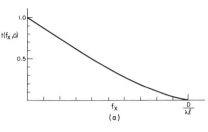

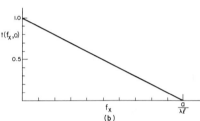

Fig. 3.29 Transfer functions of diffraction-limited imaging radiometers having (a) circular and (b) rectangular exit apertures. In (a) the exit aperture diameter is D. In (b) the rectangular aperture width is a. For both cases l is the distance from exit aperture to image plane.

In the more general situation where the system is not diffraction limited the wave front in the exit aperture deviates from a converging sphere and $W(x', y') \neq 0$. In this case the general relation between the transfer function and the exit aperture complex wave amplitude $U(x', y')$ is

$$t(f_x, f_y) = \iint |K(x, y)|^2 \exp[-i2\pi(f_x x + f_y y)]\, dx\, dy \qquad (3.127)$$

$$t(f_x, f_y) = a^{-1} \iint_{\text{exit aper.}} U^*(x', y') U(x' + \lambda l f_x, y' + \lambda l f_y)\, dx'\, dy', \qquad (3.128)$$

where the factor a^{-1} is simply a normalizing constant which ensures that $t(0,0) = 1$. Because the exit aperture constitutes the limiting aperture in a diffraction-limited imaging radiometer it is convenient to specify the aberrations $W(x', y')$ at this location also. Specification of these aberrations then permits specification of $U(x', y')$ in the exit aperture and the calculation of the transfer function from Eq. (3.128). Figure 3.29 is obtained from (3.128) when $W = 0$.

3.3.5.5   Transfer Function and Spot Diagrams

Imaging radiometers, when designed to operate over a wide field of view, are frequently limited by geometrical aberrations rather than by diffraction. The spatial distribution of radiant energy in the image of a point source can then be defined by ray tracing (Miyamoto, 1961).

In the exit aperture the ray paths define the geometrical wave front and its deviation $W(x', y')$ from a reference sphere centered on the reference image point. From Eqs. (3.128) and (3.126) it can be seen that the transfer function contains a shifted phase function. The shifted phase function can be expanded in the power series

$$\frac{2\pi}{\lambda} W(x' + \lambda l f_x, y' + \lambda l f_y) = \frac{2\pi}{\lambda}\left[W(x', y') + \lambda l\left(f_x \frac{\delta W}{\delta x'} + f_y \frac{\delta W}{\delta y'}\right) + \cdots \right].$$

(3.129)

In the geometrical optics limit the higher-order terms vanish as $\lambda \to 0$. As the wavelength approaches zero the transfer function (3.128) then becomes

$$t(f_x, f_y) = a^{-1}\iint\limits_{\text{exit aper.}} \exp[i2\pi l(f_x \delta W/\delta x' + f_y \delta W/\delta y')]\, dx'\, dy'. \quad (3.130)$$

This is the transfer function in the geometrical optics limit. Notice that unlike (3.128), (3.130) does not involve shifting the exit aperture. As a result (3.130) can be expected to be most accurate at low spatial frequencies. It is found that if the aberrations in the exit aperture exceed about two wavelengths, then (3.130) is fairly accurate.

Spot diagrams are also important in a geometrical optics evaluation (Smith, 1966). Figure 3.30 illustrates the production of a spot diagram. Rays are usually traced from the source point in such a way that they are equally spaced in the entrance or exit apertures. This represents the assumption of uniform aperture irradiance. From geometrical optics it can be shown that the transverse aberration, which is the distance between the paraxial image point and the intersection of each ray with the image plane, has x and y components given by

$$TA_x = l\,\delta W/\delta x', \qquad TA_y = l\,\delta W/\delta y'. \quad (3.131)$$

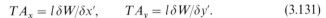

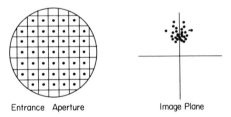

Entrance   Aperture   Image Plane

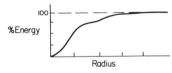

Fig. 3.30   Spot diagram (top right) produced from a uniform distribution of ray intercepts with the entrance aperture. The percent energy contained within a circle of specified radius is found from the number of encircled rays.

These are the same quantities that appear in (3.130). The geometrical transfer function can then be determined from transverse aberration data on a number of rays. In the geometrical optics limit both the spot diagram and the geometrical transfer function describe the spatial distribution of radiant energy in an imaging radiometer.

3.4   TRANSMISSION AND REFLECTION AT AN INTERFACE

In the previous sections of this chapter we have considered the spatial distribution of radiant energy on the basis of both geometrical and wave optics. However, we have not considered the losses which are associated with the passage of a wave front across the interface between different media. These losses are the subject of this section.

In many applications the media of interest are dielectrics rather than conductors. We assume in the following that the media are dielectrics and so are characterized by a real index of refraction. Figure 3.31 shows the boundary between two media of index n and n'. The incident wave front is represented by a ray incident on the interface at an angle θ with the normal to the interface. The transmitted wave front is refracted according to Snell's law ($n \sin \theta = n' \sin \theta'$) and so makes an angle θ' with the normal. The ray representing the reflected wave front is also at the angle θ with respect to the normal. We define the plane of incidence as the plane containing both the incident ray and the normal. Then the transmitted and reflected rays are also found to be in this same plane.

It is necessary to recall that the electric field which is associated with each wave front or ray and which is perpendicular to the ray can be represented by two components. One of these components denoted by the symbol $\parallel$ is parallel to the plane of incidence and the other represented by the symbol $\perp$ is perpendicular to the plane of incidence. We now define the transmission

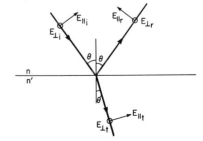

Fig. 3.31   Reflection and transmission at the interface between two dielectric media of index n and n'.

and reflection coefficients as

$$r_{\parallel} = E_{\parallel_r}/E_{\parallel_i}, \qquad r_{\perp} = E_{\perp_r}/E_{\perp_i} \tag{3.132}$$

$$t_{\parallel} = E_{\parallel_t}/E_{\parallel_i}, \qquad t_{\perp} = E_{\perp_t}/E_{\perp_i}, \tag{3.133}$$

where, for example, $E_{\parallel_r}$ represents the reflected electric field component which is parallel to the plane of incidence. From electromagnetic theory it is then found (Born and Wolf, 1964) that the reflection and transmission coefficients are given by the *Fresnel equations*

$$r_{\parallel} = (n' \cos\theta - n \cos\theta')/(n' \cos\theta + n \cos\theta'), \tag{3.134}$$

$$r_{\perp} = (n \cos\theta - n' \cos\theta')/(n \cos\theta + n' \cos\theta'), \tag{3.135}$$

$$t_{\parallel} = (2n \cos\theta)/(n' \cos\theta + n \cos\theta'), \tag{3.136}$$

$$t_{\perp} = (2n \cos\theta)/(n \cos\theta + n' \cos\theta'), \tag{3.137}$$

where n, n', θ, and θ' are defined in Fig. 3.31.

For normal incidence ($\theta = 0$) Eqs. (3.134)–(3.137) become

$$t = 2n/(n' + n) \tag{3.138}$$

and

$$r = (n' - n)/(n' + n), \tag{3.139}$$

where for the reflected wave the sign convention is such that if $n' > n$, there is a phase reversal on reflection.

The reflection and transmission coefficients (3.134)–(3.137) are related to the radiant transmittance and reflectance quantities defined in Chapter 2 by

$$\tau_{\parallel} = |t_{\parallel}|^2, \qquad \tau_{\perp} = |t_{\perp}|^2, \tag{3.140}$$

$$\rho_{\parallel} = |r_{\parallel}|^2, \qquad \rho_{\perp} = |r_{\perp}|^2. \tag{3.141}$$

These are the regular transmittance and reflectance quantities.

At normal incidence one finds from (3.134)–(3.137) that

$$\tau = [2n/(n' + n)]^2, \tag{3.142}$$

$$\rho = [(n' - n)/(n' + n)]^2. \tag{3.143}$$

Figure 3.32 shows the radiant transmittance (3.140) and reflectance (3.141) for index of refraction ratios n'/n of 1.5 and 4. These correspond to an air–glass interface in the visible part of the spectrum and an air–Ge interface at 10 μm in the infrared. The angle at which $\rho_{\parallel}$ drops to zero is the polarizing angle. At this angle the reflected beam is linearly polarized in a plane perpendicular to the plane of incidence.

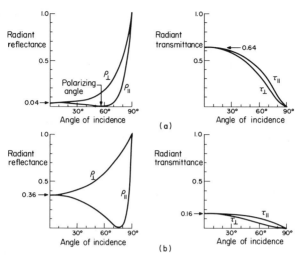

Fig. 3.32 Radiant transmittance and reflectance for dielectrics of index of refraction ratio n'/n equal to (a) 1.5 and (b) 4.

Because the basic radiance is invariant across the boundary between two dielectrics of different index of refraction as stated in (3.9), it is convenient to express the reflection losses in terms of this quantity. The radiance L' in the medium of index n' shown in Fig. 3.31 is given for each component, parallel or perpendicular to the plane of incidence, by

$$L' = (n'^2/n^2)L\tau \quad (\text{W m}^{-2}\,\text{sr}^{-1}), \tag{3.144}$$

where τ is one of the coefficients in (3.140).

3.5 ABSORPTION AND SCATTERING IN A MEDIUM

3.5.1 Attenuation Coefficients

In addition to the losses which can occur due to reflection at an interface, the radiance along a ray from source to detector can also be reduced by attenuation due to absorption and scattering within a medium. In some applications the medium also emits radiant energy and this can increase the radiance along a ray. In this section we consider only the reduction of radiance by absorption and scattering while reserving our discussion of emission effects for the next section.

The actual attenuation effects observed in real media can be very complex. A real medium may be inhomogeneous or anisotropic and may introduce multiple scattering effects. Although in principle it is possible to accommodate all aspects of attenuation into a model of a transmission medium, it is seldom practicable to do so. In the following we present an approximate description of attenuation losses which is found to provide a sound starting point for many applications.

We assume that the medium is isotropic. We also assume that:

(1)   The fractional decrease in radiance is linearly proportional to distance in the attenuating medium.
(2)   No radiant power absorbed in the medium is reradiated.
(3)   No radiant power scattered out of the beam is scattered back into the beam.

Consider Fig. 3.33a. A beam of radiant power of wavelength λ is incident from the left on an attenuating layer of thickness dl'. The fractional change in radiance under assumptions (1)–(3) is

$$dL_\lambda/L_\lambda = -\mu(\lambda, l')\,dl', \qquad (3.145)$$

where $\mu(\lambda, l')$ is the *spectral attenuation coefficient* as a function of position l' at wavelength λ. Integrating (3.145) along a path length l in the medium leads to

$$\int_{L_\lambda(0)}^{L_\lambda(l)} dL_\lambda/L_\lambda = -\int_0^l \mu(\lambda, l')\,dl' \qquad (3.146)$$

or

$$L_\lambda(l) = L_\lambda(0)\exp\left[-\int_0^l \mu(\lambda, l')\,dl'\right] \quad (\mathrm{W\,m^{-2}\,sr^{-1}\,\mu m^{-1}}). \qquad (3.147)$$

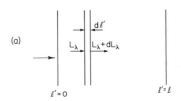

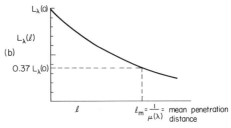

Fig. 3.33   A beam of radiant power attenuated in an absorbing and scattering layer. (a) Geometry of attenuating layer; (b) spectral radiance $L_\lambda(l)$ as a function of layer thickness L.

The radiant transmittance of the layer of thickness l is

$$\tau(\lambda) = L_\lambda(l)/L_\lambda(0) = \exp\left[-\int_0^l \mu(\lambda, l')\,dl'\right], \qquad (3.148)$$

which is *Bouguer's law*. It takes an especially simple form when the medium is homogeneous. Then $\mu(\lambda, l') = \mu(\lambda)$ and

$$\tau(\lambda) = e^{-\mu(\lambda)l}. \qquad (3.149)$$

The effects of absorption and scattering can be represented separately as

$$\mu(\lambda, l') = a(\lambda, l') + s(\lambda, l') \quad (\text{m}^{-1}), \qquad (3.150)$$

where $\mu(\lambda, l')$ is the *attenuation coefficient*, $a(\lambda, l')$ the *absorption coefficient*, and $s(\lambda, l')$ the *scattering coefficient*. Each coefficient has dimensions of reciprocal length. In some applications it is convenient to define μ, a or s with units of reciprocal centimeters or reciprocal kilometers. In general, the attenuation, absorption, and scattering coefficients will depend on wavelength and material composition. In a gaseous medium they will depend on temperature and pressure. It will be shown later that the attenuation coefficient $\mu(\lambda)$ is related to the mean penetration distance of a beam of radiant power into an attenuating medium and to the optical thickness of a layer of the medium.

The dependence of the absorption and scattering coefficients on wavelength is of fundamental importance. As a result of this dependence the radiant transmittance (3.149) integrated over a range of wavelengths $\Delta\lambda$

$$\tau = \int_{\Delta\lambda} \tau(\lambda)\,d\lambda = \int_{\Delta\lambda} e^{-\mu(\lambda)}\,d\lambda \qquad (3.151)$$

is a valid measure of transmittance only when the spectral power distribution is uniform over $\Delta\lambda$. Similar considerations apply to the inhomogeneous medium represented by Eq. (3.148).

3.5.2 Mean Penetration Distance and Optical Thickness

When the attenuation coefficient $\mu(\lambda, l')$ is constant, the fraction of the original radiance $L_\lambda(0)$ which passes through the path length l is, from (3.149),

$$L_\lambda(l)/L_\lambda(0) = \tau(\lambda) = e^{-\mu(\lambda)}. \qquad (3.152)$$

as shown in Fig. 3.33(b).

The fraction of (3.152) which is attenuated in a layer from l to $l + dl$ is

$$[L_\lambda(l) - L_\lambda(l + dl)]/L_\lambda(0) = -\{d[L_\lambda(l)/L_\lambda(0)]/dl\}\,dl = \mu(\lambda)e^{-\mu(\lambda)l}\,dl. \quad (3.153)$$

The *mean penetration distance* of the beam is then found by multiplying the distance l by the fraction attenuated at l and integrating over all path lengths.

This is

$$l_m = \int_0^\infty l\mu(\lambda)e^{-\mu(\lambda)l}\,dl = 1/\mu(\lambda) \quad \text{(m)}. \tag{3.154}$$

as indicated in Fig. 3.33b.

According to Eq. (3.154) the mean distance which the beam penetrates into the medium before absorption or scattering is the reciprocal of the attenuation coefficient $\mu(\lambda)$. The parameter l_m is a simple measure of the opacity of an attenuating medium to a beam of radiant power.

Optical thickness is defined as

$$\beta(\lambda) = \mu(\lambda)l \tag{3.155}$$

and by using Eq. (3.154) one finds that

$$\beta(\lambda) = l/l_m. \tag{3.156}$$

That is, the optical thickness is the number of mean penetration distances in the path length l. The definitions of mean penetration distance l_m and optical thickness $\beta(\lambda)$ can be generalized to cover situations where the attenuation coefficient $\mu(\lambda)$ is not independent of path length l.

3.6   EMISSION

When a beam of radiant energy passes through a transmission medium which is emitting radiant energy, the radiance along a ray can increase. In Section 3.6.1 we determine the radiance when the emission of the medium is the only source of radiant energy, while in Section 3.6.2 we consider the combined effects of an external source and emission by the medium. In both cases we include the effects of attenuation.

3.6.1   Emission and Attenuation within a Medium

Figure 3.34 shows a medium which is both emitting and attenuating radiant power. We assume that any contributions from external sources are zero. In the absence of absorption, each volume element of the medium emits an

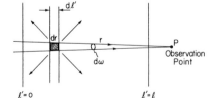

Fig. 3.34   Emitting and absorbing medium.

amount of spectral radiant power

$$d\Phi_\lambda = j_\lambda \rho \, dv \quad (\text{W} \, \mu\text{m}^{-1}),$$ (3.157)

where $d\Phi_\lambda$ is the spectral radiant power ($\text{W} \, \mu\text{m}^{-1}$), j_λ the *spectral emission coefficient* ($\text{W} \, \text{kg}^{-1} \mu\text{m}^{-1}$), ρ the material density ($\text{kg} \, \text{m}^{-3}$), and dv the volume element (m^3). The spectral irradiance at an observation point P at a distance r due to this volume element is

$$dE_\lambda = d\Phi_\lambda/4\pi r^2 \quad (\text{W} \, \text{m}^{-2} \mu\text{m}^{-1}),$$ (3.158)

under the assumption that the power is uniformly distributed over a sphere. From (3.157) and (3.158) and $dv = r^2 \, dr \, d\omega$ the spectral radiance at the observation point is

$$dL_\lambda = dE_\lambda/d\omega = j_\lambda \rho \, dr/4\pi \quad (\text{W} \, \text{m}^{-2} \, \text{sr}^{-1} \mu\text{m}^{-1}).$$ (3.159)

However, in the presence of attenuation over a distance r between the element dv and the edge of the medium, the observed spectral radiance is

$$dL_\lambda = (j_\lambda \rho \, dr/4\pi)e^{-\mu(\lambda)}.$$ (3.160)

The spectral radiance due to the total emitting and attenuating medium of thickness l is then the integral of (3.160):

$$L_\lambda = (j_\lambda \rho/4\pi) \int_0^l e^{-\mu(\lambda)} \, dr.$$ (3.161)

From (3.161),

$$L_\lambda = [j_\lambda \rho/4\pi\mu(\lambda)](1 - e^{-\mu(\lambda)l}) \quad (\text{W} \, \text{m}^{-2} \, \text{sr}^{-1} \mu\text{m}^{-1}).$$ (3.162)

Notice that the radiance (3.162) at the observation point P is also the radiance at the near edge of the volume ($l' = l$) according to the concept of radiance conservation derived in Section 3.2.2. Figure 3.35 shows Eq. (3.162) in graphical form. According to definition (3.155) the quantity $\mu(\lambda)$ is the optical depth $\beta(\lambda)$. As the optical depth increases the spectral radiance approaches the value $j_\lambda \rho/4\pi\mu(\lambda)$. The quantity has the dimensions of radiance and can be regarded as the intrinsic radiance of the emitting–attenuating

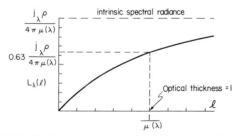

Fig. 3.35 Radiance due to an emitting and attenuating medium.

volume. Then the spectral radiance is

$$L_\lambda = L_{i\lambda}(1 - e^{-\beta(\lambda)}) \quad (\text{W m}^{-2}\,\text{sr}^{-1}\,\mu\text{m}^{-1}), \tag{3.163}$$

where $L_{i\lambda}$ is the *intrinsic spectral radiance* and $\beta(\lambda)$ the optical thickness.

If the intrinsic radiance $L_{i\lambda}$ is that of a blackbody, then the quantity $1 - e^{-\beta(\lambda)}$ can be regarded as the spectral emissivity of the emitting–attenuating volume according to the definition of emissivity given in Chapter 4.

3.6.2   External Sources: The Equation of Transfer

In the previous section only internal emission and attenuation within the volume were considered. When an external source is also present, the intervening attenuating–emitting volume affects the apparent radiance of the source as measured at point P.

With reference to Fig. 3.34 the change in spectral radiance at P due to an elementary volume of length dr is found from (3.145) and (3.159) to be

$$dL_\lambda = -L_\lambda\mu(\lambda)\,dr + (j_\lambda\rho/4\pi)\,dr, \tag{3.164}$$

where the first term on the right-hand side of the equation represents the decrease in radiance due to attenuation and the second term represents the increase in radiance due to emission. Rearranging (3.164) gives

$$(dL_\lambda/dr) + \mu(\lambda)L_\lambda = j_\lambda\rho/4\pi. \tag{3.165}$$

The solution of (3.165), the observed spectral radiance, is given (Chandrasekhar, 1950) by the *equation of transfer*

$$L_\lambda = L_{s\lambda}e^{-\beta(\lambda)} + L_{i\lambda}(1 - e^{-\beta(\lambda)}) \quad (\text{W m}^{-2}\,\text{sr}^{-1}\,\mu\text{m}^{-1}). \tag{3.166}$$

where $L_{s\lambda}$ is the spectral radiance of the source and $L_{i\lambda}$ the intrinsic spectral radiance of the volume as in (3.163) and $\beta(\lambda)$ is the optical thickness given by (3.156). The first term on the right of Eq. (3.166) represents the decreased source spectral radiance due to attenuation by the volume. The second term represents the contribution to the observed spectral radiance due to the internal emission and attenuation of the volume.

3.7   POLARIZATION

In addition to spatial, spectral, temporal, and coherence properties, beams of optical radiant energy have polarization properties. Polarization is associated with the transverse wave characteristics of electromagnetic fields. In the following we consider the elementary description of the polarization properties of radiant energy and the definition and interpretation of the Stokes parameters.

The Stokes parameters constitute a set of four measurable quantities which completely characterize the polarization properties of a beam of radiant energy. With the use of the Stokes parameters it is possible to determine the irradiance at a detector at the output of an arbitrary optical instrument. These techniques are necessary because in general the amount of radiant energy which passes through an optical instrument is sensitive to the polarization properties of the radiant energy from the source and also to the polarization characteristics of the components of the instrument. These components include the transmission medium, reflecting or transmitting materials, refracting or reflecting optical focusing elements, and detector.

In the following description of polarization properties we assume that the beam of radiant energy covers only a narrow range of wavelengths. The reason for this assumption is that the polarization properties of a beam consisting of a broad range of wavelengths may vary within this range, depending on which small portion of the range of wavelengths is examined.

3.7.1 The Polarization Ellipse

Consider a plane wave traveling out of the page as shown in Fig. 3.36 with electric field components in the x and y directions in the plane of the page given by

$$\mathscr{E}_x = A_x(t)\cos[2\pi vt - \theta_x(t)], \qquad (3.167)$$

$$\mathscr{E}_y = A_y(t)\cos[2\pi vt - \theta_y(t)], \qquad (3.168)$$

where v is the mean optical frequency, $A_x(t)$ and $A_y(t)$ the amplitudes, and $\theta_x(t)$ and $\theta_y(t)$ the phases of the wave. The electric field vector is then

$$\bar{\mathscr{E}} = \bar{x}\mathscr{E}_x + \bar{y}\mathscr{E}_y, \qquad (3.169)$$

where $\bar{x}$ and $\bar{y}$ are unit vectors in the x and y directions.

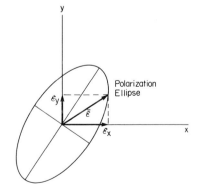

Fig. 3.36 Polarization ellipse and instantaneous electric field vector $\bar{\mathscr{E}}$.

The amplitudes $A(t)$ and phases $\theta(t)$ are functions of time with variations which are slow compared to that of the mean frequency v, being on the order of the bandwidth Δv. The amplitudes and phases are then essentially constant over a fairly large number of cycles. Over this short time the two perpendicular sinusoidal motions represented by (3.167) and (3.168) will generate an electric field vector with the locus of the tip describing an ellipse as shown in Fig. 3.36. This ellipse is the *polarization ellipse*.

As the bandwidth is narrowed the amplitudes and phases, $A(t)$ and $\theta(t)$, remain constant over longer times so that in the limit a monochromatic wave is completely described by a fixed polarization ellipse. The wave is then said to be *completely polarized*. Some of the terms used to describe special cases of elliptical polarization in monochromatic waves are as follows.

(1)   *Linearly polarized.*   When $\theta_x = \theta_y$, the wave is linearly polarized at an angle $\tan^{-1}(\mathscr{E}_y/\mathscr{E}_x)$ with the x axis.

(2)   *Circularly polarized.*   When $\theta_x - \theta_y = \pm 90°$ and $A_x = A_y$, the wave is circularly polarized. If $\theta_x - \theta_y = -90°$, the electric field vector rotates counterclockwise when the observer faces into the oncoming wave. The wave is circularly polarized. If $\theta_x - \theta_y = -90°$, the electric field vector rotates *positive helicity* in the terminology of modern physics or antenna engineering). On the other hand if $\theta_x - \theta_y = +90°$, the wave rotates clockwise and is *right circularly polarized* (*negative helicity*).

For nonmonochromatic beams of radiant energy it is possible for only the size of the ellipse to vary in such a way that $A_x(t)/A_y(t) = \text{const.}$ and $\theta_x(t) - \theta_y(t) = \text{const.}$ In any measurement this wave behaves as would a monochromatic wave. The x and y components of the electric field are said to be completely correlated or completely coherent and this wave is also *completely polarized*.

On the other hand, when the beam of radiant energy covers a small but finite range of frequencies so that $A_x(t)$, $A_y(t)$, and $\theta_x(t) - \theta_y(t)$ vary randomly in time, it is possible that there is no preferred orientation or direction of rotation of the electric field vector which can be observed in a measurement. The wave is then said to be *unpolarized*.

3.7.2   The Stokes Parameters

The preceding section deals with *completely polarized* waves where $A_x(t)$, $A_y(t)$, and $\theta_x(t) - \theta_y(t)$ are constants or are very slowly varying so that the two components of the field are completely coherent. This section deals with techniques for characterizing *partially polarized* waves. The characterization is in terms of four measurable parameters which allow us to

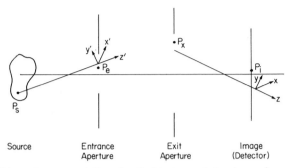

Fig. 3.37 Schematic representation of optical instrument showing surfaces and coordinate systems used to define the Stokes parameters.

determine the irradiance at the detector of any optical instrument when the input to the instrument is described by (3.167) and (3.168).

Figure 3.37 shows an optical instrument. The input is the electric field distribution which exists in the entrance aperture due to the source. Consider the electric field at a point P_e in the entrance aperture due to a small region of the source, around P_s small enough so that the field is spatially coherent over the entrance aperture in the sense of Section 3.3.4.3 or 3.3.4.5. Two coordinate systems (x', y', z') and (x, y, z) are established with z axes along the direction of the ray shown. The component of the electric field at P_e which is along the x' axis can in general have components along both the x and y axes in the image, and similarly for the component at P_e along the y' axis. The two components of the field at P_i in the image are then

$$a_x(t) = \tau_{xx'}a_{x'}(t) + \tau_{xy'}a_{y'}(t), \tag{3.170}$$

$$a_y(t) = \tau_{yx'}a_{x'}(t) + \tau_{yy'}a_{y'}(t), \tag{3.171}$$

where the τ parameters are transfer functions which depend on the location of P_e and P_i and on wavelength and are to be calculated from the geometry of the instrument and its components. The field components a_x and a_y are defined by $a_x(t) = A_x(t)\exp(-i2\pi vt)$, and so on, where the amplitude $A(t)$ is that used in (3.167) and (3.168).

The time-averaged irradiance measured by a detector at P_i is

$$E(P_i) = (\varepsilon c/2n)[\langle|a_x(t)|^2\rangle + \langle|a_y(t)|^2\rangle] \tag{3.172}$$

from Eq. (3.60) were the angle brackets represent the time average. Substitution of (3.170) and (3.171) into (3.172) will lead to

$$E(P_i) = \tfrac{1}{2}(|\tau_{xx'}|^2 + |\tau_{yx'}|^2)(S_0 + S_1) + \tfrac{1}{2}(|\tau_{xy'}|^2 + |\tau_{yy'}|^2)(S_0 - S_1)$$
$$+ R_e[(\tau_{xx'}\tau_{xy'}^* + \tau_{yx'}\tau_{yy'}^*)(S_2 + iS_3), \tag{3.173}$$

where

$$S_0 = (\varepsilon c/2n)(\langle A_x^2 \rangle + \langle A_y^2 \rangle), \tag{3.174}$$

$$S_1 = (\varepsilon c/2n)(\langle A_x^2 \rangle - \langle A_y^2 \rangle), \tag{3.175}$$

$$S_2 = (\varepsilon c/2n)\langle 2A_{x'}A_{y'}\cos(\theta_x' - \theta_y') \rangle, \tag{3.176}$$

$$S_3 = (\varepsilon c/2n)\langle 2A_{x'}A_{y'}\sin(\theta_x' - \theta_y') \rangle. \tag{3.177}$$

The quantities S_0, S_1, S_2, and S_3 are the *Stokes parameters* of the beam of radiant energy at the input (entrance aperture) of the optical instrument. They are measurable quantities and together with the measurable transfer functions τ they determine the irradiance at the detector through (3.173). In using formulas (3.173)–(3.177) it should be kept in mind that the transfer functions τ depend on the positions P_e and P_i as well as on wavelength and the choice of coordinate systems. Also, the Stokes parameters depend on the points P_s and P_e as well as the choice of coordinate system in the entrance aperture. In applying (3.173) to determine the total irradiance at P_i due to all source and entrance aperture points it is necessary to integrate (3.173) with proper attention to source coherence. When the source is incoherent the Stokes parameters from various source points add.

Measurement of the Stokes parameters is described elsewhere (Stone, 1963). Their interpretation is as follows. The parameter S_0 is simply the irradiance; S_1 the difference of the irradiances in the x and y components of the field, and, in terms of Fig. 3.36, indicates the eccentricity of the ellipse; S_2 and S_3 products of the x and y components, and indicate both the angular orientation of the ellipse and the direction of rotation of the electric field vector, clockwise or counterclockwise.

For an unpolarized wave $S_1 = S_2 = S_3 = 0$. In a partially polarized wave S_1, S_2, and S_3 are nonzero but $S_1{}^2 + S_2{}^2 + S_3{}^2 < S_0{}^2$, whereas in a completely polarized wave $S_1{}^2 + S_2{}^2 + S_3{}^2 = S_0{}^2$. We can then define the *degree of polarization* by

$$\text{degree of polarization} = (S_1{}^2 + S_2{}^2 + S_3{}^2)^{1/2}/S_0. \tag{3.178}$$

Any set of irradiance measurements on a beam of radiant energy which is made with the aid of polarizers, filters, etc., can yield no more than the Stokes parameters.

3.8 NOISE

The fundamental limit to the accuracy of all radiometric measurements is noise. In the optical portion of the electromagnetic spectrum the fundamental noise mechanisms are associated with the particle nature of photons and also with the thermal fluctuations which occur in the lower frequency electronic circuits which are designed to amplify the signal coming from a detector. We refer to the first of these as *photon shot noise* and to the second as *thermal* or *Johnson noise* (Oliver, 1965).

In addition to the photon shot noise which is identified with a beam of radiant energy there is an additional noise which is relatively small at optical frequencies in comparison with the shot noise. This additional noise is known as excess radiation noise and is characteristic of thermal (nonlaser) sources. Although it is small its characteristics are of interest because its measurement is the basis for the radiometric techniques of intensity interferometry.

In practice, other nonfundamental noise mechanisms can contribute to the reduction of radiometric accuracy. Some of these are discussed in Chapter 6 in connection with the description of detectors.

3.8.1 Photon Shot Noise

The quantum or photon properties of electromagnetic radiation play an increasingly important role as one moves from the longer to the shorter wavelength regions of the spectrum. The fundamental reason for this change is that the energy of each photon is

$$E = hv = hc/\lambda \quad (\text{J}), \tag{3.179}$$

where h is Planck's constant $= 6.626 \times 10^{-34}$ (J s), v the optical frequency (Hz), c the speed of light $= 2.998 \times 10^8$ (m s^{-1}), and λ the wavelength (m). As the wavelength decreases, a beam of fixed radiant power becomes more grainy; that is, the number of photons per unit time decreases. The number of photons per unit time or the photon flux is

$$\Phi_p = \Phi/hv \quad (\text{s}^{-1}), \tag{3.180}$$

where Φ is the radiant power (W). According to Eq. (3.180) a beam of fixed radiant power Φ has a photon flux Φ_p which is inversely proportional to the frequency. Because the number of photons per unit time decreases and because the arrival of photons at a surface is a random or statistical event, a beam of fixed radiant power becomes more noisy at higher frequencies.

The following brief derivation shows the basis for determining the statistical properties of a beam of photons. Consider a beam of photons incident on an ideal photon counter. The counter is ideal in the sense that the arrival of each photon at the counter registers as an event with no losses due to surface reflections, internal quantum efficiency, etc. The important parameters are: Δt the time interval over which photons are counted, r the probability per unit time of an event (rate of arrival of photons), and $\langle n \rangle = r \Delta t$ the mean number of photons counted.

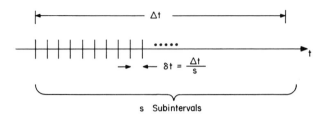

Fig. 3.38   Division of measurement time interval Δt into s subintervals of time δt.

Consider the time interval Δt to be divided into s subintervals as shown in Fig. 3.38. The length of each subinterval is then $\delta t = \Delta t/s$. Each of these subintervals is chosen to be short enough so that the probability of more than one event occurring is negligibly small. That is,

$$P(0, \delta t) + P(1, \delta t) = 1, \tag{3.181}$$

where $P(0, \delta t)$ is the probability of no event occurring during time interval δt and $P(1, \delta t)$ the probability of one event occurring during time interval δt. It is also reasonable to assume that the probability of one event occurring in the interval δt is

$$P(1, \delta t) = r \, \delta t = r \, \Delta t/s, \tag{3.182}$$

where r is the rate of arrival of photons defined earlier. Finally, it is assumed that the probability of the arrival of a photon in a time interval δt is statistically independent of the number of photons which arrived during previous intervals.

With the three assumptions represented by (3.181), (3.182), and statistical independence, the probability of n events in the time interval Δt is given by the Bernoulli distribution

$$P(n) = (p)^n (1 - p)^{s - n} \binom{s}{n}, \tag{3.183}$$

where $p = P(1, \delta t)$ and

$$\binom{s}{n} = \frac{s!}{n!(s - n)!}. \tag{3.184}$$

Equation (3.183) can then be written as

$$P(n) = \left(\frac{r \, \Delta t}{s}\right)^n \left(1 - \frac{r \, \Delta t}{s}\right)^{s - n} \frac{s!}{n!(s - n)!} \tag{3.185}$$

$$P(n) = \left(\frac{r \, \Delta t}{s}\right)^n \left(1 - \frac{r \, \Delta t}{s}\right)^{s} \left(1 - \frac{r \, \Delta t}{s}\right)^{-n} \frac{1}{n!} \frac{s!}{(s - n)!}. \tag{3.186}$$

In the limit as the number of subintervals s is allowed to increase while keeping Δt fixed,

$$P(n) = \frac{\langle n \rangle^n}{n!} e^{-\langle n \rangle}, \tag{3.187}$$

where $\langle n \rangle = r \Delta t$ is the mean number of events in Δt.

Equation (3.187) is the Poisson distribution. The fluctuations described by (3.187) are known as *shot noise*. An important property of shot noise is that the variance in the number of photons counted is

$$\sigma^2 = \langle (n - \langle n \rangle)^2 \rangle = \langle n \rangle. \tag{3.188}$$

The standard deviation is then $\sigma = \langle n \rangle^{1/2}$. If the ratio of the mean number of photons to the standard deviation in the number of photons is regarded as a signal-to-noise ratio SNR, then, using (3.188),

$$\text{SNR} = \langle n \rangle / \sigma = \langle n \rangle^{1/2} = (r \Delta t)^{1/2}. \tag{3.189}$$

According to (3.189) the SNR increases as the square root of either the photon flux r or the time interval Δt.

Figure 3.39 shows the Poisson distribution (3.187) as a function of the discrete variable n. Figure 3.40 shows (3.187) as a function of the continuous variable $r \Delta t$. It can be shown that as $\langle n \rangle$ becomes large the probability distribution (3.187) for shot noise approaches the Gaussian distribution:

$$P(n) = (2\pi \langle n \rangle)^{-1/2} \exp[-(n - \langle n \rangle)^2 / 2 \langle n \rangle]. \tag{3.190}$$

This trend can be seen in Figs. 3.39 and 3.40.

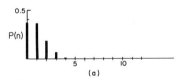

(a)

Fig. 3.39 Poisson distribution for the number n of photons counted with the mean number $\langle n \rangle$ as a parameter. (a) $\langle n \rangle = 1$; (b) $\langle n \rangle = 2$; (c) $\langle n \rangle = 5$.

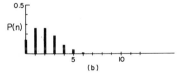

(b)

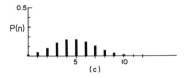

(c)

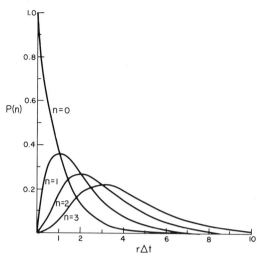

Fig. 3.40 Probability of counting n photons as a function of the continuous variable $r\,\Delta t$ using Eq. (3.187). Here r is the rate of arrival of photons and Δt the time interval.

3.8.2 Thermal Noise and Excess Radiation Noise

The spectral distribution of blackbody radiation is described in some detail in Chapter 4. It is found there that the average spectral radiant energy density in a beam of blackbody radiation is given by (4.23) as

$$\langle w_v \rangle = (8\pi v^2/c^3)[hv/(e^{hv/kT} - 1)] \quad (\mathrm{J\,m^{-3}\,Hz^{-1}}). \tag{3.191}$$

Because this spectral distribution is not independent of frequency v, the number of photons counted by a detector would not be expected to be statistically independent as smaller and smaller counting intervals are employed. For this reason the assumption of statistical independence which was used to derive the shot noise statistics represented by Eq. (3.187) is not always valid. In this section the statistics of blackbody radiation are examined in more detail to determine the conditions under which the shot noise properties (3.187)–(3.189) are valid. Some of the relationships found in this section are also the basis of the important radiometric techniques of intensity interferometry.

From the Fowler–Einstein equation of thermodynamics, the variance of the spectral energy density is

$$\langle \Delta w_v{}^2 \rangle = \langle (w_v - \langle \bar{w}_v \rangle)^2 \rangle = kT^2\, \delta \langle w_v \rangle / \delta T. \tag{3.192}$$

Using (3.191) one finds that

$$\frac{\delta \langle w_v \rangle}{\delta T} = \left(\frac{8\pi v^2}{c^3} \frac{hv}{e^{hv/kT} - 1} \right) \left(\frac{hv}{kT^2} \right) \left(\frac{e^{hv/kT}}{e^{hv/kT} - 1} \right) \tag{3.193}$$

From (3.192) and (3.193) then

$$\langle \Delta w_v^2 \rangle = hv \langle w_v \rangle \left(1 + \frac{1}{e^{hv/kT} - 1} \right). \qquad (3.194)$$

The average number of photons per unit frequency interval is $\langle n_v \rangle = \langle w_v \rangle / hv$ and the standard deviation in the number of photons per unit frequency interval is

$$\langle \Delta n_v^2 \rangle = \langle (n_v - \langle n_v \rangle)^2 \rangle = \Delta w_v^2 / (hv)^2. \qquad (3.195)$$

The combination of (3.195) and (3.194) then leads to the conclusion that

$$\langle \Delta n_v^2 \rangle = \langle n_v \rangle \left[1 + \frac{1}{e^{hv/kT} - 1} \right]. \qquad (3.196)$$

The first term on the right of (3.196) represents shot noise as described in the previous section. This term corresponds to the result in Eq. (3.188). It is the variance in the number of photons per unit frequency interval which would be obtained if the total energy were due to independent particles of energy hv. The second term on the right of (3.196) represents wave noise or *excess radiation noise*. It is associated with the wave characteristics of electromagnetic radiation.

Figure 3.41 shows the parameter hv/kT as a function of wavelength for two temperatures. Figure 3.42 then shows graphically the variance in the number of photons per unit frequency interval given by Eq. (3.196). As indicated by the figure there are two cases of interest.

(1)   Low frequencies ($hv/kT \ll 1$).   When $hv/kT \ll 1$, then

$$1/(e^{hv/kT} - 1) \cong kT/hv. \qquad (3.197)$$

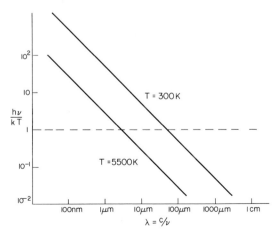

Fig. 3.41   Parameter hv/kT for various temperatures.

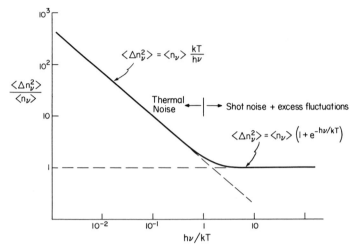

Fig. 3.42   Variance of number of photons $\langle \Delta n_v{}^2 \rangle$ as a function of the parameter hv/kT.

Using (3.197) in (3.196) leads to

$$\langle \Delta n_v{}^2 \rangle = \langle n_v \rangle [1 + (kT/hv)] \cong \langle n_v \rangle (kT/hv) \qquad (3.198)$$

or with the use also of (3.194),

$$\langle \Delta w_v{}^2 \rangle \cong \langle w_v \rangle kT. \qquad (3.199)$$

Equation (3.199) shows the interesting result that at low frequencies the fluctuations in radiant energy are linearly proportional to temperature and do not involve Planck's constant h. The quantum or photon properties are not important at low frequencies. Eq. (3.199) represents *thermal noise*.

(2)   High frequencies ($hv/kT \gg 1$).   At the higher frequencies such as in the optical region of the spectrum where typically $hv/kT \gg 1$,

$$1/(e^{hv/kT} - 1) \cong e^{-hv/kT}, \qquad (3.200)$$

so that (3.196) becomes

$$\langle \Delta n_v{}^2 \rangle = \langle n_v \rangle (1 + e^{-hv/kT}). \qquad (3.201)$$

The first term on the right of (3.201) is shot noise as in Eq. (3.188). The second term is referred to in the optical region as *excess fluctuations*. The existence of these excess fluctuations forms the basis for the optical intensity stellar interferometer experiments (Hanbury-Brown, 1974) conducted by Hanbury-Brown and Twiss at Manchester, England in 1955. The intensity interferometer greatly extends the baseline and consequently the angular resolution which can be achieved for astronomical measurements. However, for most applications in the infrared, visible, and ultraviolet regions of the spectrum, the quantity hv/kT is much larger than unity so that to a good approximation $\langle \Delta n_v^2 \rangle = \langle n_v \rangle$. This corresponds to the result obtained in the previous section in Eq. (3.188) where it was assumed that a beam of radiant power consists of a stream of independent particles of energy hv.

REFERENCES

Beckmann, P. (1963). "The Scattering of Electromagnetic Waves from Rough Surfaces." Pergamon, Oxford.

Beran, M. and Parrent, G. B., Jr. (1964). "Theory of Partial Coherence." Prentice–Hall, Englewood Cliffs, New Jersey.

Blevin, W. R. (1970). Diffraction losses in radiometry and photometry, *Metrologia* **6**, 39–44.

Born, M., and Wolf, E. (1964). "Principles of Optics," 2nd ed. Pergamon, Oxford.

Bracewell, R. (1965). "The Fourier Transform and Its Applications." McGraw–Hill, New York.

Brouwer, W. (1964). "Matrix Methods in Optical Instrument Design." Benjamin, New York.

Chandrasekhar, S. (1950). "Radiative Transfer." Oxford Univ. Press, London and New York.

Fante, R. (1975). Electromagnetic beam propagation in turbulent media, *Proc. IEEE* **63**, 1669–1692.

Françon, M., and Mallick, S. (1967). Measurement of the second order degree of coherence, *in* "Progress in Optics," Vol. VI (E. Wolf, ed.). North–Holland Publ. Amsterdam.

Goodman, J. (1968). "Introduction to Fourier Optics." McGraw–Hill, New York.

Hanbury Brown, R. (1974). "The Intensity Interferometer." Taylor and Francis, London.

Jacquinot, P. (1965). *Jpn. J. Appl. Phys.* **4**, Suppl. 1, 447.

Kingslake, R. (1978). "Lens Design Fundamentals." Academic Press, New York.

Lawrence, R., and Strohbehn, J. (1970). A survey of clear-air propagation effects relevant to optical communications, *Proc. IEEE* **58**, 1523–1545.

Marchand, E., and Wolf, E. (1974). Radiometry with sources of any state of coherence, *J. Opt. Soc. Am.* **64**, 1219–1226.

Miyamoto, K. (1961). Wave optics and geometrical optics in optical design, *in* "Progress in Optics," Vol. I (E. Wolf, ed.). North–Holland Publ., Amsterdam.

Murata, K. (1966). Instruments for the measuring of optical transfer functions, *in* "Progress in Optics," Vol. V (E. Wolf, ed.). North–Holland Publ., Amsterdam.

Nicodemus, F. (1963). Radiance, *Am. J. Phys.* **31**, 368.

Oliver, B. (1965). Thermal and quantum noise, *Proc. IEEE* **53**, 436–454.

O'Neill, E. L. (1963). "Introduction to Statistical Optics." Addison–Wesley, Reading, Massachusetts.

Palmer, J. M. (1971). "Lens Aberration Data." Amer. Elsevier, New York.

Siegel, R., and Howell, J. (1972). "Thermal Radiation Heat Transfer." McGraw–Hill, New York.

Smith, W. (1966). "Modern Optical Engineering." McGraw–Hill, New York.

Sparrow, E., and Cess, R. (1978). "Radiation Heat Transfer." augmented edition, McGraw-Hill, New York.

Steel, W. H., De, M., and Bell, J. A. (1972). Diffraction corrections in radiometry, *J. Opt. Soc. Am.* **62**, 1099–1103.

Stone, J. (1963). "Radiation and Optics." McGraw–Hill, New York.

Strohbehn, J. W., ed. (1978). "Laser Beam Propagation in the Atmosphere." Springer–Verlag, Berlin and New York.

Walther, A. (1968). Radiometry and coherence, *J. Opt. Soc. Am.* **58**, 1256–1259.

Walther, A. (1973). Radiometry and coherence, *J. Opt. Soc. Am.* **63**, 1622–1623.

Wolf, E. (1978). Coherence and radiometry, *J. Opt. Soc. Am.* **68**, 6–17.

4

Blackbody Radiation

4.1 INTRODUCTION

In this chapter we consider the angular and spectral characteristics of the radiant energy emitted by a blackbody. A *blackbody* is an ideal body which allows all incident radiant energy to pass into it so that all incident energy is absorbed and none is reflected. The definition of a blackbody requires that this be true at all wavelengths and at all angles of incidence. A blackbody is therefore a perfect absorber.

As might be expected, the emission properties of a blackbody are related to its absorption properties. The important emission properties are (1) the blackbody is a perfect emitter at each angle and at each wavelength, and (2) the total radiant energy emitted by a blackbody (over all angles and wavelengths) is a function only of its temperature. The blackbody is a perfect emitter in the sense that it emits the maximum possible radiant energy for a body at that temperature.

That these emission properties of a blackbody are a direct consequence of its absorption properties can be seen by considering Fig. 4.1 which shows a

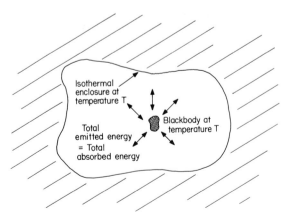

Fig. 4.1 Arrangement of blackbody inside black isothermal cavity to demonstrate emission properties of blackbody.

small blackbody of arbitrary shape inside a black isothermal enclosure. At thermal equilibrium the blackbody must be at the same temperature as the enclosure. To maintain this temperature the small blackbody must emit as much radiant energy as it absorbs, and this must remain true when the angular orientation of the blackbody is changed or when the black enclosure is arranged to radiate and absorb only within a narrow wavelength interval. Furthermore when the enclosure temperature changes, the energy absorbed and emitted by the blackbody must change in a corresponding way to maintain thermal equilibrium.

In practice, only a few surfaces such as gold black, carbon black, platinum black, and carborundum approach the blackbody in their absorbing properties. However, we will see that a small aperture in an isothermal cavity can also closely approximate the absorbing and emitting properties of a blackbody.

The angular and spectral distribution of radiant energy from an ideal blackbody constitutes an important standard of comparison for the emitting and absorbing characteristics of real objects and as a result plays a central role in radiometry. Differences between ideal and real characteristics are then described by the emissivity, absorptance, and reflectance parameters introduced in Chapter 2.

It should be kept in mind in reading this chapter that local thermal equilibrium is assumed. Although this is usually a good assumption, it does not apply to laser sources or gas discharge sources. As a result, the angular and spectral characteristics of lasers and the spectral characteristics of discharge sources can be quite different from the blackbody characteristics described here.

In this chapter we shall begin by examining the angular characteristics of blackbody radiation. After this we shall determine the spectral characteristics of blackbody radiation first derived by Planck in 1900. We shall describe a blackbody simulator which is used to approximate the angular and spectral characteristics of the ideal blackbody. Finally, we shall describe Kirchhoff's law, which relates the emissivity and absorptance of bodies which are not ideal blackbodies.

4.2   ANGULAR CHARACTERISTICS OF BLACKBODY RADIATION

4.2.1   Angular Independence of Blackbody Radiance

Radiance is the fundamental radiometric quantity. It was defined in Chapters 2 and 3 to be

$$L_\lambda(\theta,\phi) = d^2\Phi_\lambda(\theta,\phi)/dA\cos\theta\,d\omega \quad (\text{W m}^{-2}\,\text{sr}^{-1}\,\mu\text{m}^{-1}), \qquad (4.1)$$

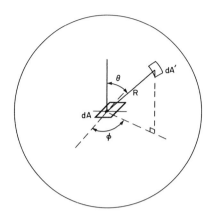

Fig. 4.2 Blackbody element of area dA surrounded by an isothermal black spherical enclosure at the same temperature.

where $d^2\Phi_\lambda$ is the radiant power, $dA \cos\theta$ the projected area in a given direction, $d\omega$ the solid angle about that direction, (θ, ϕ) the polar angles, and λ the wavelength.

We shall now show that the radiance of a blackbody is independent of angle. Consider the blackbody element of area dA shown in Fig. 4.2 to be surrounded by an isothermal spherical black enclosure at the same temperature. The radiant power at wavelength λ which is emitted by the blackbody element dA and collected by area element dA' on the sphere is found from (4.1) to be

$$d^2\Phi_\lambda(\theta, \phi) = L_\lambda(\theta, \phi)\, dA \cos\theta\, dA'/R^2, \tag{4.2}$$

where θ and ϕ are the angles shown in the figure. The power at wavelength λ emitted by the area element dA' on the sphere and collected by the black-body element dA is

$$d^2\Phi_\lambda(0, 0) = L_\lambda(0, 0)\, dA'(dA \cos\theta/R^2), \tag{4.3}$$

where the angles are $(0, 0)$ since the radiant power from dA' to dA is emitted normal to the surface of the sphere.

In thermal equilibrium the power emitted by dA must equal the power absorbed by dA. Equating (4.2) and (4.3) we then find

$$L_\lambda(\theta, \phi) = L_\lambda(0, 0). \tag{4.4}$$

That is, the radiance of the blackbody at (θ, ϕ) is the same as that at $(0, 0)$ so the radiance of the blackbody is independent of angle. Since the blackbody is a perfect absorber and emitter, then the conclusion is also valid when the blackbody is not surrounded by the isothermal enclosure. In general then the radiance of a blackbody is independent of angle.

4.2.2 Lambert's Cosine Law

The radiance defined by (4.1) and in Chapter 2 involves the projected area element $dA \cos \theta$. We can determine from (4.1) that the power emitted by a body at each wavelength per unit of actual surface area per unit solid angle is

$$d^2\Phi_\lambda(\theta, \phi)/dA \, d\omega = L_\lambda(\theta, \phi) \cos \theta. \qquad (4.5)$$

From (4.4) it then follows that for a blackbody

$$d^2\Phi_\lambda(\theta, \phi)/dA \, d\omega = L_\lambda(0, 0) \cos \theta \qquad (4.6)$$

so that the emitted power per unit area per unit solid angle is proportional to the cosine of the angle from the vertical where the angle θ is as shown in Fig. 4.2. Relation (4.6) is *Lambert's cosine law.*

4.2.3 Radiance and Radiant Exitance Relationship

There is a very simple relationship between the radiance and the radiant exitance of a blackbody. The radiant exitance was defined in Chapter 2 to be

$$M_\lambda = d\Phi_\lambda/dA = \int L_\lambda(\theta, \phi) \cos \theta \, d\omega \quad (\text{W m}^{-2}) \qquad (4.7)$$

or the radiant power per unit area. From (4.4) and the solid angle relationship $d\omega = \sin \theta \, d\phi \, d\theta$ shown in Fig. 4.3, Eq. (4.7) becomes

$$M_\lambda = L_\lambda(0, 0) \int_0^{2\pi} \int_0^{\pi/2} \cos \theta \sin \theta \, d\theta \, d\phi$$

$$= \pi L_\lambda \quad (\text{W m}^{-2}). \qquad (4.8)$$

The simple relationship found in (4.8) is that the radiant exitance of a blackbody at each wavelength is π times the radiance normal to the surface at that wavelength. In (4.8) $L_\lambda(\theta, \phi) = L_\lambda(0, 0) = L_\lambda$.

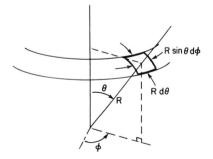

Fig. 4.3 Geometry for determining relation between radiance and radiant exitance, with $d\omega = R^2 \sin \theta \, d\phi \, d\theta/R^2$.

4.2.4 Finite Solid Angle

In some applications the radiant power per unit area contained in only a part of the hemispherical solid angle is desired. In this case the limits of integration in the equation preceding (4.8) are changed to extend from θ_1 to θ_2 and from ϕ_1 to ϕ_2 so that

$$M_\lambda(\theta_1 - \theta_2, \phi_1 - \phi_2) = L_\lambda(0,0) \int_{\phi_1}^{\phi_2} \int_{\theta_1}^{\theta_2} \cos\theta \sin\theta \, d\theta \, d\phi$$

$$= L_\lambda \tfrac{1}{2} (\sin^2\theta_2 - \sin^2\theta_1)(\phi_2 - \phi_1). \qquad (4.9)$$

It can easily be shown that (4.9) reduces to (4.8) when the entire hemisphere is covered.

4.3 SPECTRAL CHARACTERISTICS
OF BLACKBODY RADIATION

The angular characteristics of the radiant energy from a blackbody presented in Section 4.2 were essentially based on thermodynamic arguments. In contrast to this situation the spectral characteristics presented in this section cannot be obtained from thermodynamics. They require an assumption that radiant energy is quantized, an assumption which was first introduced by Planck (1901) and which became the foundation of quantum physics.

In the following we shall first determine the number of electromagnetic modes in each wavelength or frequency interval within a closed cavity. By combining this information with the assumption that energy is quantized we shall obtain Planck's spectral distribution of radiant energy within the cavity (Richtmeyer et al., 1968). On this basis it will then be possible to find the spectral radiant exitance at a small aperture which is introduced into the side of this cavity, an approach which is in fact used in the experimental realization of a blackbody simulator as described in Section 4.4.

4.3.1 Modes in a Cavity

A *mode* is a natural vibration of the electromagnetic field in a cavity or enclosure. Consider the cubic cavity shown in Fig. 4.4 with sides of length l. We know that the electromagnetic waves within the cavity must satisfy the wave equation which can be written in rectangular coordinates as

$$\frac{\partial^2 U}{\partial x^2} + \frac{\partial^2 U}{\partial y^2} + \frac{\partial^2 U}{\partial z^2} = \frac{1}{c^2} \frac{\partial^2 U}{\partial t^2}. \qquad (4.10)$$

In (4.10) the spatial coordinates are x, y, z, the time coordinate is t, U the wave amplitude, and c the speed of light. In addition to satisfying the wave equation

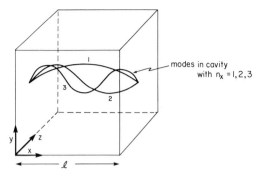

Fig. 4.4   Isothermal cavity containing radiant energy.

the wave amplitude must also be zero at the walls of the cavity to meet the boundary conditions.

The only waves which meet both of these conditions are of the form

$$U(x, y, z, t) = \sin(2\pi v_x x/c)\sin(2\pi v_y y/c)\sin(2\pi v_z z/c)\sin(2\pi vt), \quad (4.11)$$

where v is the frequency of the wave. By substituting (4.11) into (4.10) we see that

$$v = (v_x^2 + v_y^2 + v_z^2)^{1/2}. \quad (4.12)$$

The values which can be assumed by v_x, v_y, and v_z depend on the size of the cavity. Since the field amplitude $U(x, y, z, t) = 0$ at $x = 0$ and at $x = l$, for example, then $2\pi v_x l/c = n_x \pi$, where n_x is a positive integer. The permissible values of v_x are then $v_x = (c/2)(n_x/l)$. Following a similar approach for the y and z directions and substituting in (4.12) shows that the permissible frequencies v are

$$v = (c/2l)(n_x^2 + n_y^2 + n_z^2)^{1/2} = v_0(n_x^2 + n_y^2 + n_z^2)^{1/2}, \quad (4.13)$$

where $v_0 = c/2l$ is the fundamental frequency. Each combination of integers n_x, n_y, n_z represents a single mode of the electromagnetic field within the cavity.

To find the total number of modes in the cavity in each frequency interval it is helpful to construct a lattice as shown in Fig. 4.5. Each point of the lattice represents a single combination of integers n_x, n_y, n_z. To find the number of modes within a frequency interval dv about the frequency v we see from (4.13) that we can simply count the number of lattice points within a spherical shell of radius v/v_0 and thickness $d(v/v_0)$. Since n_x, n_y, and n_z are all positive the lattice points lie only in one octant of a sphere. The volume of this shell and the corresponding number of modes within the frequency range dv is then

$$\text{no. of modes} = \tfrac{1}{8} \cdot 4\pi(v/v_0)^2 \, d(v/v_0) = (4\pi v^2/c^3) \, dv \, l^3. \quad (4.14)$$

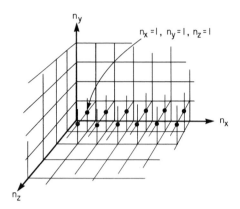

Fig. 4.5   Lattice representing possible modes in cavity.

If we now recall that electromagnetic radiant energy has two modes of polarization for each cavity mode shown in (4.14) and calculate the total number of modes per unit volume of the cavity within the frequency range dv we conclude from (4.14) that

$$N_v \, dv = (2 \times \text{no. of modes})/l^3 = (8\pi v^2/c^3) \, dv, \qquad (4.15)$$

where N_v represents the mode density. Although we assumed a cubic cavity to find (4.15), it can be shown that this result is independent of cavity shape. Notice that (4.15) depends only on frequency.

4.3.2   Average Energy per Mode

The previous section provided us with the number of modes in a unit volume and unit frequency interval in a cavity. If we knew the average energy per mode, we could then determine the average energy per unit volume and unit frequency interval. This was first done by Planck (1901) who made the fundamental assumption that energy is quantized. We now refer to these energy quanta as *photons*. The assumption of quantized energy enters in the following way.

To calculate the average energy per mode when the energy can have only discrete values Q_n we calculate

$$\bar{Q} = \sum_{n=0}^{\infty} Q_n P(Q_n) \bigg/ \sum_{n=0}^{\infty} P(Q_n) \qquad (4.16)$$

where $\bar{Q}$ is the average energy, Q_n the nth energy value, and $P(Q_n)$ a probability function which represents the relative probability of occurrence of the nth energy value. Planck assumed that the energy must be an integral multiple of hv, where v is the frequency and h Planck's constant ($h = 6.626 \times 10^{-34}$ J s). The energy values must then be $Q_n = nhv$.

For the relative probability $P(Q_n)$ we use the Boltzmann distribution (Richtmeyer *et al.*, 1968; TerHaar, 1960)

$$P(Q_n) = \exp(-Q_n/kT), \tag{4.17}$$

which is appropriate for distinguishable energy quanta and in which T is the absolute temperature (K) and k Boltzmann's constant ($k = 1.38 \times 10^{-23}$ J K^{-1}). Using (4.17) and $Q_n = nhv$ in (4.16) we find

$$\bar{Q} = hv \sum_{n=0}^{\infty} n\exp(-nhv/kT) \bigg/ \sum_{n=0}^{\infty} \exp(-nhv/kT). \tag{4.18}$$

To evaluate the sum in the denominator we expand the sum and use the series relation $1 + x + x^2 + x^3 + \cdots = (1 - x)^{-1}$ to find

$$\sum_{n=0}^{\infty} \exp(-nhv/kT) = [1 - \exp(-hv/kT)]^{-1}. \tag{4.19}$$

Also, the sum in the numerator of (4.18) is evaluated by noticing that $\sum n\exp(-n\alpha) = -d/d\alpha \sum \exp(-n\alpha)$. This leads to

$$\sum_{n=0}^{\infty} n\exp(-nhv/kT) = \exp(-hv/kT)[1 - \exp(-hv/kT)]^{-2}, \tag{4.20}$$

and, by combining (4.19) and (4.20) in (4.18), we determine that

$$\bar{Q} = hv/[\exp(hv/kT) - 1] \quad \text{(J)}. \tag{4.21}$$

This is Planck's relationship for the average energy per mode. Notice that the only variables in (4.21) are the frequency v and temperature T. It should also be noted that the energy per mode can fluctuate (TerHaar, 1960) about the average energy per mode $\bar{Q}$.

4.3.3 Spectral Radiant Exitance of Blackbody

In this section we first find the average spectral energy density in a cavity and then from this determine the average spectral radiant exitance of a small aperture in the side of the cavity. We assume that the medium is vacuum. For other media see Section 4.3.9.

The average spectral energy density in a cavity is

$$\bar{w}_v = N_v \bar{Q}, \tag{4.22}$$

where $\bar{w}_v$ is the average energy per unit volume per unit frequency interval, $\bar{Q}$ the average energy per mode, and N_v the number of modes per unit volume per unit frequency interval. From (4.15) and (4.21) we find that (4.22) is

$$\bar{w}_v = (8\pi v^2/c^3)[hv/(\exp(hv/kT) - 1)] \quad (\text{J m}^{-3}\,\text{Hz}^{-1}). \qquad (4.23)$$

Equation (4.23) is *Planck's equation* for the spectral radiant energy density in a cavity. It can be expressed in terms of wavelength rather than frequency by using the relations $\bar{w}_v dv = \bar{w}_\lambda d\lambda$, $c = v\lambda$, and $d\lambda = -(c/v^2)\,dv$, where $\bar{w}_\lambda$ is the average energy per unit volume per unit wavelength interval. It is important to note that $\bar{w}_v \neq \bar{w}_\lambda$. Using these relations with (4.23) we have

$$\bar{w}_\lambda = (8\pi hc/\lambda^5)[1/(\exp(hc/\lambda kT) - 1)] \quad (\text{J m}^{-3}\,\mu\text{m}^{-1}). \qquad (4.24)$$

Now consider a small aperture in the side of a cavity which contains radiant energy with spectral density given by (4.24). If the aperture is small enough, and we assume it is, the energy density within the cavity will not change significantly. To find the spectral radiant exitance M_λ at the aperture we use the relation

$$M_\lambda = \pi L_\lambda = \pi(c/4\pi)\bar{w}_\lambda, \qquad (4.25)$$

where L_λ is the spectral radiance at the aperture. The basis for the first relation in (4.25) is (4.8) and the basis for the $c/4\pi$ factor in the second relation is that the energy density at any point in the cavity represents the integrated radiance from 4π sr, and that in a vacuum this energy propagates a distance of one wavelength λ in one wave period T so that $\lambda/T = \lambda v = c$.

The average spectral radiant exitance at the aperture in the cavity is found from (4.25) and (4.24) to be

$$M_\lambda = (2\pi hc^2/\lambda^5)[1/(\exp(hc/\lambda kT) - 1)] \quad (\text{W m}^{-2}\,\mu\text{m}^{-1}). \qquad (4.26)$$

This is the *Planck blackbody spectral radiant exitance formula* (for vacuum) which is the basis for much of the radiometric analysis of thermal sources. Figure 4.6 shows Eq. (4.26) in graphical form as a function of wavelength λ with the temperature T as a parameter. Spectral radiance is also found from these curves through the relation $L_\lambda = M_\lambda/\pi$. Notice that the spectral radiant exitance at each wavelength increases with temperature and that the peak spectral radiant exitance shifts to shorter wavelengths with increasing temperature. For $T = 300$ K, which is approximately room temperature, M_λ peaks at approximately 9.6 μm in the infrared. For $T = 5000$–6000 K, which is the temperature of the sun's surface, the peak is in the visible region of the spectrum.

It is possible to put (4.26) in a more convenient form which does not require providing a separate curve for each temperature. To do this we divide both sides of (4.26) by T^5 and obtain the *universal Planck formula*:

$$M_\lambda/T^5 = 2\pi hc^2/(\lambda T)^5[\exp(hc/k\lambda T) - 1] \quad (\text{W m}^{-2}\,\mu\text{m}^{-1}\,\text{K}^{-5}), \qquad (4.27)$$

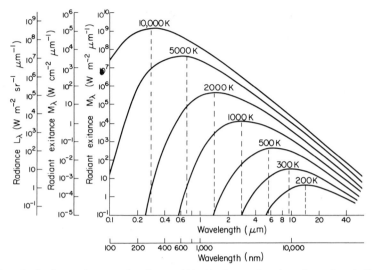

Fig. 4.6   Radiant exitance and radiance of blackbody as a function of wavelength. Dashed line indicates wavelength λ_m at which peak occurs.

which is a function only of the single variable λT. This relation is shown in graphical form (Pivovonsky and Nagel, 1961) in Fig. 4.7. This form of the blackbody spectral distribution is used in later sections of this chapter to derive other important spectral characteristics of blackbody radiation.

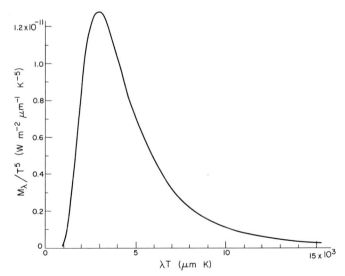

Fig. 4.7   Universal Planck curve showing the normalized quantity M_λ/T^5 as a function of λT. The Wien displacement law shows that the peak occurs at $\lambda_m T = 2.88 \times 10^3 \ \mu m \ K$.

4.3.4   Wien and Rayleigh–Jeans Approximations

There are two approximate forms of the spectral radiant exitance formula
(4.26) which are convenient because of their simplicity and are valid when
the product λT is very small or very large. Historically, both approximations
actually preceded the accurate derivation by Planck of the blackbody
radiation law.

When the quantity in the exponential of (4.26) satisfies the condition
$hc/\lambda kT \gg 1$ so that $\exp(hc/\lambda kT) \gg 1$, then

$$M_\lambda \cong (2\pi hc^2/\lambda^5)\exp(-hc/\lambda kT) \quad (\text{W m}^{-2}\,\mu\text{m}^{-1}). \tag{4.28}$$

This is *Wien's approximation* for the spectral radiant exitance of a blackbody
in vacuum (Wien, 1896). This approximation is valid at short wavelengths
and/or low temperatures. Figure 4.8 shows the region in which the Wien
formula closely approximates the correct formula based on Planck's law.

On the other hand, when $hc/\lambda kT \ll 1$ so that the exponential in (4.26) is
$\exp(hc/\lambda kT) \cong 1$, then we can use the expansion $e^x = 1 + x + x^2 + x^3 + \cdots$
to write $e^x \simeq 1 + x$ for $x \ll 1$. Then (4.26) becomes

$$M_\lambda \cong 2\pi ckT/\lambda^4 \quad (\text{W m}^{-2}\,\mu\text{m}^{-1}). \tag{4.29}$$

This is the *Rayleigh–Jeans approximation* for the spectral radiant exitance
of a blackbody in vacuum (Rayleigh, 1900; Jeans, 1905). It is valid for long
wavelengths and/or high temperatures. Within the normal range of temperatures the wavelengths which satisfy the condition $hc/\lambda kT \ll 1$ are so
long that they are in the radio part of the spectrum. The Rayleigh–Jeans
approximation is also shown in Fig. 4.8.

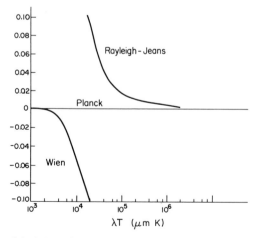

Fig. 4.8   Fractional deviation of Wien and Rayleigh–Jeans approximations from Eq. (4.26).

4.3.5   Wien Displacement Law

An important characteristic of the spectral radiant exitance formula (4.26) and the universal formula (4.27) is that the peak shifts or is displaced to shorter wavelengths with increasing temperature (Wien, 1894). This was shown in Fig. 4.6. To find the wavelength at which this peak occurs we can differentiate (4.26) or (4.27) with respect to λ and set this equal to zero. The result is

$$\lambda_m T = hc/5k = 2.88 \times 10^3 \quad (\mu\text{m K}), \tag{4.30}$$

where λ_m is the wavelength at which the peak of M_λ occurs. This is the *Wien displacement law*. Notice that if we had determined the wavelength at which M_ν expressed in terms of frequency has its peak, the result would be different. In that case

$$\lambda_m T = hc/3k = 4.80 \times 10^3 \quad (\mu\text{m K}), \tag{4.31}$$

where λ_m is now the wavelength at which M_ν reaches its peak.

By rearranging the left sides of (4.30) and (4.31) and using $c = \nu\lambda$, we can also obtain the conveniently remembered results that M_λ peaks at $h\nu/kT = 5$ and M_ν peaks at $h\nu/kT = 3$.

4.3.6   Stefan–Boltzmann Law

If we integrate the spectral radiant exitance in (4.26) over all wavelengths, we obtain the total radiant exitance of a blackbody. This is

$$M = 2\pi hc^2 \int_0^\infty \lambda^{-5}/\left[\exp(hc/\lambda kT) - 1\right] d\lambda. \tag{4.32}$$

To facilitate the integration set $x = hc/\lambda kT$ so that $\lambda = hc/xkT$ and $d\lambda = (-hc/x^2 kT)\,dx$. Then

$$M = (2\pi h/c^2)(kT/h)^4 \int_0^\infty x^3/(e^x - 1)\,d\lambda$$
$$= \sigma T^4 \quad (\text{W m}^{-2}), \tag{4.33}$$

where σ is the Stefan–Boltzmann constant ($\sigma = 5.67 \times 10^{-8}\ \text{W m}^{-2}\,\text{K}^{-4}$). The relationship $M = \sigma T^4$ is the *Stefan–Boltzmann law* (Stefan, 1879; Boltzmann, 1884). To give some idea of typical magnitudes of the total radiant exitance, (4.33) shows that at $T = 300$ K the total radiant exitance of a blackbody in vacuum is approximately 460 W m^{-2}.

4.3.7   Peak Spectral Radiant Exitance

The Stefan–Boltzmann law in the previous section applies to the total radiant exitance integrated over all wavelengths or frequencies. If we consider the dependence of the spectral radiant exitance on temperature, we find a different relationship. By substituting the Wien displacement law (4.30) into (4.27) we find that at the peak where $\lambda_m T = 2.88 \times 10^3$ (μm K) we have

$$M_\lambda(\lambda = \lambda_m) = \text{const.} \cdot T^5 \quad (\text{W m}^{-2}\,\mu\text{m}^{-1}).$$ (4.34)

That is, the peak spectral radiant exitance is proportional to the fifth power of the absolute temperature.

4.3.8 Radiant Exitance in Wavelength Interval

In many applications we are interested in finding the radiant exitance of a blackbody between two wavelengths λ_1 and λ_2. This is illustrated in Fig. 4.9. The fraction of the total radiant exitance contained between λ_1 and λ_2 is

$$f_{\lambda_1 - \lambda_2} = \int_{\lambda_1}^{\lambda_2} M_\lambda \, d\lambda \bigg/ \int_0^\infty M_\lambda \, d\lambda = (\sigma T^4)^{-1} \int_{\lambda_1}^{\lambda_2} M_\lambda \, d\lambda.$$ (4.35)

This last integral can be expressed as two integrals each starting at $\lambda = 0$ so that

$$f_{\lambda_1 - \lambda_2} = (\sigma T^4)^{-1} \left[\int_0^{\lambda_2} M_\lambda \, d\lambda - \int_0^{\lambda_1} M_\lambda \, d\lambda \right] = f_{0-\lambda_2} - f_{0-\lambda_1}.$$ (4.36)

The fraction $f_{\lambda_1 - \lambda_2}$ can then be found for any application from values of the integrated fractional radiant exitance $f_{0-\lambda}$.

To obtain a single set of values of $f_{0-\lambda}$ which will be valid for all temperatures it is convenient to rewrite (4.36) as

$$f_{\lambda_1 - \lambda_2} = (1/\sigma) \left[\int_0^{\lambda_2 T} (M_\lambda/T^5)\, d(\lambda T) - \int_0^{\lambda_1 T} (M_\lambda/T^5)\, d(\lambda T) \right]$$

$$= f_{0-\lambda_2 T} - f_{0-\lambda_1 T}.$$ (4.37)

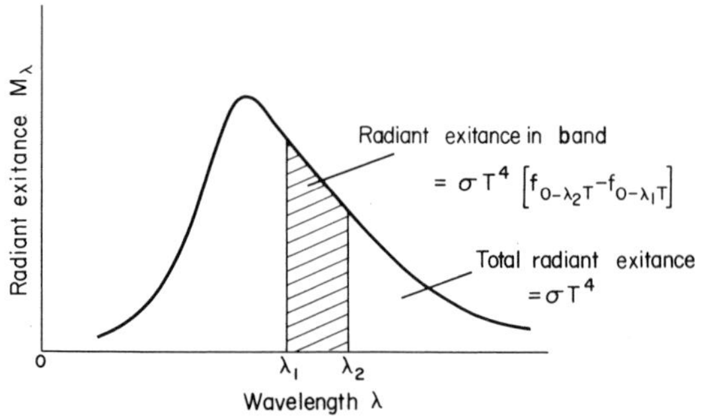

Fig. 4.9 Radiant exitance in a wavelength interval between λ_1 and λ_2.

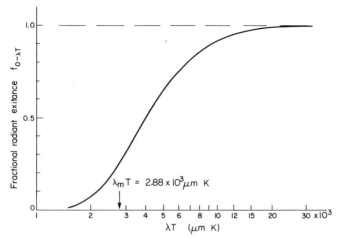

Fig. 4.10   Fractional radiant exitance in the range 0 to λT.

The form of M_λ/T^5 is known from (4.27) and can be integrated. Results of this integration are shown in Fig. 4.10 as values of $f_{0-\lambda T}$. Tabulated values of $f_{0-\lambda T}$ to higher accuracy are also available in the literature (Pivovonsky and Nagel, 1961). However, the graphical results shown in Fig. 4.10 are useful for estimating the fraction of the radiant exitance of a blackbody within a wavelength interval when the fraction is not too small. The radiant exitance in the interval is then

$$M_{\lambda_1-\lambda_2} = \sigma T^4 [f_{0-\lambda_2 T} - f_{0-\lambda_1 T}]. \quad (\text{W m}^{-2}) \qquad (4.38)$$

Figure 4.10 can also be used to determine the fraction of the total radiance between λ_1 and λ_2 since the radiance is simply related to the radiant exitance by (4.8).

4.3.9   Medium Other than Vacuum

In all preceding sections of this chapter, we have assumed that the medium is vacuum. In some cases the medium is a dielectric with index of refraction n different from $n = 1$. To find the corresponding results for this medium, we use $\lambda = \lambda_0/n$ and $c = c_0/n$ in all vacuum equations. Here λ and c are the actual wavelength and speed of light in the medium, λ_0 and c_0 the corresponding values in vacuum, and n the index in the medium.

As a result, we find that in the medium of index n the Wien displacement law (4.30) is

$$n\lambda_{\mathrm{m}} T = 2.88 \times 10^3 \quad (\mu\mathrm{m\,K}) \qquad (4.39)$$

for the peak of M_λ where λ_m is the wavelength in the medium. Similarly, the Stefan–Boltzmann law (4.33) becomes

$$M = n^2\sigma T^4 \quad (\text{W m}^{-2}).\tag{4.40}$$

Equations (4.23)–(4.29) are also modified. The modified forms are similarly found by representing the vacuum values as λ_0 and c_0 and using the relations $\lambda_0 = n\lambda$ and $c_0 = nc$.

4.4 EXPERIMENTAL REALIZATION OF A BLACKBODY SIMULATOR

In a number of measurement applications it is necessary to provide a source with angular and spectral characteristics which closely approximate those of the ideal blackbody. This source is often called a *blackbody simulator*.

Figure 4.11 shows a cutaway view of one configuration of a blackbody simulator. A copper or stainless steel cylinder with a cavity which is part cylindrical and part conical is surrounded by electrical heating coils which carry a controlled current. Close regulation of this current is important to ensure close regulation and uniformity of the cavity temperature. Insulation is used outside the heater coils to ensure that the cavity temperature is determined only by the heating coil current. The surface material inside the cavity is chosen for its thermal stability as well as for absorption characteristics. The cavity geometry itself is designed to ensure total absorption of radiant energy which is incident through the small opening in the polished surface. As indicated in the figure, incident radiant energy can undergo multiple reflections within the cavity with only a small probability of being reflected back to the opening. The opening in the cavity then acts as a

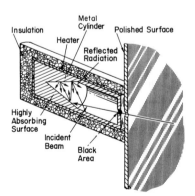

Fig. 4.11 Cutaway view of one configuration of a blackbody simulator.

blackbody. Measurements of the angular distribution of radiant power from blackbody simulators of this type have shown that they closely approximate the ideal blackbody. Achievable emissivities are typically 0.95 to 0.99.

4.5 KIRCHHOFF'S LAW

As mentioned in the introduction to this chapter, there is a close relationship between the absorption and emission properties of a body or surface. Consider again an absorbing and emitting body located in an isothermal enclosure as shown in Fig. 4.1. If the body is not perfectly black, it is characterized by an *emissivity* parameter defined as

$$\text{emissivity} = \frac{\text{radiant power emitted/area}}{\text{radiant power emitted by a blackbody/area}}. \tag{4.41}$$

More specifically, $\varepsilon(\lambda, \theta, \phi)$ is the *directional spectral emissivity* and can be written in terms of the radiance as

$$\varepsilon(\lambda, \theta, \phi) = \frac{L_{\lambda,e}(\lambda, \theta, \phi)\cos\theta\, d\omega\, d\lambda}{L_{\lambda,bb}(\lambda, \theta, \phi)\cos\theta\, d\omega\, d\lambda} = \frac{L_{\lambda,e}(\lambda, \theta, \phi)}{L_{\lambda,bb}(\lambda, \theta, \phi)}, \tag{4.42}$$

where (θ, ϕ) are the angular coordinates shown in Fig. 4.2, $L_{\lambda,e}$ the emitted spectral radiance, and $L_{\lambda,bb}$ the spectral radiance of a blackbody. Since the blackbody emits the maximum possible radiant power in each direction and at each wavelength, it can be seen that $0 \le \varepsilon(\lambda, \theta, \phi) \le 1$.

The absorbing properties of the surface are characterized by the *absorptance* defined as

$$\text{absorptance} = \frac{\text{radiant power absorbed}}{\text{radiant power incident on surface}}, \tag{4.43}$$

and specifically the *directional spectral absorptance* $\alpha(\lambda, \theta, \phi)$ is

$$\alpha(\lambda, \theta, \phi) = L_{\lambda,a}(\lambda, \theta, \phi)/L_{\lambda,i}(\lambda, \theta, \phi), \tag{4.44}$$

where $L_{\lambda,a}$ and $L_{\lambda,i}$ are the absorbed and incident spectral radiance.

For the nonblackbody in the cavity shown in Fig. 4.1 the emitted and absorbed radiance must be equal at each wavelength and in each direction just as for the blackbody, otherwise thermal equilibrium would not be maintained. Since the radiance incident on the body in the isothermal cavity is that for a blackbody, we have $\varepsilon(\lambda, \theta, \phi)L_{\lambda,bb} = \alpha(\lambda, \theta, \phi)L_{\lambda,bb}$ which leads to the important conclusion that

$$\varepsilon(\lambda, \theta, \phi) = \alpha(\lambda, \theta, \phi) \qquad (4.45)$$

in each direction (θ, ϕ) and at each wavelength λ. The fundamental relation (4.45) is known as *Kirchhoff's law*. It shows that good emitters are good absorbers. Notice that there are no restrictions on the applicability of this law other than thermal equilibrium. Equation (4.45) applies to each polarization component separately and therefore applies directly to a beam of unpolarized radiant power.

However, it should also be noted that geometrically or spectrally averaged emissivity and absorptance quantities are equal only under certain conditions. Table 4.1 defines the geometrical and spectral averages of emissivity. Similar average absorptance quantities are defined in terms of the incident and absorbed radiance shown in (4.44). The *hemispherical* quantities have been averaged over the solid angle of 2π sr while the *total* quantities have been averaged over all wavelengths λ. The conditions under which the various average emissivity and absorptance quantities are equal can be found from the following considerations.

In Table 4.1 the emitted radiance is $L_{\lambda,e} = \varepsilon(\lambda, \theta, \phi)L_{\lambda,\text{bb}}$. This relation can be used in each of the definitions of average emissivity shown in the table. Similarly, the absorbed radiance is $L_{\lambda,a} = \alpha(\lambda, \theta, \phi)L_{\lambda,i}$ and this can be used in the average absorptance definitions which involve the same angular and spectral integrations as shown for the emissivity in Table 4.1. When this is done and the average emissivity is compared with the corresponding average absorptance we find the required conditions shown in Table 4.2. Only under the conditions shown in Table 4.2 are the averaged emissivity and absorptance quantities equal.

TABLE 4.1   Directional Spectral Emissivity and Average Emissivity Quantities[a]

Quantity	Symbol	Definition
Directional spectral emissivity	$\varepsilon(\lambda, \theta, \phi)$	$\dfrac{L_{\lambda,e}(\lambda, \theta, \phi)}{L_{\lambda,\text{bb}}(\lambda, \theta, \phi)}$
Directional total emissivity	$\varepsilon(\theta, \phi)$	$\dfrac{\int_0^\infty L_{\lambda,e}(\lambda, \theta, \phi)\,d\lambda}{\int_0^\infty L_{\lambda,\text{bb}}(\lambda, \theta, \phi)\,d\lambda}$
Hemispherical spectral emissivity	$\varepsilon(\lambda)$	$\dfrac{\int_{(2\pi)} L_{\lambda,e}(\lambda, \theta, \phi)\cos\theta\,d\omega}{\int_{(2\pi)} L_{\lambda,\text{bb}}(\lambda, \theta, \phi)\cos\theta\,d\omega}$
Hemispherical total emissivity	ε	$\dfrac{\int_0^\infty \int_{(2\pi)} L_{\lambda,e}(\lambda, \theta, \phi)\cos\theta\,d\omega\,d\lambda}{\int_0^\infty \int_{(2\pi)} L_{\lambda,\text{bb}}(\lambda, \theta, \phi)\cos\theta\,d\omega\,d\lambda}$

[a] The notation (2π) indicates integration over the hemisphere.

TABLE 4.2 Required Conditions for the Equality of Average Emissivity and
Absorptance Quantities

Quantity	Equality	Required conditions
Directional spectral emissivity	$\varepsilon(\lambda, \theta, \phi) = \alpha(\lambda, \theta, \phi)$	None (except thermal equilibrium)
Directional total emissivity	$\varepsilon(\theta, \phi) = \alpha(\theta, \phi)$	(1) Spectral distribution of incident energy is proportional to that of bb, or (2) $\varepsilon(\lambda, \theta, \phi) = \alpha(\lambda, \theta, \phi)$ is independent of wavelength
Hemispherical spectral emissivity	$\varepsilon(\lambda) = \alpha(\lambda)$	(1) Incident radiant energy is independent of angle, or (2) $\varepsilon(\lambda, \theta, \phi) = \alpha(\lambda, \theta, \phi)$ is independent of angle
Hemispherical total emissivity	$\varepsilon = \alpha$	(1) Incident radiant energy is independent of angle *and* has spectral distribution proportional to that of bb, or (2) Incident radiant energy is independent of angle *and* $\varepsilon(\lambda, \theta, \phi) = \alpha(\lambda, \theta, \phi)$ is independent of wavelength, or (3) Incident radiant energy at each angle has spectral distribution proportional to that of bb *and* $\varepsilon(\lambda, \theta, \phi) = \alpha(\lambda, \theta, \phi)$ is independent of angle, or (4) $\varepsilon(\lambda, \theta, \phi) = \alpha(\lambda, \theta, \phi)$ is independent of angle and wavelength

For an opaque body it is true that the incident power is the sum of the absorbed power and the reflected power. That is, $\alpha + \rho = 1$ where ρ is the reflectance. This equality applies at all times to directional spectral as well as to average quantities. However, it is only under the conditions shown in Table 4.2 that the equality of absorptance and emissivity can be used to establish a relation between reflectance and emissivity. Thus $\alpha(\lambda, \theta, \phi) + \rho(\lambda, \theta, \phi) = \varepsilon(\lambda, \theta, \phi) + \rho(\lambda, \theta, \phi) = 1$ under all conditions but $\varepsilon(\theta, \phi) + \rho(\theta, \phi) = 1$ and $\varepsilon(\lambda) + \rho(\lambda) = 1$ and $\varepsilon + \rho = 1$ only under the stated conditions.

In the matter of terminology it should be noted that a *graybody* refers to a body or surface for which the directional spectral emissivity $\varepsilon(\lambda, \theta, \phi)$ is independent of wavelength. The spectral distribution of radiant power from a graybody is then proportional to that from a blackbody at each angle. Also, a *diffuse* emitting surface refers to a surface for which the directional spectral emissivity $\varepsilon(\lambda, \theta, \phi)$ is independent of direction. The angular distribution of radiant power from a diffuse surface is then proportional to that from a blackbody at each wavelength.

REFERENCES

Boltzmann, L. (1884). *Ann. Phys.* **2**(22), 291.

Jeans, J. (1905). *Phil. Mag.* **10**, 91.

Pivovonsky, M., and Nagel, M. (1961). "Tables of Blackbody Radiation Functions." Macmillan, New York.

Planck, M. (1901). *Ann. Phys.* **4**(3), 553.

Rayleigh, Lord (1900). *Phil. Mag.* **49**, 539.

Richtmeyer, F., Kennard, E., and Cooper, R. (1968). "Introduction to Modern Physics," 5th ed. McGraw–Hill, New York.

Sears, F. (1953). "Thermodynamics," 2nd ed. Addison–Wesley, Reading, Massachusetts.

Stefan, J. (1879). *Sitzber. Akad. Wiss. Wien* **79**(2), 391.

TerHaar, D. (1960). "Elements of Statistical Mechanics," 2nd ed. Holt, New York.

Wien, W. (1894). *Ann. Phys.* **2**(52), 132.

Wien, W. (1896). *Ann. Phys.* **3**(58), 662.

5

Radiation Sources

5.1 INTRODUCTION

This chapter is concerned with sources of optical radiant energy. The emphasis is given to those sources which are most likely to be of interest to scientists and engineers dealing with optical radiation measurements in the near UV–visible–near IR regions. Sources in the far IR region are usually either lasers or black body type sources as described in Chapter 4.

The basic concepts of optical radiation and the distinction between blackbody and graybody radiators have been presented in Chapters 3 and 4. In this chapter, we shall describe two types of radiation sources, natural and artificial. Although natural radiation sources such as sunlight, skylight, and daylight are not normally used with laboratory optical instruments, reference is often made to them. In some special applications daylight simulators are used to evaluate the optical properties of materials and particularly to measure and evaluate the color of materials. In such cases one is interested in the measurement and specification of the colors of light.

The original impetus for international agreement on methods and data for measuring and specifying color came from illuminating engineering and applications of colorimetry to light sources and signal lights. Until quite recently, all light sources consisted of incandescent substances. Methods of measurement and specification of the color of light originated in programs for measuring such sources. Because the spectral distribution of power radiated by blackbodies can be predicted from fundamental physical principles as described in Chapter 4, they are the basis for measurement and specification of the colors of light from many kinds of light sources. The spectral distributions and colors of blackbodies (Planckian sources) depend only on their temperatures. Consequently, their spectral distributions and colors are commonly specified in terms of their temperatures. By extension, the colors and, in appropriate cases, the spectral distributions of many other nonlaser light sources are also specified in terms of temperature, the concept of which has been discussed in the previous chapter. A natural source such as daylight, in its many different phases, can also be appropriately specified by its color temperature. Although the spectral distributions of the various phases of daylight are distinctly different from the spectral distributions of the most nearly similar blackbody sources, enough is known about the spectral distributions of daylight so that the spectral distribution of any particular phase of daylight can be computed from its color temperature or from measurements of its relative spectral power at a very few wavelengths.

5.2 NATURAL SOURCES

5.2.1 Sunlight, Skylight, and Daylight

When discussing natural sources of radiant energy, we have primarily in mind *sunlight*; however, we should not omit the radiation from the sky which is termed *skylight* and the combination of the two, termed *daylight*.

The relative spectral distribution and radiant power of sunlight which reaches the earth's surface are modified by the atmosphere through which the sunlight is transmitted. The attenuation of sunlight is caused by the scattering properties of molecules, water droplets, and aerosols suspended in the atmosphere, and since the scattering effect of these particles is wavelength-dependent, both the radiant power and color temperature of sunlight and daylight (sunlight plus skylight) decrease with increasing air mass or decreasing solar altitude. The radiant power and spectral distribution of skylight and daylight also depend on the abundance and particle size of water droplets and aerosols in the air. For this reason, the spectral distribution of various phases of daylight depends not only on solar altitude but also on atmospheric conditions.

Until recently very few data pertaining to the spectral energy distribution of daylight were available in the literature. It is curious, however, that while the spectral power distribution of direct sunlight has been actively studied for more than 70 years, investigation of the spectral composition of daylight has been largely neglected. The notable exception has been the book of Taylor and Kerr, who 35 years ago published the results of their measurements (Taylor and Kerr, 1941). While their data have been used in establishing some standards (American Standards Association, 1958), they do have certain limitations. The measurements of Taylor and Kerr were made with the receiver in the horizontal plane, which may not be the most representative. Furthermore, their measurements do not include the ultraviolet region of the spectrum and no data are given on the relationship between the spectral distribution of daylight and solar altitude.

More recently, extensive measurements of daylight have been made under a variety of experimental conditions (Condit and Grum, 1964; Judd *et al.*, 1964a; Henderson and Hodgkiss, 1963; Nayatani and Wyszecki, 1963; Schutze, 1970).

5.2.2 Effect of Atmospheric Conditions on Spectral Distribution of Daylight

It has been stated previously that the spectral distribution of sunlight is modified by the atmosphere through which it is transmitted. The four curves in Fig. 5.1 represent the spectral irradiance of daylight measured on a 15–0° plane for four different atmospheric conditions for the solar altitude of 40°. These are clear sky (curve 1), light hazy sky (curve 2), very light to light clouds (curve 3), and heavy overcast sky (curve 4).

Although the spectral irradiance of daylight is not that of a blackbody, reference is usually made to the color temperature of daylight when discussing its color quality. Therefore the chromaticity coordinates (u, v) are shown for all curves in these figures (MacAdam, 1937; Brussels Section of the CIE, 1960). Calculation of chromaticity coordinates is described in Section 5.4.2.

An examination of the data in Fig. 5.1 shows that a marked decrease in the blue component of daylight occurs when the sky becomes hazy. As light clouds develop, the blue component increases slightly; with heavy overcast, the spectral composition of daylight is only slightly less blue than for the clear sky condition (and actually exceeds it in the ultraviolet region).

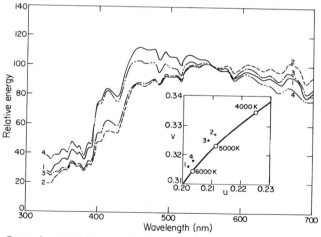

Fig. 5.1 Spectral composition of daylight as a function of atmospheric conditions. Curve 1, clear sky; curve 2, light hazy sky; curve 3, very light to light clouds; curve 4, heavy overcast sky.

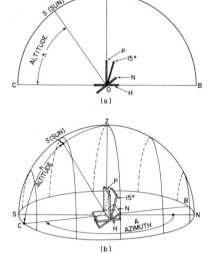

Fig. 5.2 Target orientation. (a) two-dimensional plane; (b) three-dimensional.

The orientation designated 15°–0° means the measurements were made from a plane tipped back from the perpendicular by 15°, but facing the sun in azimuth. Figure 5.2 illustrates this orientation. Part (a) represents a vertical plane through the sun and (b) illustrates the three-dimensional relationships, and together with the cross-sectional view above, serve to define the terms *azimuth A* and *solar altitude h*.

5.2.3 Effect of Solar Altitude

Spectral irradiances of daylight for clear sky condition at solar altitudes of 70, 40, 30, 20, 15, 10, and 8° are given in Fig. 5.3. The correlated color temperatures, as defined in Section 5.4.2, for the conditions in Fig. 5.3 go from about 6000 K for 70° solar altitude down to 4000 K at the altitude of 8°. The effect of solar altitude on the color temperature of daylight can be best demonstrated graphically. Figure 5.4 is a plot of color temperature as a function of solar altitude. Curves 1, 2, and 3 represent clear sky conditions on different days. Curve 4 is for light haze condition, while curve 5 represents heavy overcast.

The altitude of the sun at any hour of the day depends on the position on the earth at which it is observed, the time of the day and year, and the declination of the sun. Solar altitude at any place and time may be determined by the formula (Jones and Condit, 1948).

$$\sin \alpha = \sin \beta \sin \gamma + \cos \beta \cos \gamma \cos(\text{hour angle}),$$

where α is the solar altitude, β the latitude of the place, and γ the declination of the sun. The hour angle is the number of hours before or after local solar

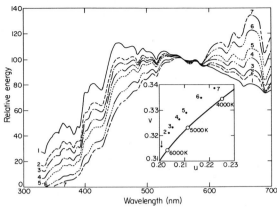

Fig. 5.3 Spectral composition of daylight as a function of solar altitude.

Curve and u, v Plot No.	Solar Altitude (Deg.)
1	70
2	40
3	30
4	20
5	15
6	10
7	8

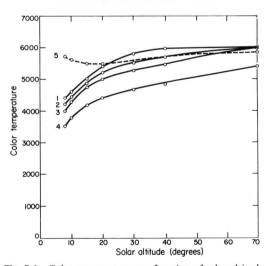

Fig. 5.4 Color temperature as a function of solar altitude.

Curve No.	Illuminant	Sky Conditions
1	Daylight	Clear
2	Daylight	Clear
3	Daylight	Clear
4	Daylight	Light haze
5	Daylight	Heavy overcast

noon. The solar altitude varies continuously from day to day, but the total variation during any one-month period is relatively small.

5.2.4   Effect of Haze and Overcast

In Fig. 5.1, curve 3 represents the spectral irradiance of daylight measured under a light haze condition. The data clearly show that the spectral composition of daylight from a light hazy day is redder than for a clear sky condition at all solar altitudes.

Spectral irradiance of daylight for heavy overcast is given in Fig. 5.5. An inspection of the figure shows that the color temperature is considerably lower than 6000 K and that the spectral composition of radiant energy from the overcast sky varies much less with solar altitude than does daylight for clear atmospheric conditions. Below a solar altitude of approximately 20° the color temperature may even increase with a decrease in the sun's angular elevation. (Fig. 5.4, curve 5).

It should be noted in the cases of clear hazy and light cloud conditions that all of the chromaticity points are above the blackbody locus, i.e., daylight for these conditions is slightly green with respect to the blackbody

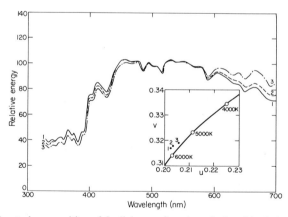

Fig. 5.5   Spectral composition of daylight as a function of solar altitude (overcast sky).

Curve and u, v Plot No.	Solar Altitude (Deg.)
1	70
2	40
3	20

nearest to it in color temperature. In the case of daylight from overcast sky, however, the spectral composition is slightly pink at very low solar altitudes.

5.2.5 Skylight

Spectral irradiance of skylight is markedly different from daylight. Such data for a clear sky are shown in Figs. 5.6–5.8. Inspection of these curves shows that the bluest sky condition does not occur with the sun at its highest elevation. In fact, there is a marked increase in color temperature with decreasing solar altitude from 70 to 20° (Fig. 5.3), after which point the color temperature decreases with lower solar altitudes. This is accounted for by the fact that for higher solar altitudes much of the radiant energy on the target comes from an extended area of the sky surrounding the sun which is far brighter and much less blue than that portion of the sky toward which the target is facing. With a decrease of solar altitude, the effect of this bright area on the spectral composition of skylight incident on the target diminishes.

Another characteristic of the spectral composition of skylight is that it varies much less with solar altitude than does daylight (sunlight plus skylight) for a clear sky, although it certainly varies more with atmospheric conditions than does daylight.

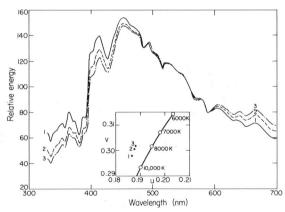

Fig. 5.6 Spectral composition of skylight as a function of solar altitude (clear sky).

Curve and u, v Plot No.	Solar Altitude (Deg.)
1	15
2	10
3	8

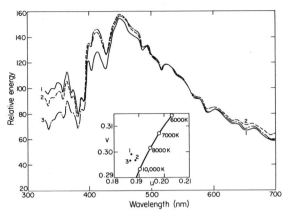

Fig. 5.7 Spectral composition of skylight as a function of solar altitude (clear sky).

Curve and u, v Plot No.	Solar Altitude (Deg.)
1	40
2	30
3	20

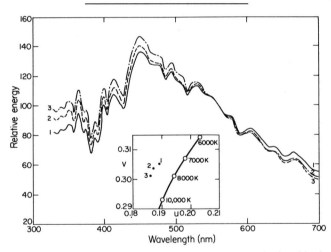

Fig. 5.8 Spectral composition of skylight as a function of solar altitude.

Curve and u, v Plot No.	Solar Altitude (Deg.)
1	70
2	60
3	50

TABLE 5.1 Color Temperatures and Chromacity Coordinates

Fig. No.	Curve No.	Solar Alt. Deg.	Orientation (Deg.)	Sky conditions	CIE coordinates x	y	u	v	Color temp. (K)
5.1	1	40	15–0	Clear	0.3228	0.3367	0.2019	0.3159	5960
	2	40	15–0	Light haze	0.3505	0.3619	0.2111	0.3269	4840
	3	40	15–0	Very light to light clouds	0.3444	0.3572	0.2088	0.3249	5040
	4	40	15–0	Clear	0.3276	0.3409	0.2036	0.3178	5040
5.3	1	70	15–0	Clear	0.3226	0.3346	0.2025	0.3152	5980
	2	40	15–0	Clear	0.3228	0.3367	0.2019	0.3159	5960
	3	30	15–0	Clear	0.3261	0.3401	0.2029	0.3174	5800
	4	20	15–0	Clear	0.3351	0.3492	0.2056	0.3213	5390
	5	15	15–0	Clear	0.3448	0.3570	0.2092	0.3248	5020
	6	10	15–0	Clear	0.3581	0.3668	0.2142	0.3292	4600
	7	8	15–0	Clear	0.3643	0.3710	0.2168	0.3311	4420
5.5	1	70	15–0	Heavy overcast	0.3250	0.3393	0.2025	0.3170	5850
	2	40	15–0	Heavy overcast	0.3276	0.3409	0.2036	0.3178	5720
	3	20	15–0	Heavy overcast	0.3325	0.3436	0.2060	0.3192	5490
5.8	1	70	15–180	Clear	0.2932	0.3155	0.1892	0.3053	7880
	2	60	15–180	Clear	0.2887	0.3123	0.1872	0.3037	8300
	3	50	15–180	Clear	0.2846	0.3071	0.1861	0.3013	8800
5.7	1	40	15–180	Clear	0.2828	0.3020	0.1867	0.2991	9150
	2	30	15–180	Clear	0.2823	0.2967	0.1884	0.2969	9420
	3	20	15–180	Clear	0.2798	0.2967	0.1865	0.2966	9660
5.6	1	15	15–180	Clear	0.2815	0.2997	0.1866	0.2981	9350
	2	10	15–180	Clear	0.2857	0.3052	0.1876	0.3006	8790
	3	8	15–180	Clear	0.2877	0.3078	0.1881	0.3018	8530

Table 5.1 gives color temperatures and chromaticity coordinates for daylight and skylight presented in Figs. 5.1, 5.3, and 5.5–5.8.

5.3   EXTRATERRESTRIAL SOLAR RADIATION

The data for the integrated irradiance and spectral distribution of extraterrestrial and terrestrial solar radiation are needed as a basis for the uniform calculation of the radiation load of objects and for the purpose of simulation. These data are required in aerospace applications for testing satellites and space vehicles in sun simulators, in space biology for testing biological objects under extraterrestrial conditions in the laboratory, in biological and medical science for irradiation of organisms in solaria under the conditions found in nature, and in atmospheric physics for determining the radiation heat balance of the earth.

The determination of the integrated irradiance of extraterrestrial solar radiation, which was called the *solar constant* by Pouillet as early as 1838, was originally based on extrapolation of measured results obtained on high mountains to extraterrestrial conditions. These were later followed by a series of spectrobolometrical measurements by the Smithsonian Institution, again on high mountains. Recently, Drummond *et al.* (1968a,b) and then Thekaekara *et al.* (1968; Thekaekara and Drummond, 1971) and his staff succeeded in determining the solar constant directly from an airplane at high altitudes. The solar constant can be calculated on the one hand from a series of astronomical observations of the chromosphere of the sun and on the other hand can be derived from measurements at high altitudes.

The annual mean of the integrated irradiance just beyond the earth's atmosphere is called the *solar constant* E_0. The value of E_0 obtained by Thekaekara *et al.* from an aircraft is

$$E_0 = 1350 \text{ W m}^{-2}.$$

This value for the solar constant is recommended by the CIE (1972).

Absolute values of the spectral irradiance of extraterrestrial solar radiation in watts per square meter per micrometer according to Thekaekara *et al.* (1968) are given in Table 5.2 (column 2). Column 3 gives the values of the irradiance related to a constant relative spectral band $\Delta\lambda/\lambda$, i.e., the spectral irradiance $E_\lambda \cdot \lambda$. Table 5.3 gives the irradiance for broad spectral bands of extraterrestrial solar radiation. Figures 5.9 and 5.10 give the spectral irradiance of the extraterrestrial solar radiation according to Thekaekara *et al.* (1968). Figure 5.10 is simply a logarithmic plot of data in Table 5.2 with Planckian radiation of 6000 K included in the figure for comparison.

TABLE 5.2   Spectral Irradiance of Extraterrestrial Solar Radiation[a]

Wave-length λ (μm)	Spectral Irradiance E_λ ($Wm^{-2}\mu m^{-1}$)	$E_\lambda\lambda$ ($Wm^{-2}\mu m^{-1}\mu m$)	Wave-length λ (μm)	Spectral Irradiance E_λ ($Wm^{-2}\mu m^{-1}$)	$E_\lambda\lambda$ ($Wm^{-2}\mu m^{-1}\mu m$)
.12	1.0 $-$ 1	1.2 $-$ 2	.43	1.639 $+$ 3	7.09 $+$ 2
.14	3.0 $-$ 2	4.2 $-$ 3	.44	1.810 $+$ 3	7.96 $+$ 2
.15	7.0 $-$ 2	1.05 $-$ 2	.45	2.006 $+$ 3	9.03 $+$ 2
.16	2.3 $-$ 1	3.68 $-$ 2	.46	2.066 $+$ 3	9.50 $+$ 2
.17	6.3 $-$ 1	1.07 $-$ 1	.47	2.033 $+$ 3	9.56 $+$ 2
.18	1.25	2.25 $-$ 1	.48	2.074 $+$ 3	9.96 $+$ 2
.19	2.71	5.15 $-$ 1	.49	1.950 $+$ 3	9.55 $+$ 2
.20	1.07 $+$ 1	2.14	.50	1.942 $+$ 3	9.71 $+$ 2
.21	2.29 $+$ 1	4.60	.51	1.882 $+$ 3	9.60 $+$ 2
.22	5.75 $+$ 1	1.27 $+$ 1	.52	1.833 $+$ 3	9.53 $+$ 2
.23	6.67 $+$ 1	1.53 $+$ 1	.53	1.842 $+$ 3	9.76 $+$ 2
.24	6.30 $+$ 1	1.51 $+$ 1	.54	1.783 $+$ 3	9.63 $+$ 2
.25	7.04 $+$ 1	1.74 $+$ 1	.55	1.725 $+$ 3	9.49 $+$ 2
.26	1.30 $+$ 2	3.38 $+$ 1	.56	1.695 $+$ 3	9.45 $+$ 2
.27	2.32 $+$ 2	6.26 $+$ 1	.57	1.712 $+$ 3	9.76 $+$ 2
.28	2.22 $+$ 2	6.22 $+$ 1	.58	1.715 $+$ 3	9.94 $+$ 2
.29	4.82 $+$ 2	1.40 $+$ 2	.59	1.700 $+$ 3	1.003 $+$ 3
.30	5.14 $+$ 2	1.54 $+$ 2	.60	1.666 $+$ 3	9.99 $+$ 2
.31	6.89 $+$ 2	2.13 $+$ 2	.61	1.635 $+$ 3	9.97 $+$ 2
.32	8.30 $+$ 2	2.65 $+$ 2	.62	1.602 $+$ 3	1.043 $+$ 3
.33	1.059 $+$ 3	3.49 $+$ 2	.63	1.570 $+$ 3	9.89 $+$ 2
.34	1.074 $+$ 3	3.60 $+$ 2	.64	1.544 $+$ 3	9.88 $+$ 2
.35	1.093 $+$ 3	3.76 $+$ 2	.65	1.511 $+$ 3	9.82 $+$ 2
.36	1.068 $+$ 3	3.84 $+$ 2	.66	1.486 $+$ 3	9.81 $+$ 2
.37	1.181 $+$ 3	4.41 $+$ 2	.67	1.456 $+$ 3	9.76 $+$ 2
.38	1.120 $+$ 3	4.29 $+$ 2	.68	1.427 $+$ 3	9.70 $+$ 2
.39	1.098 $+$ 3	4.28 $+$ 2	.69	1.402 $+$ 3	9.67 $+$ 2
.40	1.429 $+$ 3	5.71 $+$ 2	.70	1.369 $+$ 3	9.58 $+$ 2
.41	1.751 $+$ 3	7.18 $+$ 2	.71	1.344 $+$ 3	9.56 $+$ 2
.42	1.747 $+$ 3	7.34 $+$ 2	.72	1.314 $+$ 3	9.44 $+$ 2

[a] Altitude of the sun is 90°. Radiation near the earth are for $\Delta\lambda$ = const. and for $\Delta\lambda/\lambda$ = const. Values are extracted from the tables of Thekaekara *et al.* (1968). The numbers beside the values of spectral irradiance are exponents of 10; e.g., -2 means 10^{-2}.

TABLE 5.2 *(Continued)*

Wave-length λ (μm)	Spectral Irradiance E_λ ($Wm^{-2}\,\mu m^{-1}$)	Spectral Irradiance $E_\lambda\lambda$ ($Wm^{-2}\mu m^{-1}\mu m$)	Wave-length λ (μm)	Spectral Irradiance E_λ ($Wm^{-2}\mu m^{-1}$)	Spectral Irradiance $E_\lambda\lambda$ ($Wm^{-2}\mu m^{-1}\mu m$)
.73	1.290 + 3	9.42 + 2	2.8	3.90 + 1	1.09 + 2
.74	1.260 + 3	9.32 + 2	2.9	3.50 + 1	1.02 + 2
.75	1.235 + 3	9.26 + 2	3.0	3.10 + 1	9.3 + 1
.80	1.107 + 3	8.86 + 2	3.1	2.60 + 1	8.06 + 1
.90	8.89 + 2	8.00 + 2	3.2	2.26 + 1	7.23 + 1
1.0	7.46 + 2	7.46 + 2	3.3	1.92 + 1	6.34 + 1
1.1	5.92 + 2	6.51 + 2	3.4	1.66 + 1	5.64 + 1
1.2	4.84 + 2	5.81 + 2	3.5	1.46 + 1	5.11 + 1
1.3	3.96 + 2	5.15 + 2	3.6	1.35 + 1	4.86 + 1
1.4	3.36 + 2	4.71 + 2	3.7	1.23 + 1	4.55 + 1
1.5	2.87 + 2	4.30 + 2	3.8	1.11 + 1	4.22 + 1
1.6	2.44 + 2	3.90 + 2	3.9	1.03 + 1	4.02 + 1
1.7	2.02 + 2	3.43 + 2	4.0	9.5	3.80 + 1
1.8	1.59 + 2	2.86 + 2	5.0	3.83	1.92 + 1
1.9	1.26 + 2	2.39 + 2	6.0	1.75	1.05 + 1
2.0	1.03 + 2	2.06 + 2	7.0	9.90 − 1	6.93
2.1	9.0 + 1	1.89 + 2	8.0	6.00 − 1	4.80
2.2	7.9 + 1	1.74 + 2	9.0	3.80 − 1	3.42
2.3	6.8 + 1	1.56 + 2	10.0	2.50 − 1	2.50
2.4	6.4 + 1	1.44 + 2	15.0	4.9 − 2	7.35 − 1
2.5	5.4 + 1	1.35 + 2	20.0	1.6 − 2	3.20 − 1
2.6	4.8 + 1	1.25 + 2	50.0	3.8 − 4	1.90 − 2
2.7	4.3 + 1	1.16 + 2	100.0	3.0 − 5	3.0 − 3

TABLE 5.3 Irradiance of Extraterrestrial Solar Radiation[a]

Spectral bands	Wavelength (μm)	Irradiance (W m^{-2})		Percentages of solar constant (%)	
Ultraviolet					
UV–C	<0.28	8		0.6	
UV–B	0.28–0.315	18	118	1.3	8.7
UV–A	0.315–0.40[b]	92		6.8	
Visible	0.38–0.78[b]	638	638	47.3	47.3
Infrared					
IR–A	0.78–1.4	406		30.1	
IR–B	1.4–3.0	183	618	13.5	45.7
IR–C	>3.0	29		2.1	

[a] According to Thekaekara *et al.* (1968) for $E_0 = 1.35$ kW m^{-2} perpendicular incidence.

[b] Within the region 0.38–0.4 μm the irradiance and the percentages are counted twice due to overlapping as defined. (21 W m^{-2} represents 1.7%).

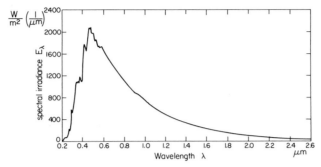

Fig. 5.9 Spectral irradiance of extraterrestrial solar radiation according to Thekaekara (1968), where $\Delta\lambda$ = const. Area under the curve is equal to the solar constant $E_0 = 1.35\ \text{kW m}^{-2}$.

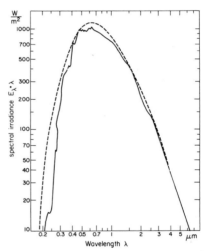

Fig. 5.10 Spectral irradiance of extra-terrestrial solar radiation $E_\lambda\lambda$ according to Thekaekara (solid line) and of Planckian radiation 6000 K, with $E_0 = 1.35$ kW m^{-2} (dashed line) and $\Delta\lambda/\lambda$ = const.

5.4 ARTIFICIAL RADIATION SOURCES

5.4.1 Incandescent Lamps

From the large variety of artificial radiation sources used in practice we shall limit our discussion to those commonly used in the measurement of

optical radiant energy. Also discussed here are some of those sources that
are often used in specific applications where knowledge of their spectral
characteristics is essential, i.e., sources used in color reproduction and
color-matching systems and sources used in instrumentation designed for
measurement of photometric and radiometric characteristics of fluorescent
materials.

The most common artificial source is a tungsten-filament incandescent
lamp (a thermal source). An incandescent lamp is defined as a lamp in which
light is produced by a filament heated to incandescence. This is accomplished
by a flow of electric current. Incandescence is the self-emission of radiant
energy, primarily in the visible part of the spectrum, due to the thermal
excitation of atoms or molecules (U.S.A. Std., 1967).

Such a lamp consists of a glass bulb that houses the filament and an inert
gas. The gas retards rapid evaporation of the filament and early blackening
of the bulb. The filament is a tungsten ribbon or coil. The electric current is
conducted to the filament via lead-in wires. There is also a support wire that
holds the filament in place. The support wire is usually made of molybdenum
or tungsten.

5.4.2 Color of Light

Although the relative spectral distribution from an incandescent-filament
lamp is often assumed to be the same as the relative spectral distribution
from a blackbody at some related temperature (usually higher or lower than
the actual temperature of the filament), the actual distribution is usually
somewhat different from that of a blackbody. An example of that for
tungsten at a temperature of 3000 K is shown in Fig. 5.11. The curve marked
Planckian radiator is for a blackbody at the actual temperature of the tungsten. The curve shown for the tungsten is everywhere lower by an amount
that is almost constant in the visible region (0.4–0.7 μm), but which deviates
much more at longer wavelengths. The ratio of the exitance of the tungsten
to the exitance of the blackbody, which is different for different wavelengths, as shown in Fig. 5.12, is called the *emissivity* of tungsten. As shown,
the emissivity is also dependent on temperature. A radiator whose emissivity
is independent of wavelength (but not necessarily independent of temperature) is called a graybody, as indicated on the dashed curve of Fig. 5.11,
which has exactly the same shape as the curve for the Planckian radiator
(blackbody).

Although the temperature of the tungsten will be different from the temperature of the blackbody or the graybody that has the most similar shape
in the visible region (vertically displaced by as nearly as possible equal
amounts from 0.4 to 0.7 μm in Fig. 5.11), the tungsten is said to have the

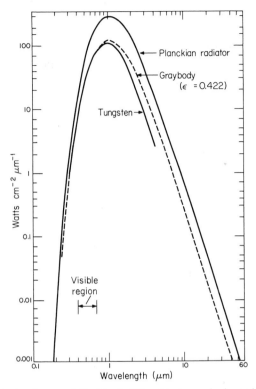

Fig. 5.11 Spectral exitance of blackbody (Planckian), graybody, and tungsten ribbon. operating at 3000 K.

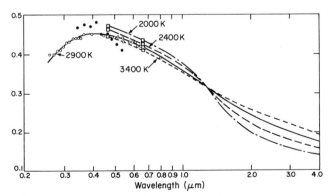

Fig. 5.12 Emissivities of tungsten ribbon, at indicated temperatures. Data are from (○) Hoffman and Willenberg (1932); (●) Hulburt (1948); and (□) Worthing (1941).

same color temperature as the blackbody or the graybody. Until recently, color temperature was determined visually, by adjusting the temperature of a blackbody until the light from it appeared to have the same color as the light from the source that was being measured. Simultaneously, the luminances of either or both the blackbody and the source being measured were adjusted, without change of color, to obtain a brightness match as well as a color match.

TABLE 5.4 Chromaticity Coordinates (x, y) of Points on Blackbody (Planckian) Locus and Reciprocal Slopes of Iso-Color–Temperature Loci[a]

$(MK)^{-1}$	T (K)	x	y	1/Slope	$(MK)^{-1}$	T (K)	x	y	1/Slope
0	∞	0.2399	0.2413	−1.4711	300	3333	0.4149	0.3953	0.4090
10	100,000	0.2426	0.2381	−1.3579	310	3226	0.4216	0.3982	0.4328
20	50,000	0.2456	0.2425	−1.2790	320	3125	0.4282	0.4009	0.4552
30	33,333	0.2489	0.2472	−1.1830	330	3030	0.4247	0.4033	0.4770
40	25,000	0.2525	0.2523	−1.0858	340	2941	0.4410	0.4055	0.4976
50	20,000	0.2565	0.2577	−0.9834	350	2857	0.4473	0.4074	0.5181
60	16,667	0.2607	0.2634	−0.8863	360	2778	0.4535	0.4091	0.5377
70	14,286	0.2653	0.2693	−0.7874	370	2703	0.4595	0.4105	0.5565
80	12,500	0.2701	0.2755	−0.6934	380	2632	0.4654	0.4118	0.5740
90	11,111	0.2752	0.2818	−0.6022	390	2564	0.4712	0.4128	0.5917
100	10,000	0.2806	0.2883	−0.5140	400	2500	0.4769	0.4137	0.6080
110	9091	0.2863	0.2949	−0.4260	410	2439	0.4824	0.4143	0.6248
120	8333	0.2921	0.3015	−0.3607	420	2381	0.4878	0.4148	0.6401
130	7692	0.2982	0.3081	−0.2881	430	2326	0.4931	0.4151	0.6554
140	7143	0.3045	0.3146	−0.2227	440	2273	0.4982	0.4153	0.6706
150	6667	0.3110	0.3211	−0.1615	450	2222	0.5032	0.4153	0.6847
160	6250	0.3176	0.3275	−0.1044	460	2174	0.5082	0.4151	0.6984
170	5882	0.3243	0.3338	−0.0512	470	2128	0.5129	0.4148	0.7124
180	5556	0.3311	0.3399	−0.0004	480	2083	0.5176	0.4145	0.7257
190	5263	0.3380	0.3459	+0.0458	490	2041	0.5221	0.4140	0.7382
200	5000	0.3450	0.3516	0.0889	500	2000	0.5266	0.4133	0.7508
210	4762	0.3521	0.3571	0.1294	510	1961	0.5309	0.4126	0.7630
220	4545	0.3591	0.3624	0.1676	520	1923	0.5350	0.4118	0.7745
230	4348	0.3662	0.3674	0.2045	530	1887	0.5391	0.4109	0.7865
240	4167	0.3733	0.3722	0.2386	540	1852	0.5431	0.4099	0.7981
250	4000	0.3804	0.3767	0.2712	550	1818	0.5470	0.4089	0.8095
260	3846	0.3874	0.3810	0.3011	560	1786	0.5508	0.4078	0.8204
270	3704	0.3944	0.3850	0.3303	570	1754	0.5545	0.4066	0.8307
280	3571	0.4013	0.3887	0.3578	580	1724	0.5581	0.4054	0.8411
290	3448	0.4081	0.3921	0.3836	590	1695	0.5616	0.4041	0.8511
					600	1667	0.5650	0.4028	0.8613

[a] Based on $c_2 = 14,388$ μm K and orthogonals on CIE 1960 (u, v) diagram.

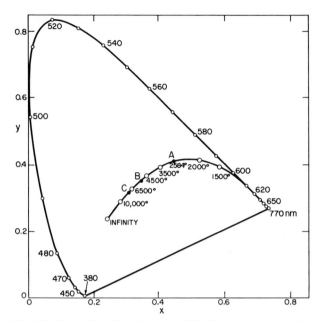

Fig. 5.13 Blackbody (Planckian) locus on CIE 1931 x, y chromacity diagram.

Today it is a common practice to determine the color temperature from the spectroradiometric measurements. Consequently, the chromaticity of light from the source being tested can be computed by use of the CIE (1971a) color mixture data given in Appendix 5.2. The color temperature is now most usually determined by interpolation of the temperature of the black-body whose chromaticity corresponds most closely to the chromaticity of the source. The interpolation is done using a table or a graph of similarly computed chromaticities of blackbodies at various temperatures. The chromaticity coordinates for blackbodies at a number of temperatures are given in Table 5.4. Figure 5.13 represents those chromaticities by points on the CIE 1931 chromaticity diagram. The curve shown through those points is called the *Planckian locus*.

Because practical light sources rarely have spectral distributions exactly proportional to the spectral distribution of a blackbody at any specified temperature, the point that represents the chromaticity of a practical light source usually does not coincide with any point on the Planckian locus. If it is above the locus, the light from the source may appear slightly greenish compared to the most nearly similar blackbody. If the point is below the Planckian locus the light from that source may appear slightly pinkish compared to the most nearly similar blackbody, but it should not appear either bluer or yellower than a blackbody at the correct color temperature.

When color temperature is determined by interpolation of a computed chromaticity, in a table or a chromaticity diagram, the question of what constitutes the visually most nearly similar chromaticity, or point on the Planckian locus, is moot when the chromaticity of the source being evaluated is represented by a point that is not exactly on that locus. There is a method that is recommended (Wyszecki and Stiles, 1967; CIE, 1971a) by the International Commission on Illumination (CIE) for determining the equivalent or *correlated color temperature* of a source whose chromaticity is represented by a point that is not exactly on the blackbody locus.

The CIE recommends plotting the Planckian locus on the CIE 1960 (u,v) chromaticity diagram, as shown in Fig. 5.14, and then drawing the normal of that curve to pass through the off-locus point representing the chromaticity of the source for which the correlated color temperature is desired. The correlated color temperature is defined by the CIE as the temperature of the blackbody whose chromaticity is represented by the point on the Planckian locus at the foot of that normal. Lines in the CIE 1931 x,y diagram that correspond to a number of such normals for the indicated color temperatures are shown in Fig. 5.15. The correlated color temperature of a source whose chromaticity is represented by a point that is not too far away from the Planckian locus may be determined by interpolation between the two

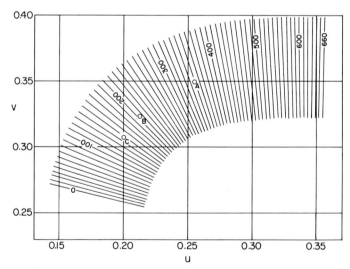

Fig. 5.14　Planckian locus and isotemperature loci at 10 MK^{-1} intervals on u, v diagram for $C_2 = 14{,}388\ \mu\text{m K}$.

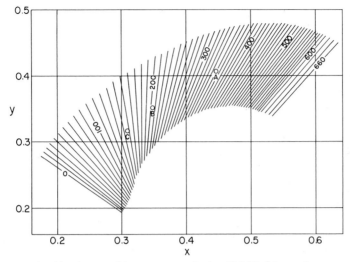

Fig. 5.15   Planckian locus and isotemperature loci at 10 MK^{-1} intervals on x, y diagram for $C_2 = 14{,}388$ μm K.

nearest of those lines. Their slopes and intercepts with the Planckian locus are given as functions of color temperature in Table 5.4.

From Wien's formula, a simple formula can be derived for the transmittance of a light filter that can be used to modify the spectral distribution of a blackbody at one temperature T_1 to be proportional to the spectral distribution at some other desired temperature T_2. The color temperature of the source modified by the use of that filter is T_2, even though the absolute spectral power may not be equal to that from a blackbody at T_2. When T_2 is less than T_1 (and T is less than 3400 K), then, within the visible and ultraviolet regions of the spectrum, the spectral transmittance of the filter needed to convert the spectral distribution of a blackbody at T_1 into the (absolute) spectral distribution of a blackbody at T_2 at λ (μm) is

$$\tau(\lambda) = \exp\left[\frac{14{,}388}{\lambda}\left(\frac{1}{T_1} - \frac{1}{T_2}\right)\right]. \tag{5.1}$$

At one wavelength λ the spectral density of such a filter is proportional to $(1/T_1) - (1/T_2)$, which was formerly expressed in terms of microreciprocal degrees (abbreviated "mired"). However, besides having clumsy and misleading syntax (the prefix "micro" was not intended to modify "reciprocal" but the quantity "reciprocal degrees,") whose significance is not clearly indicated by those words, the term incorporates the word "degrees," which is forbidden in the International Metric (SI) system of nomenclature. A replacement term "reciprocal megakelvins" (MK^{-1}) has been suggested as being compatible with the SI nomenclature; this was adopted by the CIE (CIE, 1975).

For example, a dye which when used in a light filter in a certain concentration would modify the spectral distribution of a blackbody at 3333.3 K, i.e., 300 MK^{-1}, to be the same as that of a blackbody at 3125 K, i.e., 320 MK^{-1}, a difference of 20 MK^{-1}, if used with twice that concentration would produce a difference of 40 MK^{-1} and would modify the spectral distribution from the 3333.3 K blackbody to be the same as that of a blackbody at 2960 K, i.e., 340 MK^{-1}. In terms of reciprocal megakelvins (MK^{-1}), the power of a light filter to modify the spectral distribution ($\lambda \leq 0.7$ μm) from a blackbody ($T \leq 3400$) is independent of the temperature T of the blackbody and is proportional to the concentration of the dye with which the filter is made. In other words, the shape of the curve that shows the logarithm of the spectral density of any such blackbody distribution–conversion filter is independent of the temperature of either the input source or the equivalent output source, provided that the former temperature is higher than the latter and not more than about 3400 K.

Although the spectral transmittance of such conversion filters ($T_2 \leq T_1 \leq$ 3400 K) is less than 1.0 at all wavelengths less than or equal to 0.7 μm, and therefore the filter could conceivably be made, the maximum transmittance (0.7 μm) will be quite low, e.g., $e^{-4.1108} = 0.0164$ for a 20 MK^{-1} filter and 0.00027 for a 40 MK^{-1} filter. Therefore, such filters are not made for absolute value conversion of blackbody distributions, but are designed to have high transmittance while preserving the desired conversion power (i.e., shape of the requisite spectral density curve). Consequently, the curves of logarithms of the spectral densities of practical filters that have different conversion powers do not have the same shapes nor can filters with greater conversion powers be made by merely increasing the concentrations of the dyes used for lower conversion powers. Such filters are called color–temperature conversion filters; their color–temperature conversion powers, in terms of reciprocal megakelvins (MK^{-1}) are independent of the color temperature of the input source, but they do yield the same absolute spectral distribution as a blackbody at the output color temperature. If the input distribution is from a blackbody and the conversion is to a lower color temperature, the output spectral distribution can be greater than that from a blackbody at the lower temperature; it usually is greater if the conversion power of the filter is 10 MK^{-1} or more.

The same kind of discussion holds for light filters intended for conversion from a lower to a higher color temperature; in such cases the conversion power is negative; the wavelength of maximum transmittance is at the short-wavelength end of the visible spectrum ($\lambda = 0.4$ μm); the output spectral distribution is always less than from a blackbody at the output temperature T_2.

Strictly speaking, T_2 should not exceed 3400 K. However, the need for spectral distributions colorimetrically equivalent to temperatures higher than 3400 K, e.g., 5000 K for artificial sunlight or 6500 K for artificial daylight, or 7500–12,000 K for artificial north skylight, is so frequent and urgent that color–temperature conversion filters are commonly made for such applications. For the most critical applications, the Wien formula is not adequate for the design of such filters; Planck's law is used. The conversion powers of such filters are often expressed in "mireds." They are numerically equal to reciprocal megakelvins. The conversion power (in reciprocal megakelvins) of a filter for converting tungsten-quality light ($T_c \sim 3000$ K) to sunlight or daylight ($T_c \sim 5500$ K) is not precisely invariant with input color temperature, but may be taken as approximately constant for variations of a few hundred kelvin input color temperature. The corresponding variations of output color temperature are about four times as great.

Because color–temperature conversion filters having such great powers, e.g., $333.3 - 136.7 = 166.6$ MK^{-1} are very difficult to make precisely; the chromaticity of the output light rarely falls precisely on the blackbody locus. The color temperature of the output light is then determined in the manner indicated in Figs. 5.14 and 5.15. The power of a filter to alter color temperatures, thus determined, is nearly invariant for moderate changes of the input color temperature, e.g., from 2500 to 3400 K, and is also nearly independent of the extents of the departures of the input and output chromaticities from the blackbody locus, within the extent of the transverse lines in Figs. 5.14 and 5.15.

The noticeable differences of color temperatures are nearly constant from about 1000 to 20,000 K when the color temperatures (and their differences) are expressed in terms of reciprocal megakelvins. Under the most favorable conditions of comparison (e.g., brightness-matched and sharply juxtaposed or patterned photometric fields, subtending four or more degrees of visual angle), two color temperatures whose reciprocals differ by about 1 MK^{-1} are just noticeably different.

Blackbody, or Planckian, simulators cannot be built to operate at temperatures higher than about 4000 K, and blackbody simulators are so cumbersome and inconvenient to operate at any temperatures that they are rarely used. Incandescent-filament lamps, carbon-arc lamps, and some high-temperature, high-pressure vapor lamps that provide continuous spectra qualitatively similar to light from blackbody sources are much more commonly used. Their spectral distributions cannot be computed from fundamental formulas comparable to Planck's or Wien's formulas. Rather, their spectral distributions are measured by the techniques of *spectroradiometry*.

The chromaticities of practical sources rarely correspond to points that fall precisely on the blackbody locus. Nevertheless, the color quality of the light for such sources is commonly expressed in terms of color temperature, which, when precision is desired, is determined in the manner indicated in Figs. 5.14 and 5.15.

5.4.3   Tungsten Halogen

An addition of a halogen to the gas of an incandescent lamp will, under proper conditions, increase the life and efficiency of the lamp. When internal lamp temperatures are high enough, a halogen (iodine or bromine) added to the filling gases in a lamp vaporizes and combines chemically with the evaporated tungsten—the resulting tungsten–halogen gas migrates back to the filament, where the very high temperature decomposes it. Tungsten is redeposited on the filament and the halogen repeats the cycle. For this reason these types of lamps are usually called "incandescent tungsten lamps with halogen regenerating cycle." This action makes the lamp capable of giving full radiant power throughout the life of the lamp.

Due to the high temperature required (over 250°C) within the lamp, the lamp envelope must be able to withstand such temperatures. For this reason the envelopes of tungsten–halogen cycle lamps are made of high-silica materials. These materials have low coefficients of expansion, are optically clear, and have a high thermal shock resistance. The current-carrying wires must also be able to stand this high temperature, otherwise they would fail due to oxidation and hence give a short lamp life.

These types of lamps provide a concentrated source of radiant energy in a relatively small lamp size. When higher desired voltage of these lamps is required, the filament tubes are used in a glass outer envelope filled with inert gas. This avoids failure due to oxidation and allows greater latitude in the filament design.

In view of the above described characteristics, the tungsten–halogen lamps are now widely used as standards of spectral irradiance for the spectral region 0.25–2.60 μm (Stair *et al.*, 1971). This topic and this application of tungsten–halogen lamps shall be discussed in a later chapter.

5.4.4   Discharge Sources

Discharge sources (lamps) operate on the principle of passing an electric current through an ionized vapor of an element. Such a lamp consists of two electrodes sealed in a glass or quartz tube. The tube is filled with a gas at a pressure usually of a few millimeters of mercury. When ions and electrons are excited, the gas becomes conducting if only a low voltage is applied. The application of voltage produces an electric field that pulls the positive ions to the cathode and the electrons to the anode and also accelerates the charge carriers. Due to the collision of carriers with gas molecules and the atoms of the electrodes, the outer electrons are detached from the neutral molecules and atoms, hence, the number of free electrons in the gas increases due to this ionization by a collision process. Even some of the inner electrons are raised by this process to a higher energy level of the atom. Then, when the electrons of the excited atoms return to the lower energy level, the atoms will emit photons of the wavelengths of light.

The most common discharge sources of concern in optical radiation measurement (particularly in radiometry and spectroradiometry) are: hydrogen lamps, mercury vapor lamps, various glow lamps, and fluorescent lamps. Hydrogen or deuterium lamps are used in ultraviolet spectrophotometry and also as standards of irradiance for spectroradiometry (National Bureau of Standards, 1976). Mercury vapor and glow lamps are often used in calibrating the wavelength scale of radiometers and spectrophotometers. The hydrogen sources can be regarded as high-pressure lamps at a pressure of a few hundred pascals. A discharge can pass though hydrogen, and a continuum spectrum is produced which penetrates far into the ultraviolet. The spectral power distribution of a low-wattage hydrogen lamp is shown in Fig. 5.16. This spectrum is a molecular spectrum of hydrogen, so one must use special electrodes to ensure that the atomic hydrogen which is produced during the discharge can quickly combine to form molecules. Heavy hydrogen (deuterium) is often used instead of ordinary hydrogen with resulting increase in radiance by a factor of 2 to 3.

Mercury-vapor lamps are also used in ultraviolet spectrophotometry for specialized purposes. For more details on this type of lamps the reader should consult the lamp manufacturers' manuals.

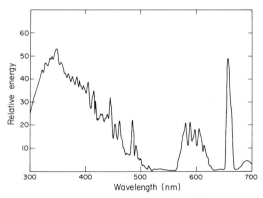

Fig. 5.16   Spectral irradiance of a hydrogen lamp.

Another discharge source of interest in radiometry is the corona-discharge source. Corona-discharge devices are used in a variety of research and development applications. Examples of such applications are the electrostatic charging of surfaces, the treatment of plastic materials, and in electrophotography the enhancing of surface adhesion in film coating. The corona discharge is also known to produce a considerable amount of electromagnetic radiation, mainly in the UV region of the spectrum (Loeb, 1965; Grum and Costa, 1976). The mechanism involved is presumed to be the ionization and radiative recombination of the gases surrounding the high-voltage terminal of the device. The spectral power distributions of the corona discharge, generated by using a high-voltage coil of the type CENCO BD 10 at a potential of 20 kV at its terminal, are shown in Figs. 5.17–5.21. These figures show the emission of corona discharges in air and in the indicated

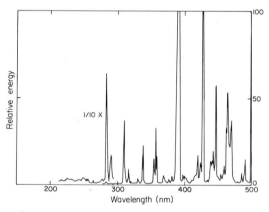

Fig. 5.17 Spectral emission of a corona discharge in helium atmosphere.

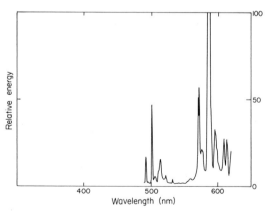

Fig. 5.18 Spectral emission of a corona discharge in helium atmosphere.

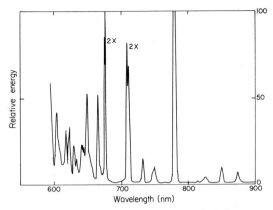

Fig. 5.19 Spectral emission of a corona discharge in helium atmosphere.

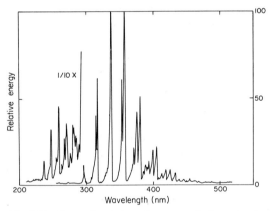

Fig. 5.20 Spectral emission of a corona discharge in air.

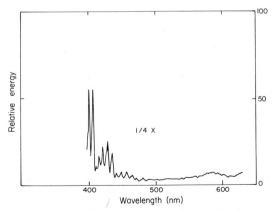

Fig. 5.21 Emission of a corona discharge in air.

atmospheres. A continuous concentrated arc from the terminal coil to an iron ground wire was used to generate the emission to be measured.

5.4.5   Fluorescent Sources

Fluorescent lamps are of increasing interest in science and industry. These sources are usually not used instrumentally, i.e., in conjunction with spectrometers, but are used as a promoter of photochemical reactions and in visual assessments of materials. Also, fluorescent lamps are now widely used in illumination and in the simulation of various phases of daylight.

The fluorescent lamp is an electric-discharge source which produces light by the radiation of fluorescent materials (powders) activated by ultraviolet energy generated by a low-pressure mercury arc. The chemical composition of the phosphor determines the color of the light produced. Fluorescent lamps have a negative resistance characteristic and because of that they must be operated from some current-limiting device. This device, commonly called a *ballast*, limits the lamp current to the designed value. It also provides the required starting and operating voltages.

Fluorescent lamps are noted for their high luminous efficiency (up to 80 lm/W), long life, and the small change in the light output with variations in the line voltage.

There are several white fluorescent lamps available, each having a characteristic spectral distribution. There are also red, gold, green, and blue fluorescent lamps. Also there is a black-light fluorescent lamp which has its tube made of dark blue ultraviolet-transmitting glass to minimize the visible output of phosphor and mercury radiation.

Figures 5.22–5.25 show typical spectral distributions of three 15-W fluorescent lamps. The relative and absolute spectral outputs greatly depend on

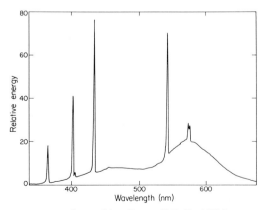

Fig. 5.22   Spectral irradiance, G.E. Cool White.

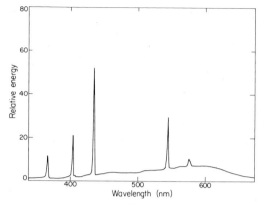

Fig. 5.23 Spectral irradiance, Westinghouse Deluxe Cool White.

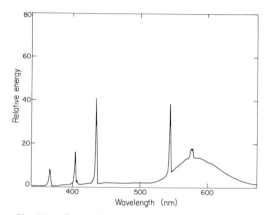

Fig. 5.24 Spectral irradiance, Sylvania Warm White.

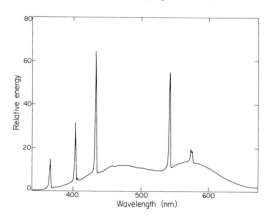

Fig. 5.25 Spectral irradiance, G.E. Daylight.

such factors as gas pressure, ambient temperature, and discharge voltage. All these may cause considerable variation in the spectral output so that an average curve is of little value in obtaining the characteristics of a particular lamp. The line spectra, especially their intensity and width, are also dependent on the bandwidth of measuring equipment. The proper setup for measuring such spectral power distributions is well presented by Nonaka *et al.* (1972).

The color-rendering properties of fluorescent lamps are normally expressed in terms of the color-rendering index (CIE, 1974a). The calculation of the color-rending index is based on the spectral energy distribution of the source and computation of chromaticity coordinates of the source itself and of eight selected Munsell reflecting test samples under both sample and reference illuminants. The value of the general color-rendering index R_a is then determined using

$$R_a = 100 - 4.6\,\overline{\Delta E}_a, \qquad (5.2)$$

where

$$\overline{\Delta E}_a = \sum_{i=1}^{8} \Delta E_i/8 \qquad (5.3)$$

and

$$\Delta E_i = 800\{[(u_{k,i} - u_k) - (u_{o,i} - u_o)]^2 + [(v_{k,i} - v_k) - (v_{o,i} - v_o)]^2\}^{1/2}, \qquad (5.4)$$

where $u_{k,i}$ and $v_{k,i}$ are the UCS chromaticity coordinates of test samples under sample light source k, $u_{o,i}$ and $v_{o,i}$ the UCS chromaticity coordinates of test samples under reference illuminant o, u_k and v_k the CIE UCS coordinates of sample light source k, u_o and v_o the CIE UCS coordinates of reference illuminant o. Alternatively,

$$\Delta E_i = 800[(u_{o,i} - u'_{k,i})^2 + (v_{o,i} - v'_{k,i})^2]^{1/2}, \qquad (5.5)$$

where

$$u'_{k,i} = u_{k,i} + (u_o - u_k), \qquad v'_{k,i} = v_{k,i} + (v_o - v_k) \qquad (5.6)$$

and

$$R_a = 100 - 3.7(10^3)\overline{\Delta E}_{u,v}, \qquad \overline{\Delta E}_{u,v} = \overline{\Delta E}_a/800 \qquad (5.7)$$

where $\overline{\Delta E}_{u,v}$ is the average value of $\Delta E_{u,v}$ vectors for test samples 1–8 and $\overline{\Delta E}_a$ represents the same values after adjustment to provide units in which a 1% difference corresponds on the average to one just perceptible color difference. The chromaticity coordinates and the correlated color temperature can be determined by standard computation. The recommended way to do so is by using Robertson's (1968) approach. A listing of a computer program in FORTRAN IV to do so is appended.

5.4.6   Light-Emitting Diodes

Light-emitting diodes (LEDs) are solid-state sources consisting of a semi-conducting crystal containing a p–n junction. When dc voltage is applied to the crystal in the forward direction, optical radiant energy is produced. These solid-state sources convert electrical energy into radiant flux. Most common LEDs are gallium arsenide (with λ_{max} emittance of about 900 nm) and silicon carbide (λ_{max} of 580 nm). With proper mixing of these compounds the λ_{max} of emission can be shifted between 540 and 900 nm.

LEDs are usually operated at very low voltage and they may draw a current up to 200 mA. These lamps produce relatively narrow spectral emission bands as shown in Fig. 5.26. They are small and have found wide application in computer systems and in various photographic devices.

5.4.7   Lasers

Although this chapter concerns itself primarily with noncoherent sources, it seems appropriate to provide a brief survey of those laser sources

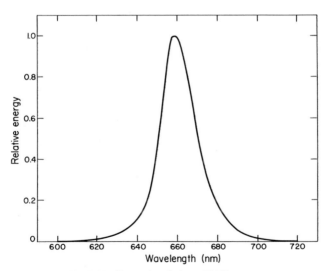

Fig. 5.26   Spectral emission of LED array.

that have found a wide application in science and engineering. The word laser is an acronym for *L*ight *A*mplification by *S*timulated *E*mission of *R*adiation. Laser radiation differs from radiation by other sources in the degree to which it is monochromatic, its very high intensity, and its coherence in time and space. The principle of laser radiation is based on a process in which a photon of light stimulates or induces an existing excited-state molecule or ion to emit a second photon in phase with the initiating photon.

The lasing material can be a gas, a liquid, or a solid, and is activated, i.e., put into an electronically excited state, by the absorption of large amounts of electromagnetic energy at wavelengths other than the wavelength of stimulated emission. When a sufficient population of lasing material is in the excited state, an outside source of photons of exactly the same energy as that to be emitted will stimulate the population of excited-state material to give up its excitation energy in phase, thus greatly amplifying the number of photons over the number which stimulated the emission. Flash tubes are normally used as external sources to supply electromagnetic energy for pumping, although this can also be accomplished by the use of an electric

TABLE 5.5   Laser Classifications[a]

Pump Energy	Active Medium					
	Solid			Liquid	Gas	
	Crystal		Glass		Molecular	Atomic	Ionic
Optical	Host lattice	Activate	Doped with neodymium, erbium, and other	Neodymium solution Cyanide Dyes Chelates			
	Ruby	Chromium (Cr)					
	YAG[b]	Neodymium (Nd) Erbium (Er) Holmium (Ho)					
	YIG[b]} FAP[b]}	Neodymium (Nd)					
Electric discharge					CO_2—He—N_2[d] CO	He—Ne[d] N_2	Argon Krypton He—Cd[d]
Electric current	Semiconductor junction, e.g., gallium arsenide (GaAs)						
Chemical					DF—CO_2[c]		
Nuclear							Argon, Boron Xenon
Electron beam	Semiconductor e.g., gallium arsenide						

[a] From Thomas (1973), p. 132.

[b] YAG = yttrium–aluminum–garnet; YIG = yttrium–iron–garnet; FAP = fluorapatite.

[c] DF = deuterium fluoride.

[d] CO_2—He—N_2 = carbon dioxide–helium–nitrogen; He—Ne = helium–neon; He—Cd = helium–cadmium.

discharge through the emitting gas. The manner and/or method of pumping classifies the laser types. Table 5.5 lists some of the most common lasers.

Laser energy is monochromatic because the electromagnetic wave is generated by anionic (atomic or molecular) transition between two specific energy states which are stable and very reproducible. Some lasers may be operated at several different wavelengths depending on the transitions between different energy states. The list of most important laser wavelengths is given in Table 5.6, and the typical values of power for some common lasers are given in Table 5.7. It has also been found that some organic dyes can be used for continuously tuning the output wavelength (Kagan *et al.*, 1968; Colles and Pigeon, 1975), so that tunable coherent radiant energy is now available from one source or another throughout most of the visible and near-infrared spectral regions. These tunable devices have made it possible for revolutionary advances in various areas of spectroscopy (especially time-resolved spectroscopy and resonance Raman spectroscopy) and in such studies as isotope separation and pollution detection.

An operational distinction concerning lasers that should be mentioned is that they can be operated in a continuous-wave (CW) or a pulsed mode. The pulsed mode, which produces the highest peak power, can be operated in

TABLE 5.6   Important Laser Wavelengths[a]

	Wavelength (μm)	Laser Medium[b]	Operation[c]	Wavelength (μm)	Laser Medium[b]	Operation[c]
Ultraviolet	0.2358	Ne*	P	0.6150	Hg*-He	P
	0.3250	He-Cd*		0.6328	He-Ne	
	0.3324	Ne*	P	0.6471	Kr*	
	0.3371	N_2	P	0.6764	Kr*	
	0.3511	Ar*		0.6943	Cr*Al$_2$O$_3$ (ruby)	
Visible	0.4280	SiC		0.85–0.9 (temp. depend.)	GaAs	
	0.4416	He-Cd*				
	0.4579	Ar*	P	1.0648	Nd*-YAG	
	0.4658	Ar*			-FAP	
	0.4680	Kr*			-glass	
	0.4727	Ar*		1.15	He-Ne	
	0.4762	Kr*		1.6602	Er*-YAG	
	0.4765	Ar*		2.0261	Xe	
	0.4825	Kr*		2.0975	Ho*-YAG	
	0.4880	Ar*		3.3912	He-Ne	
	0.4965	Ar*		3.507	Xe	
	0.5017	Ar*		5.6	CO	
	0.5145	Kr*		10.6	CO$_2$-He-N$_2$	
	0.5208	Ar*		27.9	H$_2$O	P
	0.5324	Freq. double Nd*		118.6	H$_2$O	P
	0.5682	Kr*				

[a] From Thomas (1973), p. 133.

[b] An asterisk indicates the ionic state.

[c] P indicates pulsed operation only. A blank means that the line can go continuously or pulsed.

TABLE 5.7 Typical Performance of Practical Lasers[a]

Type	Wavelength (μm)	Beampower (S) = TEM_{00} mode	Comments
He-Cd	0.325	10 mW (S)	Reliable, long life, CW,
	0.442	50 mW (S)	ultraviolet source
N_2	0.337	200 kW peak	Requires very high voltage pulses
Ar (ion)	0.458–0.514	4 W (S) total all lines	Power of ionic type of gas lasers scales as: Power $\propto$ (current)^4 $\times$ (length)^2
Kr (ion)	0.568–0.647	750 mW (S) total all lines	0.568-μm line of Kr is strong yellow
He-Ne	0.633	100 mW (S)	Most common laser: low cost,
	1.15	100 mW (S)	readily available, red source
	3.39	20 mW (S)	Very high gain
Ruby	0.694	1 mW, CW	High peak power,
		500 kW, 1 msec normal pulse at 60 ppm	high repetition rate
		150 MW, 20 nsec Q switch at 60 ppm	
GaAs (arrays)	0.85–0.94 (temp. sens.)	10 W, CW at 77 K 10 kW peak	High efficiency; requires cooling or short pulse, low duty cycle
Nd-YAG frequency doubled	0.532	10 W, CW (S)	Very efficient frequency doubling with barium-sodium-niobate
Nd-YAG	1.06	20 W, CW 10 MW, 8–10 nsec Q switch at 1–50 ppm	Lower peak power but higher repetition rate than glass
Nd-glass	1.06	100 W, CW	~6% efficiency, high power
		500 kW at 6 ppm normal pulse	
		200 MW, 10–30 nsec Q switch at 4 ppm	
CO_2	10.6	150 W, CW (as high as 8.8 kW, CW)	High efficiency (~ 20%), high power; power scales as: Power $\propto$ length
		80 W (S), 100 kW, 50 nsec Q switch at 400 pps	

[a] From Thomas (1973), p. 134.

three modes: normal mode, Q-switched mode, and mode-locked. These different modes produce pulse durations in the milliseconds, microseconds, and nanoseconds–picoseconds range, respectively. In the normal mode, a short-duration electric field is applied to either a flash lamp or the plasma tube (gas laser). In the Q-switch mode the quality factor (Q) of the resonant cavity is "spoiled" until a large amount of energy is stored in the active medium, and when the Q is returned, a high power pulse of very short duration results. This Q-switching can be achieved by placing appropriate absorbers in the laser cavity. Mode locking is achieved by the internal modulation of energy to obtain a selective buildup of bursts that have high power and very short duration.

For further details on laser types and modes and for various application requirements the reader can turn to the avalanche of papers in this field.

5.5 D-ILLUMINANTS AND SIMULATORS

5.5.1 CIE Illuminants

In 1966 the CIE recommended that the CIE standard illuminants A, B, and C be supplemented by a new standard illuminant D_{65} representing daylight (Judd *et al.*, 1964; CIE, 1974b). This new illuminant representing daylight over the spectral range 300–830 nm has a correlated color temperature of 6500 K and is defined by its relative spectral irradiance distribution.

The rules concerning a numerical procedure for defining any daylight illuminant D were also recommended (CIE, 1974b). By this procedure the spectral distribution of a new illuminant D can be computed when the correlated color temperature of the desired phase of daylight is given. The recommendation lists the spectral distribution not only of D_{65} but also of illuminants D_{55} and D_{75} (CIE, 1971b).

The CIE 1931 chromaticity coordinates (x, y) should be the following:

$$D_{65}: \quad x = 0.3127, \qquad y = 0.3291,$$
$$D_{75}: \quad x = 0.2990, \qquad y = 0.3150,$$
$$D_{55}: \quad x = 0.3324, \qquad y = 0.3475.$$

The numerical data for these illuminants and for D_{5000}, D_{6000}, and D_{7000} are given at 10 nm intervals in Table 5.8.

Although no fundamental formula can be given with which to compute the spectral distribution of daylight for any color temperature, the CIE recommended a procedure (Wyszecki and Stiles, 1967; Judd and Wyszecki, 1975; CIE, 1971b) by which representative distributions can be computed, so that there is a common basis for interlaboratory communication of color data for any desired phase of natural daylight between 4000 and 25,000 K. For use in these computations, color temperatures were based on the use of $c_2 = hc/k = 14,388$ in Planck's formula. These recommendations came about after an extensive body of data on natural daylight became available (Judd *et al.*, 1964; British Standard, 1967).

TABLE 5.8 Relative Spectral Irradiances of CIE Illuminants D_{55}, D_{65}, and D_{75}

λ (nm)	D_{55}	D_{65}	D_{75}
300	0.02	0.03	0.04
10	2.1	3.3	5.1
20	11.2	20.2	29.8
30	20.6	37.1	54.9
40	23.9	39.9	57.3
350	27.8	44.9	62.7
60	30.6	46.6	63.0
70	34.3	52.1	70.3
80	32.6	50.0	66.7
90	38.1	54.6	70.0
400	61.0	82.8	101.9
10	68.6	91.5	111.9
20	71.6	93.4	112.8
30	67.9	86.7	103.1
40	85.6	104.9	121.2
450	98.0	117.0	133.0
60	100.5	117.8	132.4
70	99.9	114.9	127.3
80	102.7	115.9	126.8
90	98.1	108.8	117.8
500	100.7	109.4	116.6
10	100.7	107.8	113.7
20	100.0	104.8	108.7
30	104.2	107.7	110.4
40	102.1	104.4	106.3
550	103.0	104.0	104.9
60	100.0	100.0	100.0
70	97.2	96.3	95.6
80	97.7	95.8	94.2
90	91.4	88.7	87.0
600	94.4	90.0	87.2
10	95.1	89.6	86.1
20	94.2	87.7	83.6
30	90.4	83.3	78.7
40	92.3	83.7	78.4
650	88.9	80.0	74.8
60	90.3	80.2	74.3
70	93.9	82.3	75.4
80	90.0	78.3	71.6
90	79.7	69.7	63.9
700	82.8	71.6	65.1
10	84.8	74.3	68.1
20	70.2	61.6	56.4
30	79.3	69.9	64.2
40	85.0	75.1	69.2
750	71.9	63.6	58.6
60	52.8	46.4	42.6
70	75.9	66.8	61.4
80	71.8	63.4	58.3
90	72.9	64.3	59.1
800	67.3	59.5	54.7
10	58.7	52.0	47.9
20	65.8	57.4	52.9
30	68.3	60.3	55.5

5.5.2   D-Simulators

In view of the CIE recommendations described in Section 5.5.1 and because of the wide international acceptance of these recommendations, the search for sources that will simulate these various D-illuminants has been widely undertaken.[†] The main applications of such artificial sources (simulators) are visual matching of object colors (particularly of fluorescent materials) and appraisal of color-rendering properties of other light sources. Daylight simulators are also widely used in various physical measurements and characterization of photometric and radiometric properties of materials.

Grum *et al.* (1970) described a series of D_{65}-simulating sources intended for use with spectroradiometers and other optical instruments. Examples of these based on a fluorescent lamp, a xenon arc lamp, and a filtered tungsten lamp are shown in Fig. 5.27–5.29. A typical application requiring such simulators is described elsewhere (Wyszecki, 1970).

Relative spectral irradiance distributions for 11 filtered and unfiltered xenon arcs, filtered tungsten–halogen lamps, and 16 fluorescent lamps were assembled by the CIE Colorimetry Committee E-1.3.1. The data on these

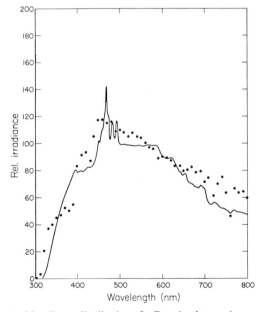

Fig. 5.27   Spectral irradiance distribution of a D_{65} simulator using a xenon arc lamp.

[†] A distinction is made between "illuminant" and "source." The term "source" refers to a physical emitter of light such as a lamp or the sun and sky. The term "illuminant" refers to a specific spectral irradiance distribution given in a table or by a curve.

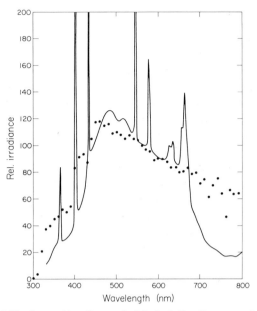

Fig. 5.28 Spectral irradiance of a Macbeth D_{65} fluorescent lamp.

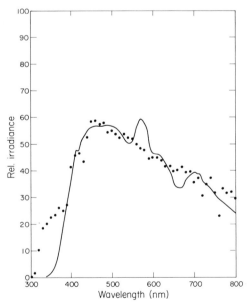

Fig. 5.29 Spectral irradiance of 1000-W halogen lamp plus Pittsburgh 2043 and Corning 5900 filters.

were collected for the purpose of studying the state-of-the-art of reproducing CIE standard illuminants D_{55}, D_{65}, and D_{75} by means of artificial light sources.

The question now arises as to how good a fit is required between such simulators and the standard illuminants. The criteria for the goodness-of-fit, however, must be weighted for a particular application. In visual work and with nonfluorescent samples it is adequate that the spectral match is acceptable within the visible region of the spectrum. When, however, a simulator is used in the measurement or in visual appraisals of fluorescent materials it is imperative that the goodness-of-fit extends spectrally into the ultraviolet region of the spectrum since the ultraviolet content of the source is very critical in the excitation of fluorescent species (Grum, 1972). Fluorescent materials can have very noticeable and peculiar tints in illumination that differ very slightly or not at all, visually, from the standard light for which the color was adjusted to a maximum. Such a peculiar sensitivity of visually trivial variations of quality of illumination is often as objectionable as the actual tints that result from nonstandard lighting.

The subject and criteria for goodness-of-fit have been discussed in the literature (Wyszecki, 1970; Stiles and Wyszecki, 1962) and are presently being studied by the CIE Subcommittee on Standard Sources (CIE, 1974; Berger and Strocka, 1975).

5.6 PLOTTING RADIANT ENERGY DATA

5.6.1 Continuous Sources

Spectral irradiance data for a source should be presented in the most meaningful manner. When the spectral irradiance of a tungsten source or any continuous source is plotted on an arithmetic scale, it is almost impossible to read off the lower ordinate with accuracy. This is illustrated in Fig. 5.30

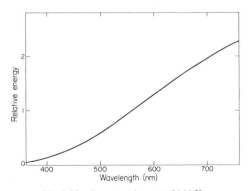

Fig. 5.30 Tungsten lamp at 2844 K.

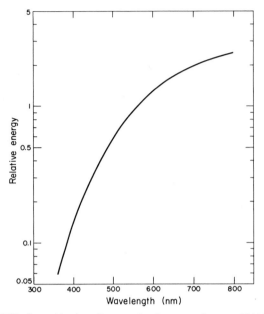

Fig. 5.31   Logarithmic ordinate scale of tungsten lamp at 2844 K.

where an arithmetic plot of the spectral irradiance of a tungsten source for a
temperature of 2844 K for λ interval 360–760 nm is shown. In Fig. 5.31 the
same data are plotted on a logarithmic ordinate scale. The values may be
read from this plot at short wavelengths with the same precision as at
longer wavelengths. Other useful relationships of the logarithmic plots have
been discussed by Moon (1936).

5.6.2   Line Sources

A basic point of view here is that lines of a mixed source shall be reported
separately from any continuum. Justification for this separate treatment is
twofold:

(1)   Composite functions, if reported as such, cannot be correctly deci-
phered as to line–power content; "lines" shown graphically as rectangles are
not usually well enough identified as to their power scales.
(2)   Since a bright line is due to its atomic origin and is emitted at a single
wavelength, its power should be reported at this wavelength. The desirability
of separately reporting lines and continuum is simply that this method has a

proper physical basis. One observes that the line "sits" on the continuum of a mixed source just as a pure line would sit on a zero baseline if there were no continuum.

The optimum measuring procedure is to scan the wavelength of the measuring radiometer slowly and carefully across all lines. Digital readout every 1 nm where bandwidth is, say, 5 nm, is a good arrangement. Spectral irradiance calculated by reference to a standard source is to be designated a tentative function, since account must be taken of the fact that the monochromator has spread a line into a nearly triangular shape above the continuum. Line power is essentially equal to the area within the spread function triangle. Further discussion of this topic is contained in Chapter 8.

On the measured data, the continuum is in proper form except that it must be interpolated whenever line spreading exists. As to the line irradiance, one way to estimate the area of the triangle is to multiply the peak tentative irradiance of a line above the interpolated continuum by the half-bandwidth in nanometers.

When plotting the spectral irradiance of a mixed source one can label the left-hand ordinate with spectral irradiance units for the continuum and the right-hand ordinates with line irradiance units. Figure 5.32 is an example of this mode of plotting.

There is not much information available in the literature for the treatment of mixed sources. The notable exceptions are the work of Nonaka and that of NBS (Grum and Costa, 1976; National Bureau of Standards, 1970). We shall treat this topic further in Chapters 8 and 9.

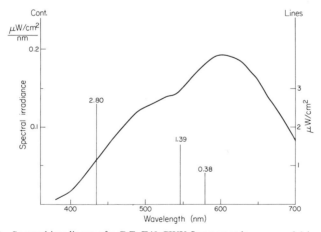

Fig. 5.32   Spectral irradiance of a G.E. F40 CWX fluorescent lamp; area 3.0 in. × 3.0 in.

APPENDIX 5.1 MAINLINE PROGRAM FOR CORRELATED COLOR TEMPERATURE

```
// JOB

LOG DRIVE    CART SPEC    CART AVAIL   PHY DRIVE
   0000         738B         738B         0000

V2 M10    ACTUAL  8K   CONFIG  8K

// DUP

*DELETE                  CCT
CART ID 738B    DB ADDR  4DC0    DB CNT    0220

// FOR
*IOCS(CARD,1132 PRINTER)
*IOCS(DISK)
*IOCS(TYPEWRITER)
*ONE WORD INTEGERS
*EXTENDED PRECISION
*LIST SOURCE PROGRAM
C      IBM 1130 COMPUTER PROGRAM 'CCT'
C      REVISED PROGRAM DATED-- 13 SEPT 1973 (D.PAINE)
C      SPECTRAL IRRADIANCE DATA FORMAT FROM 1130 PROGRAM 'SRDS'
       INTEGER FILT
       REAL LUM,M(30),MIRED
       DIMENSION XBAR(161),YBAR(161),ZBAR(161),P(161),PX(161),PY(161),
      1PZ(161),DATE(4),UCT(30),VCT(30),D(30),T(30),RATD(30),SAMP(8),
      2TTIT(8),TF(161)
       EQUIVALENCE (XBAR(1),PX(1)),(YBAR(1),PY(1)),(ZBAR(1),PZ(1),TF(1))
       DEFINE FILE 1(483,3,U,NUMBR)
       DEFINE FILE 2(195,3,U,NUMBR)
       DATA UCT(1),VCT(1),M(1)/0.18065,0.26589,-0.2548/,T(1)/100000./
       DATA UCT(2),VCT(2),M(2)/0.18132,0.26845,-0.2687/,T(2)/50000./
       DATA UCT(3),VCT(3),M(3)/0.18208,0.27118,-0.2854/,T(3)/33333./
       DATA UCT(4),VCT(4),M(4)/0.18293,0.27407,-0.3047/,T(4)/25000./
       DATA UCT(5),VCT(5),M(5)/0.18388,0.27708,-0.3267/,T(5)/20000./
       DATA UCT(6),VCT(6),M(6)/0.18494,0.28020,-0.3515/,T(6)/16667./
       DATA UCT(7),VCT(7),M(7)/0.18611,0.28340,-0.3790/,T(7)/14286./
       DATA UCT(8),VCT(8),M(8)/0.18739,0.28666,-0.4094/,T(8)/12500./
       DATA UCT(9),VCT(9),M(9)/0.18879,0.28995,-0.4426/,T(9)/11111./
       DATA UCT(10),VCT(10),M(10)/0.19031,0.29325,-0.4787/,T(10)/10000./
       DATA UCT(11),VCT(11),M(11)/0.19461,0.30139,-0.5817/,T(11)/8000./
       DATA UCT(12),VCT(12),M(12)/0.19960,0.30918,-0.7043/,T(12)/6667./
       DATA UCT(13),VCT(13),M(13)/0.20523,0.31645,-0.8484/,T(13)/5714./
       DATA UCT(14),VCT(14),M(14)/0.21140,0.32309,-1.0170/,T(14)/5000./
       DATA UCT(15),VCT(15),M(15)/0.21804,0.32906,-1.2160/,T(15)/4444./
       DATA UCT(16),VCT(16),M(16)/0.22507,0.33436,-1.4500/,T(16)/4000./
       DATA UCT(17),VCT(17),M(17)/0.23243,0.33901,-1.7280/,T(17)/3636./
       DATA UCT(18),VCT(18),M(18)/0.24005,0.34305,-2.0610/,T(18)/3333./
       DATA UCT(19),VCT(19),M(19)/0.24787,0.34653,-2.4650/,T(19)/3077./
       DATA UCT(20),VCT(20),M(20)/0.25585,0.34948,-2.9600/,T(20)/2857./
       DATA UCT(21),VCT(21),M(21)/0.26394,0.35198,-3.5760/,T(21)/2667./
       DATA UCT(22),VCT(22),M(22)/0.27210,0.35405,-4.3550/,T(22)/2500./
       DATA UCT(23),VCT(23),M(23)/0.28032,0.35575,-5.3650/,T(23)/2353./
       DATA UCT(24),VCT(24),M(24)/0.28854,0.35713,-6.7110/,T(24)/2222./
       DATA UCT(25),VCT(25),M(25)/0.29676,0.35822,-8.5720/,T(25)/2105./
       DATA UCT(26),VCT(26),M(26)/0.30496,0.35906,-11.2900/,T(26)/2000./
       DATA UCT(27),VCT(27),M(27)/0.31310,0.35968,-15.5600/,T(27)/1905./
       DATA UCT(28),VCT(28),M(28)/0.32119,0.36011,-23.2000/,T(28)/1818./
       DATA UCT(29),VCT(29),M(29)/0.32920,0.36038,-40.4100/,T(29)/1739./
```

```
       DATA UCT(30),VCT(30),M(30)/0.33713,0.36051,-113.8000/,T(30)/1667./
       READ (2,100) DATE
  100 FORMAT (4A4)
       DO 8888 N=1,1000
       READ (2,101) SAMP,KIND,FILT,KFILT
  101 FORMAT (8A6/3I1)
C     KIND=1-- PRE-CALCULATED X,Y CO-ORDINATES
C     KIND=2-- SPECTRAL RADIANCE DATA FROM 380 TO 700 NM, EVERY 2 NM
C     KIND=3-- SPECTRAL RADIANCE DATA FROM 380 TO 700 NM, EVERY 5 NM
C     KIND=4-- SPECTRAL RADIANCE DATA BETWEEN 380 AND 700 NM, EVERY 10 NM.
C     KFILT=0 DENOTES GENERAL FILTER DATA-- 380 TO 700 NM
C  (DATA MUST BE IN ONE-TO-ONE CORRESPONDENCE WITH SPECTRAL IRRADIANCE DATA.)
C     KFILT=1 DENOTES G.E.-RECORDED FILTER DATA-- 400 TO 700 NM, EVERY 5 NM
C  FOR KIND=4, ENDPOINTS MUST BE 400 NM AND 700 NM.
       WRITE (3,2019) SAMP,DATE
 2019 FORMAT (1H1,8A6/1X,4A4)
       IF (KIND-4) 1100,1100,9999
 1100 IF (KIND) 9999,9999,20000
20000 LINDX=1
       LINC=1
       LSTRT=380
       LEND=700
       GO TO (2100,2200,2300,2400), KIND
 2100 READ (2,102) X,Y
  102 FORMAT (F6.4,F7.4)
 2150 IF (FILT) 10000,10000,9999
 2200 NPTS=161
       DELT=2.
       IF (KFILT) 30000,30000,9999
 2300 NPTS=65
       NFT=NPTS
       DELT=5.
       GO TO 30000
 2400 READ (2,199) LSTRT,LEND
  199 FORMAT (I3,1X,I3)
       IF (LSTRT-380) 9999,2500,2500
 2500 IF (LEND-700) 2550,2600,9999
 2550 IF (LEND-LSTRT) 9999,9999,2600
 2600 DO 2700 JJ=1,65
       LCHK=375+5*JJ
       IF (LSTRT-LCHK) 9999,2650,2700
 2650 LINDX=JJ
       LINC=2
       NPTS=(LEND-LSTRT)/10+1
       DELT=10.
       WRITE (1,198) LSTRT,LCHK,LEND,NPTS,LINDX
  198 FORMAT (/3I4,5X,2I4)
       IF (FILT) 9999,30000,2800
 2800 IF (KFILT) 9999,2810,2900
 2810 NFT=NPTS
       INCFT=1
       GO TO 30000
 2900 IF (LSTRT-400) 9999,2901,9999
 2901 IF (LEND-700) 9999,2902,9999
 2902 NFT=61
       INCFT=2
 2700 CONTINUE
30000 INC=IFIX(DELT)
       READ (2,103) (P(J),J=1,NPTS)
```

```
  103 FORMAT (2X,E14.7,2X,E14.7,2X,E14.7,2X,E14.7,2X,E14.7)
      WRITE (3,2029) LSTRT,LEND,INC,(P(J),J=1,NPTS)
 2029 FORMAT (1H0,5X,'SPECTRAL IRRADIANCE MEASUREMENTS  ',
     1'(',I3,' NM TO ',I3,' NM AT ',I2,'-NM INTERVAL)--'/
     2(2X,E14.7,2X,E14.7,2X,E14.7,2X,E14.7,2X,E14.7))
      IF (FILT) 3100,3100,3200
 3100 IF (KIND-1) 9999,2150,10000
 3200 IF (KFILT) 3400,3400,3250
 3250 READ (2,105) TTIT,(TF(J),J=1,NFT)
  105 FORMAT (8A6/(32X,16F3.3))
      GO TO 3450
 3400 READ (2,104) TTIT,(TF(J),J=1,NFT)
  104 FORMAT (8A6/(F6.4,10F7.4))
 3450 WRITE (3,2039) LSTRT,LEND,INC,TTIT,(TF(J),J=1,NFT)
 2039 FORMAT (1H0,5X,'FILTER TRANSMITTANCE FUNCTION  ',
     1'(',I3,' NM TO ',I3,' NM AT ',I2,'-NM INTERVAL)--'/
     21X,8A6/(11F7.4))
      DO 3500 J=1,NPTS
      K=1+(J-1)*INCFT
 3500 P(J)=P(J)*TF(K)
      WRITE (3,2049) LSTRT,LEND,INC,(P(J),J=1,NFT)
 2049 FORMAT (1H0,5X,'FILTER-MODIFIED SPECTRAL IRRADIANCE DATA  ',
     1'(',I3,' NM TO ',I3,' NM AT ',I2,'-NM INTERVAL)--'/
     2(2X,E14.7,2X,E14.7,2X,E14.7,2X,E14.7,2X,E14.7))
10000 NUMBR=1
      GO TO (5000,3000,2000,2000), KIND
 2000 READ (2'NUMBR) (XBAR(J),J=1,65)
      NUMBR=66
      READ (2'NUMBR) (YBAR(J),J=1,65)
      NUMBR=131
      READ (2'NUMBR) (ZBAR(J),J=1,65)
      IF (KIND-4) 2001,2002,9999
 2001 NPTS=65
      GO TO 3150
 2002 NPTS=(LEND-LSTRT)/10+1
      GO TO 3150
 3000 READ (1'NUMBR) (XBAR(J),J=1,161)
      NUMBR=162
      READ (1'NUMBR) (YBAR(J),J=1,161)
      NUMBR=323
      READ (1'NUMBR) (ZBAR(J),J=1,161)
      NPTS=161
 3150 DO 4100 J=1,NPTS
      K=LINDX+(J-1)*LINC
      PX(J)=P(J)*XBAR(K)
      PY(J)=P(J)*YBAR(K)
 4100 PZ(J)=P(J)*ZBAR(K)
      CALL EQSF(DELT,PX,PX,NPTS)
      CALL EQSF(DELT,PY,PY,NPTS)
      CALL EQSF(DELT,PZ,PZ,NPTS)
      LUM=PY(NPTS)
      XYZ=PX(NPTS)+PY(NPTS)+PZ(NPTS)
      X=PX(NPTS)/XYZ
      Y=PY(NPTS)/XYZ
 5000 UTST=4.*X/(-2.*X+12.*Y+3.)
      VTST=6.*Y/(-2.*X+12.*Y+3.)
      DO 6000 J=1,30
 6000 D(J)=(VTST-VCT(J)-M(J)*(UTST-UCT(J)))/SQRT(1.+M(J)*M(J))
      DO 6100 J=1,29
```

```
6100 RATD(J)=D(J)/D(J+1)
     DO 6500 J=1,29
     IF (RATD(J)) 6200,6500,6500
6200 JPOS=J
     GO TO 7000
6500 CONTINUE
7000 TCOR=1./T(JPOS)+(D(JPOS)/(D(JPOS)-D(JPOS+1)))
    1*(1./T(JPOS+1)-1./T(JPOS))
     TCOR=1./TCOR+0.05
     X=X+0.00005
     Y=Y+0.00005
     UTST=UTST+0.00005
     VTST=VTST+0.00005
     IF (FILT) 7200,7200,7100
7100 WRITE (3,204) SAMP,TTIT,DATE
 204 FORMAT (1H1,8A6/1X,8A6/5X,4A4,20X,'(CONTINUED)')
7200 WRITE (3,201) X,Y,UTST,VTST
 201 FORMAT (1H0,5X,'X =',F7.4,5X,'Y =',F7.4/6X,'U =',F7.4,5X,'V =',
    1F7.4//1X,'ROBERTSON ANALYSIS--'/2X,'LINE',4X,'MIRED',5X,'T(K)',
    211X,'D',12X,'D(I)/D(I+1)')
     DO 8000 J=1,30
     MIRED=1.0E+06/T(J)+0.5
     IF (J-30) 7500,7600,8000
7500 WRITE (3,202) J,MIRED,T(J),D(J),RATD(J)
 202 FORMAT (1H ,2X,I2,4X,F5.0,4X,F7.0,2(4X,E13.6))
     GO TO 8000
7600 WRITE (3,202) J,MIRED,T(J),D(J)
8000 CONTINUE
     WRITE (3,203) TCOR
 203 FORMAT (1H0,'CORRELATED COLOR TEMPERATURE =',F9.1,' KELVINS')
     IF (KIND-1) 9999,8888,8100
8100 WRITE (3,205) LUM
 205 FORMAT (1H ,4X,'TOTAL LUMINOSITY =',E14.7)
8888 CONTINUE
9999 CALL EXIT
     END

FEATURES SUPPORTED
 ONE WORD INTEGERS
 EXTENDED PRECISION
 IOCS

CORE REQUIREMENTS FOR
 COMMON        0  VARIABLES   2604  PROGRAM   1630

END OF COMPILATION

// DUP

*STORE        WS  UA  CCT
CART ID 738B    DB ADDR   570F    DB CNT    0090
```

Subroutine EQSF of Mainline Program

```
// JOB

LOG DRIVE    CART SPEC    CART AVAIL   PHY DRIVE
  0000          7701         7701        0000

V2 M12   ACTUAL  8K   CONFIG  8K

// FOR
*EXTENDED PRECISION
*ONE WORD INTEGERS
*LIST SOURCE PROGRAM
      SUBROUTINE EQSF(H,Y,Z,NDIM)
C     EXTENDED PRECISION VERSION OF SUBROUTINE QSF
C     INTEGRATION OF AN EQUIDISTANTLY TABULATED FUNCTION BY SIMPSON'S RULE
C     DESCRIPTION OF PARAMETERS--
C       H     - THE INCREMENT OF ARGUMENT VALUES.
C       Y     - THE INPUT VECTOR OF FUNCTION VALUES.
C       Z     - THE RESULTING VECTOR OF INTEGRAL VALUES.
C             Z MAY BE IDENTICAL WITH Y.
C       NDIM  - THE DIMENSION OF VECTORS Y AND Z.
C
      DIMENSION Y(161),Z(161)
      HT=.3333333*H
      IF (NDIM-5) 7,8,1
C
C     NDIM IS GREATER THAN 5.  PREPARATIONS OF INTEGRATION LOOP
    1 SUM1=Y(2)+Y(2)
      SUM1=SUM1+SUM1
      SUM1=HT*(Y(1)+SUM1+Y(3))
      AUX1=Y(4)+Y(4)
      AUX1=AUX1+AUX1
      AUX1=SUM1+HT*(Y(3)+AUX1+Y(5))
      AUX2=HT*(Y(1)+3.875*(Y(2)+Y(5))+2.625*(Y(3)+Y(4))+Y(6))
      SUM2=Y(5)+Y(5)
      SUM2=SUM2+SUM2
      SUM2=AUX2-HT*(Y(4)+SUM2+Y(6))
      Z(1)=0.
      AUX=Y(3)+Y(3)
      AUX=AUX+AUX
      Z(2)=SUM2-HT*(Y(2)+AUX+Y(4))
      Z(3)=SUM1
      Z(4)=SUM2
      IF (NDIM-6) 5,5,2
C
C     INTEGRATION LOOP
    2 DO 4 I=7,NDIM,2
      SUM1=AUX1
      SUM2=AUX2
      AUX1=Y(I-1)+Y(I-1)
      AUX1=AUX1+AUX1
      AUX1=SUM1+HT*(Y(I-2)+AUX1+Y(I))
      Z(I-2)=SUM1
      IF (I-NDIM) 3,6,6
    3 AUX2=Y(I)+Y(I)
      AUX2=AUX2+AUX2
      AUX2=SUM2+HT*(Y(I-1)+AUX2+Y(I+1))
    4 Z(I-1)=SUM2
    5 Z(NDIM-1)=AUX1
      Z(NDIM)=AUX2
```

```
          RETURN
        6 Z(NDIM-1)=SUM2
          Z(NDIM)=AUX1
          RETURN
C         END OF INTEGRATION LOOP
C
        7 IF (NDIM-3) 12,11,8
C
C         NDIM IS EQUAL TO 4 OR 5
        8 SUM2=1.125*HT*(Y(1)+Y(2)+Y(2)+Y(2)+Y(3)+Y(3)+Y(3)+Y(4))
          SUM1=Y(2)+Y(2)
          SUM1=SUM1+SUM1
          SUM1=HT*(Y(1)+SUM1+Y(3))
          Z(1)=0.
          AUX1=Y(3)+Y(3)
          AUX1=AUX1+AUX1
          Z(2)=SUM2-HT*(Y(2)+AUX1+Y(4))
          IF (NDIM-5) 10,9,9
        9 AUX1=Y(4)+Y(4)
          AUX1=AUX1+AUX1
          Z(5)=SUM1+HT*(Y(3)+AUX1+Y(5))
       10 Z(3)=SUM1
          Z(4)=SUM2
          RETURN
C
C         NDIM IS EQUAL TO 3
       11 SUM1=HT*(1.25*Y(1)+Y(2)+Y(2)-.25*Y(3))
          SUM2=Y(2)+Y(2)
          SUM2=SUM2+SUM2
          Z(3)=HT*(Y(1)+SUM2+Y(3))
          Z(1)=0.
          Z(2)=SUM1
       12 RETURN
          END

FEATURES SUPPORTED
 ONE WORD INTEGERS
 EXTENDED PRECISION

CORE REQUIREMENTS FOR EQSF
 COMMON        0  VARIABLES     28  PROGRAM     698

RELATIVE ENTRY POINT ADDRESS IS 0036 (HEX)

END OF COMPILATION
```

APPENDIX 5.2   CIE SPECTRAL TRISTIMULUS VALUES AND CHROMATICITY COORDINATES*

λ (nm)	SPECTRAL TRISTIMULUS VALUES			CHROMATICITY COORDINATES		
	$\bar{x}(\lambda)$	$\bar{y}(\lambda)$	$\bar{z}(\lambda)$	$x(\lambda)$	$y(\lambda)$	$z(\lambda)$
360	0.000 129 900 0	0.000 003 917 000	0.000 606 100 0	0.175 56	0.005 29	0.819 15
61	0.000 145 847 0	0.000 004 393 581	0.000 680 879 2	0.175 48	0.005 29	0.819 23
62	0.000 163 802 1	0.000 004 929 604	0.000 765 145 6	0.175 40	0.005 28	0.819 32
63	0.000 184 003 7	0.000 005 532 136	0.000 860 012 4	0.175 32	0.005 27	0.819 41
64	0.000 206 690 2	0.000 006 208 245	0.000 966 592 8	0.175 24	0.005 26	0.819 50
365	0.000 232 100 0	0.000 006 965 000	0.001 086 000	0.175 16	0.005 26	0.819 58
66	0.000 260 728 0	0.000 007 813 219	0.001 220 586	0.175 09	0.005 25	0.819 66
67	0.000 293 075 0	0.000 008 767 336	0.001 372 729	0.175 01	0.005 24	0.819 75
68	0.000 329 388 0	0.000 009 839 844	0.001 543 579	0.174 94	0.005 23	0.819 83
69	0.000 369 914 0	0.000 011 043 23	0.001 734 286	0.174 88	0.005 22	0.819 90
370	0.000 414 900 0	0.000 012 390 00	0.001 946 000	0.174 82	0.005 22	0.819 96
71	0.000 464 158 7	0.000 013 886 41	0.002 177 777	0.174 77	0.005 23	0.820 00
72	0.000 518 986 0	0.000 015 557 28	0.002 435 809	0.174 72	0.005 24	0.820 04
73	0.000 581 854 0	0.000 017 442 96	0.002 731 953	0.174 66	0.005 24	0.820 10
74	0.000 655 234 7	0.000 019 583 75	0.003 078 064	0.174 59	0.005 22	0.820 19
375	0.000 741 600 0	0.000 022 020 00	0.003 486 000	0.174 51	0.005 18	0.820 31
76	0.000 845 029 6	0.000 024 839 65	0.003 975 227	0.174 41	0.005 13	0.820 46
77	0.000 964 526 8	0.000 028 041 26	0.004 540 880	0.174 31	0.005 07	0.820 62
78	0.001 094 949	0.000 031 531 04	0.005 158 320	0.174 22	0.005 02	0.820 76
79	0.001 231 154	0.000 035 215 21	0.005 802 907	0.174 16	0.004 98	0.820 86
380	0.001 368 000	0.000 039 000 00	0.006 450 001	0.174 11	0.004 96	0.820 93
81	0.001 502 050	0.000 042 826 40	0.007 083 216	0.174 09	0.004 96	0.820 95
82	0.001 642 328	0.000 046 914 60	0.007 745 488	0.174 07	0.004 97	0.820 96
83	0.001 802 382	0.000 051 589 60	0.008 501 152	0.174 06	0.004 98	0.820 96
84	0.001 995 757	0.000 057 176 40	0.009 414 544	0.174 04	0.004 98	0.820 98
385	0.002 236 000	0.000 064 000 00	0.010 549 99	0.174 01	0.004 98	0.821 01
86	0.002 535 385	0.000 072 344 21	0.011 965 80	0.173 97	0.004 97	0.821 06
87	0.002 892 603	0.000 082 212 24	0.013 655 87	0.173 93	0.004 94	0.821 13
88	0.003 300 829	0.000 093 508 16	0.015 588 05	0.173 89	0.004 93	0.821 18
89	0.003 753 236	0.000 106 136 1	0.017 730 15	0.173 84	0.004 92	0.821 24
390	0.004 243 000	0.000 120 000 0	0.020 050 01	0.173 80	0.004 92	0.821 28
91	0.004 762 389	0.000 134 984 0	0.022 511 36	0.173 76	0.004 92	0.821 32
92	0.005 330 048	0.000 151 492 0	0.025 202 88	0.173 70	0.004 94	0.821 36
93	0.005 978 712	0.000 170 208 0	0.028 279 72	0.173 66	0.004 94	0.821 40
94	0.006 741 117	0.000 191 816 0	0.031 897 04	0.173 61	0.004 94	0.821 45
395	0.007 650 000	0.000 217 000 0	0.036 210 00	0.173 56	0.004 92	0.821 52
96	0.008 751 373	0.000 246 906 7	0.041 437 71	0.173 51	0.004 90	0.821 59
97	0.010 028 88	0.000 281 240 0	0.047 503 72	0.173 47	0.004 86	0.821 67
98	0.011 421 70	0.000 318 520 0	0.054 119 88	0.173 42	0.004 84	0.821 74
99	0.012 869 01	0.000 357 266 7	0.060 998 03	0.173 38	0.004 81	0.821 81
400	0.014 310 00	0.000 396 000 0	0.067 850 01	0.173 34	0.004 80	0.821 86
01	0.015 704 43	0.000 433 714 7	0.074 486 32	0.173 29	0.004 79	0.821 92
02	0.017 147 44	0.000 473 024 0	0.081 361 56	0.173 24	0.004 78	0.821 98
03	0.018 781 22	0.000 517 876 0	0.089 153 64	0.173 17	0.004 78	0.822 05
04	0.020 748 01	0.000 572 218 7	0.098 540 48	0.173 10	0.004 77	0.822 13
405	0.023 190 00	0.000 640 000 0	0.110 200 0	0.173 02	0.004 78	0.822 20
06	0.026 207 36	0.000 724 560 0	0.124 613 3	0.172 93	0.004 78	0.822 29
07	0.029 782 48	0.000 825 500 0	0.141 701 7	0.172 84	0.004 79	0.822 37
08	0.033 880 92	0.000 941 160 0	0.161 303 5	0.172 75	0.004 80	0.822 45
09	0.038 468 24	0.001 069 880	0.183 256 8	0.172 66	0.004 80	0.822 54

* From CIE (1971a).

λ (nm)	SPECTRAL TRISTIMULUS VALUES			CHROMATICITY COORDINATES		
	$\bar{x}(\lambda)$	$\bar{y}(\lambda)$	$\bar{z}(\lambda)$	$x(\lambda)$	$y(\lambda)$	$z(\lambda)$
410	0.043 510 00	0.001 210 000	0.207 400 0	0.172 58	0.004 80	0.822 62
11	0.048 995 60	0.001 362 091	0.233 692 1	0.172 49	0.004 80	0.822 71
12	0.055 022 60	0.001 530 752	0.262 611 4	0.172 39	0.004 80	0.822 81
13	0.061 718 80	0.001 720 368	0.294 774 6	0.172 30	0.004 80	0.822 90
14	0.069 212 00	0.001 935 323	0.330 798 5	0.172 19	0.004 82	0.822 99
415	0.077 630 00	0.002 180 000	0.371 300 0	0.172 09	0.004 83	0.823 08
16	0.086 958 11	0.002 454 800	0.416 209 1	0.171 98	0.004 86	0.823 16
17	0.097 176 72	0.002 764 000	0.465 464 2	0.171 87	0.004 89	0.823 24
18	0.108 406 3	0.003 117 800	0.519 694 8	0.171 74	0.004 94	0.823 32
19	0.120 767 2	0.003 526 400	0.579 530 3	0.171 59	0.005 01	0.823 40
420	0.134 380 0	0.004 000 000	0.645 600 0	0.171 41	0.005 10	0.823 49
21	0.149 358 2	0.004 546 240	0.718 483 8	0.171 21	0.005 21	0.823 58
22	0.165 395 7	0.005 159 320	0.796 713 3	0.170 99	0.005 33	0.823 68
23	0.181 983 1	0.005 829 280	0.877 845 9	0.170 77	0.005 47	0.823 76
24	0.198 611 0	0.006 546 160	0.959 439 0	0.170 54	0.005 62	0.823 84
425	0.214 770 0	0.007 300 000	1.039 050 1	0.170 30	0.005 79	0.823 91
26	0.230 186 8	0.008 086 507	1.115 367 3	0.170 05	0.005 97	0.823 98
27	0.244 879 7	0.008 908 720	1.188 497 1	0.169 78	0.006 18	0.824 04
28	0.258 777 3	0.009 767 680	1.258 123 3	0.169 50	0.006 40	0.824 10
29	0.271 807 9	0.010 664 43	1.323 929 6	0.169 20	0.006 64	0.824 16
430	0.283 900 0	0.011 600 00	1.385 600 0	0.168 88	0.006 90	0.824 22
31	0.294 943 8	0.012 573 17	1.442 635 2	0.168 53	0.007 18	0.824 29
32	0.304 896 5	0.013 582 72	1.494 803 5	0.168 15	0.007 49	0.824 36
33	0.313 787 3	0.014 629 68	1.542 190 3	0.167 75	0.007 82	0.824 43
34	0.321 645 4	0.015 715 09	1.584 880 7	0.167 33	0.008 17	0.824 50
435	0.328 500 0	0.016 840 00	1.622 960 0	0.166 90	0.008 55	0.824 55
36	0.334 351 3	0.018 007 36	1.656 404 8	0.166 45	0.008 96	0.824 59
37	0.339 210 1	0.019 214 48	1.685 295 9	0.165 98	0.009 40	0.824 62
38	0.343 121 3	0.020 453 92	1.709 874 5	0.165 48	0.009 87	0.824 65
39	0.346 129 6	0.021 718 24	1.730 382 1	0.164 96	0.010 35	0.824 69
440	0.348 280 0	0.023 000 00	1.747 060 0	0.164 41	0.010 86	0.824 73
41	0.349 599 9	0.024 294 61	1.760 044 6	0.163 83	0.011 38	0.824 79
42	0.350 147 4	0.025 610 24	1.769 623 3	0.163 21	0.011 94	0.824 85
43	0.350 013 0	0.026 958 57	1.776 263 7	0.162 55	0.012 52	0.824 93
44	0.349 287 0	0.028 351 25	1.780 433 4	0.161 85	0.013 14	0.825 01
445	0.348 060 0	0.029 800 00	1.782 600 0	0.161 11	0.013 79	0.825 10
46	0.346 373 3	0.031 310 83	1.782 968 2	0.160 31	0.014 49	0.825 20
47	0.344 262 4	0.032 883 68	1.781 699 8	0.159 47	0.015 23	0.825 30
48	0.341 808 8	0.034 521 12	1.779 198 2	0.158 57	0.016 02	0.825 41
49	0.339 094 1	0.036 225 71	1.775 867 1	0.157 63	0.016 84	0.825 53
450	0.336 200 0	0.038 000 00	1.772 110 0	0.156 64	0.017 71	0.825 65
51	0.333 197 7	0.039 846 67	1.768 258 9	0.155 60	0.018 61	0.825 79
52	0.330 041 1	0.041 768 00	1.764 039 0	0.154 52	0.019 56	0.825 92
53	0.326 635 7	0.043 766 00	1.758 943 8	0.153 40	0.020 55	0.826 05
54	0.322 886 8	0.045 842 67	1.752 466 3	0.152 22	0.021 61	0.826 17
455	0.318 700 0	0.048 000 00	1.744 100 0	0.150 99	0.022 74	0.826 27
56	0.314 025 1	0.050 243 68	1.733 559 5	0.149 69	0.023 95	0.826 36
57	0.308 884 0	0.052 573 04	1.720 858 1	0.148 34	0.025 25	0.826 41
58	0.303 290 4	0.054 980 56	1.705 936 9	0.146 93	0.026 63	0.826 44
59	0.297 257 9	0.057 458 72	1.688 737 2	0.145 47	0.028 12	0.826 41

λ (nm)	SPECTRAL TRISTIMULUS VALUES			CHROMATICITY COORDINATES		
	$\bar{x}(\lambda)$	$\bar{y}(\lambda)$	$\bar{z}(\lambda)$	$x(\lambda)$	$y(\lambda)$	$z(\lambda)$
460	0.290 800 0	0.060 000 00	1.669 200 0	0.143 96	0.029 70	0.826 34
61	0.283 970 1	0.062 601 97	1.647 528 7	0.142 41	0.031 39	0.826 20
62	0.276 721 4	0.065 277 52	1.623 412 7	0.140 80	0.033 21	0.825 99
63	0.268 917 8	0.068 042 08	1.596 022 3	0.139 12	0.035 20	0.825 68
64	0.260 422 7	0.070 911 09	1.564 528 0	0.137 37	0.037 40	0.825 23
465	0.251 100 0	0.073 900 00	1.528 100 0	0.135 50	0.039 88	0.824 62
66	0.240 847 5	0.077 016 00	1.486 111 4	0.133 51	0.042 69	0.823 80
67	0.229 851 2	0.080 266 40	1.439 521 5	0.131 37	0.045 88	0.822 75
68	0.218 407 2	0.083 666 80	1.389 879 9	0.129 09	0.049 45	0.821 46
69	0.206 811 5	0.087 232 80	1.338 736 2	0.126 66	0.053 43	0.819 91
470	0.195 360 0	0.090 980 00	1.287 640 0	0.124 12	0.057 80	0.818 08
71	0.184 213 6	0.094 917 55	1.237 422 3	0.121 47	0.062 59	0.815 94
72	0.173 327 3	0.099 045 84	1.187 824 3	0.118 70	0.067 83	0.813 47
73	0.162 688 1	0.103 367 4	1.138 761 1	0.115 81	0.073 58	0.810 61
74	0.152 283 3	0.107 884 6	1.090 148 0	0.112 78	0.079 89	0.807 33
475	0.142 100 0	0.112 600 0	1.041 900 0	0.109 60	0.086 84	0.803 56
76	0.132 178 6	0.117 532 0	0.994 197 6	0.106 26	0.094 49	0.799 25
77	0.122 569 6	0.122 674 4	0.947 347 3	0.102 78	0.102 86	0.794 36
78	0.113 275 2	0.127 992 8	0.901 453 1	0.099 13	0.112 01	0.788 86
79	0.104 297 9	0.133 452 8	0.856 619 3	0.095 31	0.121 94	0.782 75
480	0.095 640 00	0.139 020 0	0.812 950 1	0.091 29	0.132 70	0.776 01
81	0.087 299 55	0.144 676 4	0.770 517 3	0.087 08	0.144 32	0.768 60
82	0.079 308 04	0.150 469 3	0.729 444 8	0.082 68	0.156 87	0.760 45
83	0.071 717 76	0.156 461 9	0.689 913 6	0.078 12	0.170 42	0.751 46
84	0.064 580 99	0.162 717 7	0.652 104 9	0.073 44	0.185 03	0.741 53
485	0.057 950 01	0.169 300 0	0.616 200 0	0.068 71	0.200 72	0.730 57
86	0.051 862 11	0.176 243 1	0.582 328 6	0.063 99	0.217 47	0.718 54
87	0.046 281 52	0.183 558 1	0.550 416 2	0.059 32	0.235 25	0.705 43
88	0.041 150 88	0.191 273 5	0.520 337 6	0.054 67	0.254 09	0.691 24
89	0.036 412 83	0.199 418 0	0.491 967 3	0.050 03	0.274 00	0.675 97
490	0.032 010 00	0.208 020 0	0.465 180 0	0.045 39	0.294 98	0.659 63
91	0.027 917 20	0.217 119 9	0.439 924 6	0.040 76	0.316 98	0.642 26
92	0.024 144 40	0.226 734 5	0.416 183 6	0.036 20	0.339 90	0.623 90
93	0.020 687 00	0.236 857 1	0.393 882 2	0.031 76	0.363 60	0.604 64
94	0.017 540 40	0.247 481 2	0.372 945 9	0.027 49	0.387 92	0.584 59
495	0.014 700 00	0.258 600 0	0.353 300 0	0.023 46	0.412 70	0.563 84
96	0.012 161 79	0.270 184 9	0.334 857 8	0.019 70	0.437 76	0.542 54
97	0.009 919 960	0.282 293 9	0.317 552 1	0.016 27	0.462 95	0.520 78
98	0.007 967 240	0.295 050 5	0.301 337 5	0.013 18	0.488 21	0.498 61
99	0.006 296 346	0.308 578 0	0.286 168 6	0.010 48	0.513 40	0.476 12
500	0.004 900 000	0.323 000 0	0.272 000 0	0.008 17	0.538 42	0.453 41
01	0.003 777 173	0.338 402 1	0.258 817 1	0.006 28	0.563 07	0.430 65
02	0.002 945 320	0.354 685 8	0.246 483 8	0.004 87	0.587 12	0.408 01
03	0.002 424 880	0.371 698 6	0.234 771 8	0.003 98	0.610 45	0.385 57
04	0.002 236 293	0.389 287 5	0.223 453 3	0.003 64	0.633 01	0.363 35
505	0.002 400 000	0.407 300 0	0.212 300 0	0.003 86	0.654 82	0.341 32
06	0.002 925 520	0.425 629 9	0.201 169 2	0.004 64	0.675 90	0.319 46
07	0.003 836 560	0.444 309 6	0.190 119 6	0.006 01	0.696 12	0.297 87
08	0.005 174 840	0.463 394 4	0.179 225 4	0.007 99	0.715 34	0.276 67
09	0.006 982 080	0.482 939 5	0.168 560 8	0.010 60	0.733 41	0.255 99

λ (nm)	SPECTRAL TRISTIMULUS VALUES			CHROMATICITY COORDINATES		
	$\bar{x}(\lambda)$	$\bar{y}(\lambda)$	$\bar{z}(\lambda)$	$x(\lambda)$	$y(\lambda)$	$z(\lambda)$
510	0.009 300 000	0.503 000 0	0.158 200 0	0.013 87	0.750 19	0.235 94
11	0.012 149 49	0.523 569 3	0.148 138 3	0.017 77	0.765 61	0.216 62
12	0.015 535 88	0.544 512 0	0.138 375 8	0.022 24	0.779 63	0.198 13
13	0.019 477 52	0.565 690 0	0.128 994 2	0.027 27	0.792 11	0.180 62
14	0.023 992 77	0.586 965 3	0.120 075 1	0.032 82	0.802 93	0.164 25
515	0.029 100 00	0.608 200 0	0.111 700 0	0.038 85	0.812 02	0.149 13
16	0.034 814 85	0.629 345 6	0.103 904 8	0.045 33	0.819 39	0.135 28
17	0.041 120 16	0.650 306 8	0.096 667 48	0.052 18	0.825 16	0.122 66
18	0.047 985 04	0.670 875 2	0.089 982 72	0.059 32	0.829 43	0.111 25
19	0.055 378 61	0.690 842 4	0.083 845 31	0.066 72	0.832 27	0.101 01
520	0.063 270 00	0.710 000 0	0.078 249 99	0.074 30	0.833 80	0.091 90
21	0.071 635 01	0.728 185 2	0.073 208 99	0.082 05	0.834 09	0.083 86
22	0.080 462 24	0.745 463 6	0.068 678 16	0.089 94	0.833 29	0.076 77
23	0.089 739 96	0.761 969 4	0.064 567 84	0.097 94	0.831 59	0.070 47
24	0.099 456 45	0.777 836 8	0.060 788 35	0.106 02	0.829 18	0.064 80
525	0.109 600 0	0.793 200 0	0.057 250 01	0.114 16	0.826 21	0.059 63
26	0.120 167 4	0.808 110 4	0.053 904 35	0.122 35	0.822 77	0.054 88
27	0.131 114 5	0.822 496 2	0.050 746 64	0.130 55	0.818 93	0.050 52
28	0.142 367 9	0.836 306 8	0.047 752 76	0.138 70	0.814 78	0.046 52
29	0.153 854 2	0.849 491 6	0.044 898 59	0.146 77	0.810 40	0.042 83
530	0.165 500 0	0.862 000 0	0.042 160 00	0.154 72	0.805 86	0.039 42
31	0.177 257 1	0.873 810 8	0.039 507 28	0.162 53	0.801 24	0.036 23
32	0.189 140 0	0.884 962 4	0.036 935 64	0.170 24	0.796 52	0.033 24
33	0.201 169 4	0.895 493 6	0.034 458 36	0.177 85	0.791 69	0.030 46
34	0.213 365 8	0.905 443 2	0.032 088 72	0.185 39	0.786 73	0.027 88
535	0.225 749 9	0.914 850 1	0.029 840 00	0.192 88	0.781 63	0.025 49
36	0.238 320 9	0.923 734 8	0.027 711 81	0.200 31	0.776 40	0.023 29
37	0.251 066 8	0.932 092 4	0.025 694 44	0.207 69	0.771 05	0.021 26
38	0.263 992 2	0.939 922 6	0.023 787 16	0.215 03	0.765 59	0.019 38
39	0.277 101 7	0.947 225 2	0.021 989 25	0.222 34	0.760 02	0.017 64
540	0.290 400 0	0.954 000 0	0.020 300 00	0.229 62	0.754 33	0.016 05
41	0.303 891 2	0.960 256 1	0.018 718 05	0.236 89	0.748 52	0.014 59
42	0.317 572 6	0.966 007 4	0.017 240 36	0.244 13	0.742 62	0.013 25
43	0.331 438 4	0.971 260 6	0.015 863 64	0.251 36	0.736 61	0.012 03
44	0.345 482 8	0.976 022 5	0.014 584 61	0.258 58	0.730 51	0.010 91
545	0.359 700 0	0.980 300 0	0.013 400 00	0.265 78	0.724 32	0.009 90
46	0.374 083 9	0.984 092 4	0.012 307 23	0.272 96	0.718 06	0.008 98
47	0.388 639 6	0.987 418 2	0.011 301 88	0.280 13	0.711 72	0.008 15
48	0.403 378 4	0.990 312 8	0.010 377 92	0.287 29	0.705 32	0.007 39
49	0.418 311 5	0.992 811 6	0.009 529 306	0.294 45	0.698 84	0.006 71
550	0.433 449 9	0.994 950 1	0.008 749 999	0.301 60	0.692 31	0.006 09
51	0.448 795 3	0.996 710 8	0.008 035 200	0.308 76	0.685 71	0.005 53
52	0.464 336 0	0.998 098 3	0.007 381 600	0.315 92	0.679 06	0.005 02
53	0.480 064 0	0.999 112 0	0.006 785 400	0.323 06	0.672 37	0.004 57
54	0.495 971 3	0.999 748 2	0.006 242 800	0.330 21	0.665 63	0.004 16
555	0.512 050 1	1.000 000 0	0.005 749 999	0.337 36	0.658 85	0.003 79
56	0.528 295 9	0.999 856 7	0.005 303 600	0.344 51	0.652 03	0.003 46
57	0.544 691 6	0.999 304 6	0.004 899 800	0.351 67	0.645 17	0.003 16
58	0.561 209 4	0.998 325 5	0.004 534 200	0.358 81	0.638 29	0.002 90
59	0.577 821 5	0.996 898 7	0.004 202 400	0.365 96	0.631 38	0.002 66

	SPECTRAL TRISTIMULUS VALUES			CHROMATICITY COORDINATES		
λ (nm)	$\bar{x}(\lambda)$	$\bar{y}(\lambda)$	$\bar{z}(\lambda)$	$x(\lambda)$	$y(\lambda)$	$z(\lambda)$
560	0.594 500 0	0.995 000 0	0.003 900 000	0.373 10	0.624 45	0.002 45
61	0.611 220 9	0.992 600 5	0.003 623 200	0.380 24	0.617 50	0.002 26
62	0.627 975 8	0.989 742 6	0.003 370 600	0.387 38	0.610 54	0.002 08
63	0.644 760 2	0.986 444 4	0.003 141 400	0.394 51	0.603 57	0.001 92
64	0.661 569 7	0.982 724 1	0.002 934 800	0.401 63	0.596 59	0.001 78
565	0.678 400 0	0.978 600 0	0.002 749 999	0.408 73	0.589 61	0.001 66
66	0.695 239 2	0.974 083 7	0.002 585 200	0.415 83	0.582 62	0.001 55
67	0.712 058 6	0.969 171 2	0.002 438 600	0.422 92	0.575 63	0.001 45
68	0.728 828 4	0.963 856 8	0.002 309 400	0.429 99	0.568 65	0.001 36
69	0.745 518 8	0.958 134 9	0.002 196 800	0.437 04	0.561 67	0.001 29
570	0.762 100 0	0.952 000 0	0.002 100 000	0.444 06	0.554 72	0.001 22
71	0.778 543 2	0.945 450 4	0.002 017 733	0.451 06	0.547 77	0.001 17
72	0.794 825 6	0.938 499 2	0.001 948 200	0.458 04	0.540 84	0.001 12
73	0.810 926 4	0.931 162 8	0.001 889 800	0.464 99	0.533 93	0.001 08
74	0.826 824 8	0.923 457 6	0.001 840 933	0.471 90	0.527 05	0.001 05
575	0.842 500 0	0.915 400 0	0.001 800 000	0.478 78	0.520 20	0.001 02
76	0.857 932 5	0.907 006 4	0.001 766 267	0.485 61	0.513 39	0.001 00
77	0.873 081 6	0.898 277 2	0.001 737 800	0.492 41	0.506 61	0.000 98
78	0.887 894 4	0.889 204 8	0.001 711 200	0.499 15	0.499 89	0.000 96
79	0.902 318 1	0.879 781 6	0.001 683 067	0.505 85	0.493 21	0.000 94
580	0.916 300 0	0.870 000 0	0.001 650 001	0.512 49	0.486 59	0.000 92
81	0.929 799 5	0.859 861 3	0.001 610 133	0.519 07	0.480 03	0.000 90
82	0.942 798 4	0.849 392 0	0.001 564 400	0.525 60	0.473 53	0.000 87
83	0.955 277 6	0.838 622 0	0.001 513 600	0.532 07	0.467 09	0.000 84
84	0.967 217 9	0.827 581 3	0.001 458 533	0.538 46	0.460 73	0.000 81
585	0.978 600 0	0.816 300 0	0.001 400 000	0.544 79	0.454 43	0.000 78
86	0.989 385 6	0.804 794 7	0.001 336 667	0.551 03	0.448 23	0.000 74
87	0.999 548 8	0.793 082 0	0.001 270 000	0.557 19	0.442 10	0.000 71
88	1.009 089 2	0.781 192 0	0.001 205 000	0.563 27	0.436 06	0.000 67
89	1.018 006 4	0.769 154 7	0.001 146 667	0.569 26	0.430 10	0.000 64
590	1.026 300 0	0.757 000 0	0.001 100 000	0.575 15	0.424 23	0.000 62
91	1.033 982 7	0.744 754 1	0.001 068 800	0.580 94	0.418 46	0.000 60
92	1.040 986 0	0.732 422 4	0.001 049 400	0.586 65	0.412 76	0.000 59
93	1.047 188 0	0.720 003 6	0.001 035 600	0.592 22	0.407 19	0.000 59
94	1.052 466 7	0.707 496 5	0.001 021 200	0.597 66	0.401 76	0.000 58
595	1.056 700 0	0.694 900 0	0.001 000 000	0.602 93	0.396 50	0.000 57
96	1.059 794 4	0.682 219 2	0.000 968 640 0	0.608 03	0.391 41	0.000 56
97	1.061 799 2	0.669 471 6	0.000 929 920 0	0.612 98	0.386 48	0.000 54
98	1.062 806 8	0.656 674 4	0.000 886 880 0	0.617 78	0.381 71	0.000 51
99	1.062 909 6	0.643 844 8	0.000 842 560 0	0.622 46	0.377 05	0.000 49
600	1.062 200 0	0.631 000 0	0.000 800 000 0	0.627 04	0.372 49	0.000 47
01	1.060 735 2	0.618 155 5	0.000 760 960 0	0.631 52	0.368 03	0.000 45
02	1.058 443 6	0.605 314 4	0.000 723 680 0	0.635 90	0.363 67	0.000 43
03	1.055 224 4	0.592 475 6	0.000 685 920 0	0.640 16	0.359 43	0.000 41
04	1.050 976 8	0.579 637 9	0.000 645 440 0	0.644 27	0.355 33	0.000 40
605	1.045 600 0	0.566 800 0	0.000 600 000 0	0.648 23	0.351 40	0.000 37
06	1.039 036 9	0.553 961 1	0.000 547 866 7	0.652 03	0.347 63	0.000 34
07	1.031 360 8	0.541 137 2	0.000 491 600 0	0.655 67	0.344 02	0.000 31
08	1.022 666 2	0.528 352 8	0.000 435 400 0	0.659 17	0.340 55	0.000 28
09	1.013 047 7	0.515 632 3	0.000 383 466 7	0.662 53	0.337 22	0.000 25

λ (nm)	SPECTRAL TRISTIMULUS VALUES			CHROMATICITY COORDINATES		
	$\bar{x}(\lambda)$	$\bar{y}(\lambda)$	$\bar{z}(\lambda)$	$x(\lambda)$	$y(\lambda)$	$z(\lambda)$
610	1.002 600 0	0.503 000 0	0.000 340 000 0	0.665 76	0.334 01	0.000 23
11	0.991 367 5	0.490 468 8	0.000 307 253 3	0.668 87	0.330 92	0.000 21
12	0.979 331 4	0.478 030 4	0.000 283 160 0	0.671 86	0.327 95	0.000 19
13	0.966 491 6	0.465 677 6	0.000 265 440 0	0.674 72	0.325 09	0.000 19
14	0.952 847 9	0.453 403 2	0.000 251 813 3	0.677 46	0.322 36	0.000 18
615	0.938 400 0	0.441 200 0	0.000 240 000 0	0.680 08	0.319 75	0.000 17
16	0.923 194 0	0.429 080 0	0.000 229 546 7	0.682 58	0.317 25	0.000 17
17	0.907 244 0	0.417 036 0	0.000 220 640 0	0.684 97	0.314 86	0.000 17
18	0.890 502 0	0.405 032 0	0.000 211 960 0	0.687 25	0.312 59	0.000 16
19	0.872 920 0	0.393 032 0	0.000 202 186 7	0.689 43	0.310 41	0.000 16
620	0.854 449 9	0.381 000 0	0.000 190 000 0	0.691 51	0.308 34	0.000 15
21	0.835 084 0	0.368 918 4	0.000 174 213 3	0.693 49	0.306 37	0.000 14
22	0.814 946 0	0.356 827 2	0.000 155 640 0	0.695 39	0.304 48	0.000 13
23	0.794 186 0	0.344 776 8	0.000 135 960 0	0.697 21	0.302 67	0.000 12
24	0.772 954 0	0.332 817 6	0.000 116 853 3	0.698 94	0.300 95	0.000 11
625	0.751 400 0	0.321 000 0	0.000 100 000 0	0.700 61	0.299 30	0.000 09
26	0.729 583 6	0.309 338 1	0.000 086 133 33	0.702 19	0.297 73	0.000 08
27	0.707 588 8	0.297 850 4	0.000 074 600 00	0.703 71	0.296 22	0.000 07
28	0.685 602 2	0.286 593 6	0.000 065 000 00	0.705 16	0.294 77	0.000 07
29	0.663 810 4	0.275 624 5	0.000 056 933 33	0.706 56	0.293 38	0.000 06
630	0.642 400 0	0.265 000 0	0.000 049 999 99	0.707 92	0.292 03	0.000 05
31	0.621 514 9	0.254 763 2	0.000 044 160 00	0.709 23	0.290 72	0.000 05
32	0.601 113 8	0.244 889 6	0.000 039 480 00	0.710 50	0.289 45	0.000 05
33	0.581 105 2	0.235 334 4	0.000 035 720 00	0.711 73	0.288 23	0.000 04
34	0.561 397 7	0.226 052 8	0.000 032 640 00	0.712 90	0.287 06	0.000 04
635	0.541 900 0	0.217 000 0	0.000 030 000 00	0.714 03	0.285 93	0.000 04
36	0.522 599 5	0.208 161 6	0.000 027 653 33	0.715 12	0.284 84	0.000 04
37	0.503 546 4	0.199 548 8	0.000 025 560 00	0.716 16	0.283 80	0.000 04
38	0.484 743 6	0.191 155 2	0.000 023 640 00	0.717 16	0.282 81	0.000 03
39	0.466 193 9	0.182 974 4	0.000 021 813 33	0.718 12	0.281 85	0.000 03
640	0.447 900 0	0.175 000 0	0.000 020 000 00	0.719 03	0.280 94	0.000 03
41	0.429 861 3	0.167 223 5	0.000 018 133 33	0.719 91	0.280 06	0.000 03
42	0.412 098 0	0.159 646 4	0.000 016 200 00	0.720 75	0.279 22	0.000 03
43	0.394 644 0	0.152 277 6	0.000 014 200 00	0.721 55	0.278 42	0.000 03
44	0.377 533 3	0.145 125 9	0.000 012 133 33	0.722 32	0.277 66	0.000 02
645	0.360 800 0	0.138 200 0	0.000 010 000 00	0.723 03	0.276 95	0.000 02
46	0.344 456 3	0.131 500 3	0.000 007 733 333	0.723 70	0.276 28	0.000 02
47	0.328 516 8	0.125 024 8	0.000 005 400 000	0.724 33	0.275 66	0.000 01
48	0.313 019 2	0.118 779 2	0.000 003 200 000	0.724 91	0.275 08	0.000 01
49	0.298 001 1	0.112 769 1	0.000 001 333 333	0.725 47	0.274 53	0.000 00
650	0.283 500 0	0.107 000 0	0.000 000 000 000	0.725 99	0.274 01	0.000 00
51	0.269 544 8	0.101 476 2		0.726 49	0.273 51	
52	0.256 118 4	0.096 188 64		0.726 98	0.273 02	
53	0.243 189 6	0.091 122 96		0.727 43	0.272 57	
54	0.230 727 2	0.086 264 85		0.727 86	0.272 14	

	SPECTRAL TRISTIMULUS VALUES			CHROMATICITY COORDINATES		
λ (nm)	$\bar{x}(\lambda)$	$\bar{y}(\lambda)$	$\bar{z}(\lambda)$	$x(\lambda)$	$y(\lambda)$	$z(\lambda)$
655	0.218 700 0	0.081 600 00	0.000 000 0	0.728 27	0.271 73	0.000 00
56	0.207 097 1	0.077 120 64		0.728 66	0.271 34	
57	0.195 923 2	0.072 825 52		0.729 02	0.270 98	
58	0.185 170 8	0.068 710 08		0.729 36	0.270 64	
59	0.174 832 3	0.064 769 76		0.729 68	0.270 32	
660	0.164 900 0	0.061 000 00		0.729 97	0.270 03	
61	0.155 366 7	0.057 396 21		0.730 23	0.269 77	
62	0.146 230 0	0.053 955 04		0.730 47	0.269 53	
63	0.137 490 0	0.050 673 76		0.730 69	0.269 31	
64	0.129 146 7	0.047 549 65		0.730 90	0.269 10	
665	0.121 200 0	0.044 580 00		0.731 09	0.268 91	
66	0.113 639 7	0.041 758 72		0.731 28	0.268 72	
67	0.106 465 0	0.039 084 96		0.731 47	0.268 53	
68	0.099 690 44	0.036 563 84		0.731 65	0.268 35	
69	0.093 330 61	0.034 200 48		0.731 83	0.268 17	
670	0.087 400 00	0.032 000 00		0.731 99	0.268 01	
71	0.081 900 96	0.029 962 61		0.732 15	0.267 85	
72	0.076 804 28	0.028 076 64		0.732 30	0.267 70	
73	0.072 077 12	0.026 329 36		0.732 44	0.267 56	
74	0.067 686 64	0.024 708 05		0.732 58	0.267 42	
675	0.063 600 00	0.023 200 00		0.732 72	0.267 28	
76	0.059 806 85	0.021 800 77		0.732 86	0.267 14	
77	0.056 282 16	0.020 501 12		0.733 00	0.267 00	
78	0.052 971 04	0.019 281 08		0.733 14	0.266 86	
79	0.049 818 61	0.018 120 69		0.733 28	0.266 72	
680	0.046 770 00	0.017 000 00		0.733 42	0.266 58	
81	0.043 784 05	0.015 903 79		0.733 55	0.266 45	
82	0.040 875 36	0.014 837 18		0.733 68	0.266 32	
83	0.038 072 64	0.013 810 68		0.733 81	0.266 19	
84	0.035 404 61	0.012 834 78		0.733 94	0.266 06	
685	0.032 900 00	0.011 920 00		0.734 05	0.265 95	
86	0.030 564 19	0.011 068 31		0.734 14	0.265 86	
87	0.028 380 56	0.010 273 39		0.734 22	0.265 78	
88	0.026 344 84	0.009 533 311		0.734 29	0.265 71	
89	0.024 452 75	0.008 846 157		0.734 34	0.265 66	
690	0.022 700 00	0.008 210 000		0.734 39	0.265 61	
91	0.021 084 29	0.007 623 781		0.734 44	0.265 56	
92	0.019 599 88	0.007 085 424		0.734 48	0.265 52	
93	0.018 237 32	0.006 591 476		0.734 52	0.265 48	
94	0.016 987 17	0.006 138 485		0.734 56	0.265 44	
695	0.015 840 00	0.005 723 000		0.734 59	0.265 41	
96	0.014 790 64	0.005 343 059		0.734 62	0.265 38	
97	0.013 831 32	0.004 995 796		0.734 65	0.265 35	
98	0.012 948 68	0.004 676 404		0.734 67	0.265 33	
99	0.012 129 20	0.004 380 075		0.734 69	0.265 31	
700	0.011 359 16	0.004 102 000		0.734 69	0.265 31	
01	0.010 629 35	0.003 838 453		0.734 69	0.265 31	
02	0.009 938 846	0.003 589 099		0.734 69	0.265 31	
03	0.009 288 422	0.003 354 219		0.734 69	0.265 31	
04	0.008 678 854	0.003 134 093		0.734 69	0.265 31	

λ (nm)	SPECTRAL TRISTIMULUS VALUES			CHROMATICITY COORDINATES		
	$\bar{x}(\lambda)$	$\bar{y}(\lambda)$	$\bar{z}(\lambda)$	$x(\lambda)$	$y(\lambda)$	$z(\lambda)$
705	0.008 110 916	0.002 929 000	0.000 000 0	0.734 69	0.265 31	0.000 00
06	0.007 582 388	0.002 738 139		0.734 69	0.265 31	
07	0.007 088 746	0.002 559 876		0.734 69	0.265 31	
08	0.006 627 313	0.002 393 244		0.734 69	0.265 31	
09	0.006 195 408	0.002 237 275		0.734 69	0.265 31	
710	0.005 790 346	0.002 091 000		0.734 69	0.265 31	
11	0.005 409 826	0.001 953 587		0.734 69	0.265 31	
12	0.005 052 583	0.001 824 580		0.734 69	0.265 31	
13	0.004 717 512	0.001 703 580		0.734 69	0.265 31	
14	0.004 403 507	0.001 590 187		0.734 69	0.265 31	
715	0.004 109 457	0.001 484 000		0.734 69	0.265 31	
16	0.003 833 913	0.001 384 496		0.734 69	0.265 31	
17	0.003 575 748	0.001 291 268		0.734 69	0.265 31	
18	0.003 334 342	0.001 204 092		0.734 69	0.265 31	
19	0.003 109 075	0.001 122 744		0.734 69	0.265 31	
720	0.002 899 327	0.001 047 000		0.734 69	0.265 31	
21	0.002 704 348	0.000 976 589 6		0.734 69	0.265 31	
22	0.002 523 020	0.000 911 108 8		0.734 69	0.265 31	
23	0.002 354 168	0.000 850 133 2		0.734 69	0.265 31	
24	0.002 196 616	0.000 793 238 4		0.734 69	0.265 31	
725	0.002 049 190	0.000 740 000 0		0.734 69	0.265 31	
26	0.001 910 960	0.000 690 082 7		0.734 69	0.265 31	
27	0.001 781 438	0.000 643 310 0		0.734 69	0.265 31	
28	0.001 660 110	0.000 599 496 0		0.734 69	0.265 31	
29	0.001 546 459	0.000 558 454 7		0.734 69	0.265 31	
730	0.001 439 971	0.000 520 000 0		0.734 69	0.265 31	
31	0.001 340 042	0.000 483 913 6		0.734 69	0.265 31	
32	0.001 246 275	0.000 450 052 8		0.734 69	0.265 31	
33	0.001 158 471	0.000 418 345 2		0.734 69	0.265 31	
34	0.001 076 430	0.000 388 718 4		0.734 69	0.265 31	
735	0.000 999 949 3	0.000 361 100 0		0.734 69	0.265 31	
36	0.000 928 735 8	0.000 335 383 5		0.734 69	0.265 31	
37	0.000 862 433 2	0.000 311 440 4		0.734 69	0.265 31	
38	0.000 800 750 3	0.000 289 165 6		0.734 69	0.265 31	
39	0.000 743 396 0	0.000 268 453 9		0.734 69	0.265 31	
740	0.000 690 078 6	0.000 249 200 0		0.734 69	0.265 31	
41	0.000 640 515 6	0.000 231 301 9		0.734 69	0.265 31	
42	0.000 594 502 1	0.000 214 685 6		0.734 69	0.265 31	
43	0.000 551 864 6	0.000 199 288 4		0.734 69	0.265 31	
44	0.000 512 429 0	0.000 185 047 5		0.734 69	0.265 31	
745	0.000 476 021 3	0.000 171 900 0		0.734 69	0.265 31	
46	0.000 442 453 6	0.000 159 778 1		0.734 69	0.265 31	
47	0.000 411 511 7	0.000 148 604 4		0.734 69	0.265 31	
48	0.000 382 981 4	0.000 138 301 6		0.734 69	0.265 31	
49	0.000 356 649 1	0.000 128 792 5		0.734 69	0.265 31	
750	0.000 332 301 1	0.000 120 000 0		0.734 69	0.265 31	
51	0.000 309 758 6	0.000 111 859 5		0.734 69	0.265 31	
52	0.000 288 887 1	0.000 104 322 4		0.734 69	0.265 31	
53	0.000 269 539 4	0.000 097 335 60		0.734 69	0.265 31	
54	0.000 251 568 2	0.000 090 845 87		0.734 69	0.265 31	

λ (nm)	SPECTRAL TRISTIMULUS VALUES			CHROMATICITY COORDINATES		
	$\bar{x}(\lambda)$	$\bar{y}(\lambda)$	$\bar{z}(\lambda)$	$x(\lambda)$	$y(\lambda)$	$z(\lambda)$
755	0.000 234 826 1	0.000 084 800 00	0.000 000 0	0.734 69	0.265 31	0.000 00
56	0.000 219 171 0	0.000 079 146 67		0.734 69	0.265 31	
57	0.000 204 525 8	0.000 073 858 00		0.734 69	0.265 31	
58	0.000 190 840 5	0.000 068 916 00		0.734 69	0.265 31	
59	0.000 178 065 4	0.000 064 302 67		0.734 69	0.265 31	
760	0.000 166 150 5	0.000 060 000 00		0.734 69	0.265 31	
61	0.000 155 023 6	0.000 055 981 87		0.734 69	0.265 31	
62	0.000 144 621 9	0.000 052 225 60		0.734 69	0.265 31	
63	0.000 134 909 8	0.000 048 718 40		0.734 69	0.265 31	
64	0.000 125 852 0	0.000 045 447 47		0.734 69	0.265 31	
765	0.000 117 413 0	0.000 042 400 00		0.734 69	0.265 31	
66	0.000 109 551 5	0.000 039 561 04		0.734 69	0.265 31	
67	0.000 102 224 5	0.000 036 915 12		0.734 69	0.265 31	
68	0.000 095 394 45	0.000 034 448 68		0.734 69	0.265 31	
69	0.000 089 023 90	0.000 032 148 16		0.734 69	0.265 31	
770	0.000 083 075 27	0.000 030 000 00		0.734 69	0.265 31	
71	0.000 077 512 69	0.000 027 991 25		0.734 69	0.265 31	
72	0.000 072 313 04	0.000 026 113 56		0.734 69	0.265 31	
73	0.000 067 457 78	0.000 024 360 24		0.734 69	0.265 31	
74	0.000 062 928 44	0.000 022 724 61		0.734 69	0.265 31	
775	0.000 058 706 52	0.000 021 200 00		0.734 69	0.265 31	
76	0.000 054 770 28	0.000 019 778 55		0.734 69	0.265 31	
77	0.000 051 099 18	0.000 018 452 85		0.734 69	0.265 31	
78	0.000 047 676 54	0.000 017 216 87		0.734 69	0.265 31	
79	0.000 044 485 67	0.000 016 064 59		0.734 69	0.265 31	
780	0.000 041 509 94	0.000 014 990 00		0.734 69	0.265 31	
81	0.000 038 733 24	0.000 013 987 28		0.734 69	0.265 31	
82	0.000 036 142 03	0.000 013 051 55		0.734 69	0.265 31	
83	0.000 033 723 52	0.000 012 178 18		0.734 69	0.265 31	
84	0.000 031 464 87	0.000 011 362 54		0.734 69	0.265 31	
785	0.000 029 353 26	0.000 010 600 00		0.734 69	0.265 31	
86	0.000 027 375 73	0.000 009 885 877		0.734 69	0.265 31	
87	0.000 025 524 33	0.000 009 217 304		0.734 69	0.265 31	
88	0.000 023 793 76	0.000 008 592 362		0.734 69	0.265 31	
89	0.000 022 178 70	0.000 008 009 133		0.734 69	0.265 31	
790	0.000 020 673 83	0.000 007 465 700		0.734 69	0.265 31	
91	0.000 019 272 26	0.000 006 959 567		0.734 69	0.265 31	
92	0.000 017 966 40	0.000 006 487 995		0.734 69	0.265 31	
93	0.000 016 749 91	0.000 006 048 699		0.734 69	0.265 31	
94	0.000 015 616 48	0.000 005 639 396		0.734 69	0.265 31	
795	0.000 014 559 77	0.000 005 257 800		0.734 69	0.265 31	
96	0.000 013 573 87	0.000 004 901 771		0.734 69	0.265 31	
97	0.000 012 654 36	0.000 004 569 720		0.734 69	0.265 31	
98	0.000 011 797 23	0.000 004 260 194		0.734 69	0.265 31	
99	0.000 010 998 44	0.000 003 971 739		0.734 69	0.265 31	
800	0.000 010 253 98	0.000 003 702 900		0.734 69	0.265 31	
01	0.000 009 559 646	0.000 003 452 163		0.734 69	0.265 31	
02	0.000 008 912 044	0.000 003 218 302		0.734 69	0.265 31	
03	0.000 008 308 358	0.000 003 000 300		0.734 69	0.265 31	
04	0.000 007 745 769	0.000 002 797 139		0.734 69	0.265 31	

λ (nm)	SPECTRAL TRISTIMULUS VALUES			CHROMATICITY COORDINATES		
	$\bar{x}(\lambda)$	$\bar{y}(\lambda)$	$\bar{z}(\lambda)$	$x(\lambda)$	$y(\lambda)$	$z(\lambda)$
805	0.000 007 221 456	0.000 002 607 800	0.000 000 0	0.734 69	0.265 31	0.000 00
06	0.000 006 732 475	0.000 002 431 220		0.734 69	0.265 31	
07	0.000 006 276 423	0.000 002 266 531		0.734 69	0.265 31	
08	0.000 005 851 304	0.000 002 113 013		0.734 69	0.265 31	
09	0.000 005 455 118	0.000 001 969 943		0.734 69	0.265 31	
810	0.000 005 085 868	0.000 001 836 600		0.734 69	0.265 31	
11	0.000 004 741 466	0.000 001 712 230		0.734 69	0.265 31	
12	0.000 004 420 236	0.000 001 596 228		0.734 69	0.265 31	
13	0.000 004 120 783	0.000 001 488 090		0.734 69	0.265 31	
14	0.000 003 841 716	0.000 001 387 314		0.734 69	0.265 31	
815	0.000 003 581 652	0.000 001 293 400		0.734 69	0.265 31	
16	0.000 003 339 127	0.000 001 205 820		0.734 69	0.265 31	
17	0.000 003 112 949	0.000 001 124 143		0.734 69	0.265 31	
18	0.000 002 902 121	0.000 001 048 009		0.734 69	0.265 31	
19	0.000 002 705 645	0.000 000 977 057 8		0.734 69	0.265 31	
820	0.000 002 522 525	0.000 000 910 930 0		0.734 69	0.265 31	
21	0.000 002 351 726	0.000 000 849 251 3		0.734 69	0.265 31	
22	0.000 002 192 415	0.000 000 791 721 2		0.734 69	0.265 31	
23	0.000 002 043 902	0.000 000 738 090 4		0.734 69	0.265 31	
24	0.000 001 905 497	0.000 000 688 109 8		0.734 69	0.265 31	
825	0.000 001 776 509	0.000 000 641 530 0		0.734 69	0.265 31	
26	0.000 001 656 215	0.000 000 598 089 5		0.734 69	0.265 31	
27	0.000 001 544 022	0.000 000 557 574 6		0.734 69	0.265 31	
28	0.000 001 439 440	0.000 000 519 808 0		0.734 69	0.265 31	
29	0.000 001 341 977	0.000 000 484 612 3		0.734 69	0.265 31	
830	0.000 001 251 141	0.000 000 451 810 0		0.734 69	0.265 31	

$$\Sigma\bar{x}(\lambda) \;=\; 106.865\ 469\ 489\ 595$$
$$\Sigma\bar{y}(\lambda) \;=\; 106.856\ 917\ 101\ 172$$
$$\Sigma\bar{z}(\lambda) \;=\; 106.892\ 251\ 278\ 636$$

REFERENCES

American Standards Association (1958). PH2.11-1958.
Berger, A., and Strocka, D. (1975). *Appl. Opt.* **14**, 726.
British Standard 950 (1967). "Illuminant for Color Matching and Color Appraisal," Part I.
Brussels Session of the CIE (1960). *J. Opt. Soc. Am.* **50**, 89.
CIE (1971a). Document on Colorimetry, Publ. CIE No. 15, E-1.3.1.
CIE (1971b). Publ. No. 15.
CIE (1972). Technical Committee TC-2.2, Publ. CIE No. 20. Bureau Central, 52, Boulevard
    Malesherbes, 75008 Paris, France.
CIE (1974). Colorimetry Committee TC-1.3, Subcommittee on Standard Sources.
CIE (1974a). Publication No. 13.2, TC-3.2.
CIE Committees E-1.3.1 and E-1.3.2, also CIE Publ. No. 13.2 (1974b) and No. 15 (1971).
CIE (1975). Colorimetry Committee, TC-1.3.
Colles, M. J., and Pigeon, C. R. (1975). 329 tunable lasers, *Rep. Phys.* **38**, 329–460.
Condit, H. R., and Grum, F. (1964). *J. Opt. Soc. Am.* **54**, 937.
Drummond, A. J. *et al.* (1968a). *Science* **161**, 888.
Drummond, A. J. *et al.* (1968b). *Nature (London)* **218**, 259.
Grum, F. (1972). *J. Color Appearance* **1**, 18.
Grum, F., and Costa, L. F. (1976). *Appl. Opt.* **15**, 76.
Grum, F., Saunders, S. B., and Wightman, T. E. (1970). *TAPPI* **53**, 1264.
Henderson, S. T., and Hodgkiss, D. (1963). *J. Appl. Phys.* **14**, 125.
Hoffman, F., and Willenberg, H. (1934). *Phys. Z.* **25**, 713.
Hulburt, E. O. (1917). *Astrophys. J.* **45**, 149.
Jones, L. A., and Condit, H. R. (1948). *J. Opt. Soc. Am.* **38**, 123.
Judd, D. B., and Wyszecki, G. (1975). "Color in Business, Science, and Industry," p. 117.
    Wiley, New York.
Judd, D. B., MacAdam, D. L., and Wyszecki, G. (1964). *J. Opt. Soc. Am.* **54**, 1031.
Kagan, M. R., Farmer, G. I., and Huth, B. G. (1968). *Laser Focus* **4**, No. 17, 26.
Loeb, L. B. (1965). "Electrical Coronas," p. 235. Univ. of California Press, Berkeley, California.
MacAdam, D. L. (1937). *J. Opt. Soc. Am.* **27**, 294.
Moon, P. (1936). "Scientific Basis of Illuminating Engineering," pp. 117–120. McGraw–Hill,
    New York.
National Bureau of Standards (1970). Technical Note 559, November.
National Bureau Standards (1976). Optical Radiation News, No. 14, March.
Nayatani, Y., and Wyszecki, G. (1963). *J. Opt. Soc. Am.* **53**, 626.
Nonaka, M., Kinhmeri, K., and Ishbai, M. (1972). *Metrologia* **8**, 133.
Robertson, A. R. (1968). *J. Opt. Soc. Am.* **58**, 1528.
Schutze, R. (1970). *Meteorol. Rund.* **23**, 2, 56.
Stair, R., Schneider, W. E., and Jackson, J. K. (1971). National Bureau Standard Special Publ.
    300, Vol. 7, p. 274.
Stiles, W. S., and Wyszicki, (1962). *J. Opt. Soc. Am.* **52**, 313.
Taylor, A. H., and Kerr, G. P. (1941). *J. Opt. Soc. Am.* **31**, 3.
Thekaekara, M. P., and Drummond, A. J. (1971). *Nature (London) Phys. Sci.* **229**, 6.
Thekaekara, M. P. *et al.* (1968). The solar constant and solar spectrum measured from a research
    aircraft at 38,000 ft, Goddard Space Center, Greenbelt, Maryland, X-322-68-308, NASA
    TR-R-351.
Thomas, W. (1973). "SPSE Handbook of Photo. Sci. Engineer.," Wiley, New York.
U.S.A. Standard (1967). Z 7.1-1967.
Worthing, A. G. (1941). In "Temperature, Its Measurement and Control in Science and
    Industry," pp. 1164–1187. Reinhold, New York.
Wyszecki, G. (1970). *Farbe* **19**, 43.
Wyszecki, G., and Stiles, W. S. (1967). Color Science," Table 1.6, pp. 16–28. Wiley, New York.

6

Detectors

6.1 INTRODUCTION

Detectors used in radiometry produce an electronic signal in response to radiant energy. The two primary detection mechanisms of importance are the photoelectric effect and the thermal effect, with the corresponding detectors known as *photon detectors* and *thermal detectors*. Figure 6.1a shows the photoelectric effect in which an absorbed photon excites an electron to a higher energy level. Figure 6.1b shows the thermal effect in which absorbed radiant energy raises the detector temperature from T to $T + \Delta T$. Both primary detection mechanisms subsequently produce an electronic signal.

Our approach in this chapter is first to review the physics of semiconductor materials which are used in most photon detectors. We then define the important detection parameters and describe the noise mechanisms which limit detector performance. With this background we then proceed to

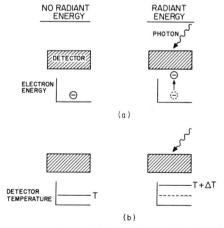

Fig. 6.1 Basic photon (a) and thermal (b) detection mechanisms showing the increase of electron energy or detector temperature.

examine individual photon and thermal detectors. We will not examine in any detail the electronic circuits which are used with detectors. For an excellent introduction to electronic circuits we refer the reader to Senturia and Wedlock (1975).

6.2   REVIEW OF SEMICONDUCTOR PHYSICS

Semiconductor materials (Smith, 1961; Richtmeyer *et al.*, 1968) are used in most photon detectors. Figure 6.2 shows a diagram of the electronic energy level structure in a semiconductor. In pure semiconductor materials the negatively charged electrons can occupy energy levels only within certain well-defined bands (hatched areas in Fig. 6.2) separated by forbidden bands.

To contribute to the electronic signal current an electron must be excited up to an energy level above the bottom of the *conduction band*. If it is originally in the *valence band*, then the required energy is greater than e_G. If the electron is at a *donor* level corresponding to the energy of an electron in a donor impurity atom which has been intentionally introduced into the semiconductor, then the required minimum energy is only e_d as shown in the figure. *Acceptor* impurities can also produce energy levels within the forbidden band. In this case the acceptor impurity atom can accept an electron which is excited from the valence band with minimum required energy equal to e_a. The effect is to produce a *hole*, the absence of an electron, in the valence band. This hole can also move through the semiconductor detector material as a positive charge carrier and contribute to the signal current. Materials in which the electrons are the majority carrier are known as *n-type* while those in which holes are the majority carrier are known as *p-type*. Pure

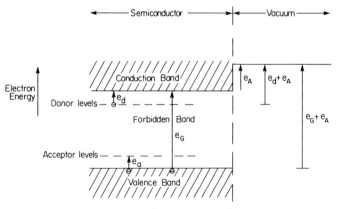

Fig. 6.2 Simplified diagram of energy band structure of semiconductors used in photon detectors. (See text for explanation.)

semiconductors in which the dominant mechanism of detection is the excitation of electrons from the valence band to the conduction band are *intrinsic* semiconductors, while those in which the dominant mechanism is excitation from donor levels or to acceptor levels are described as *extrinsic* semiconductors.

Figure 6.2 also shows the vacuum region outside the semiconductor surface. In this region an electrostatic attraction exists between an electron and the material surface. The energy required to overcome this attraction is known as the *work function* or electron affinity energy e_A of the material. If an absorbed photon from a beam of radiant energy contains energy e greater than $e_G + e_A$, then an electron can actually escape from the surface of an intrinsic semiconductor material and can be collected externally as a signal current. This is the *external* photoelectric effect or photoemission. In an extrinsic material the energy required for photoemission can be as small as $e_d + e_A$ as shown in the figure. On the other hand, if an electron is excited to the conduction band but remains within the material to contribute to the electronic signal current collected at a contact, then the effect is the *internal* photoelectric effect. For an intrinsic or extrinsic material the required photon energies for the internal photoelectric effect are shown in Fig. 6.2 to be only $e > e_G$ or $e > e_d$, respectively. As a result the internal effect can be used to detect lower energy photons than the external effect. The implication of this is as follows.

The energy of a photon is $e = hv = hc/\lambda$, where h is Planck's constant, v the optical frequency, c the velocity of light, and λ the wavelength. For the external photoelectric effect the detection requirement that $e > e_G + e_A$ or $e > e_d + e_A$ implies that

$$\lambda < hc/(e_G + e_A) \qquad \text{or} \qquad \lambda < hc/(e_d + e_A) \qquad (6.1)$$

for intrinsic and extrinsic (*n*-type) materials, respectively. By contrast the corresponding requirement for detectors using the internal photoelectric effect is only that

$$\lambda < hc/e_G \qquad \text{or} \qquad \lambda < hc/e_d. \qquad (6.2)$$

The long wavelength limit for detection by the external photoelectric effect is determined by the electron affinity energy e_A shown in (6.1), and at present is approximately 1.1 μm in the near infrared. Internal photoelectric effect detectors, both intrinsic and extrinsic, are used in the ultraviolet and visible and well out into the infrared beyond 15 μm by choosing the bandgap e_G or the donor or acceptor energy (e_d or e_a) to yield the desired cutoff wavelength. By contrast, since thermal detectors are essentially spectrally neutral and detect radiant energy only on the basis of the total absorbed energy or power, they are used at all wavelengths throughout the ultraviolet, visible, and infrared regions of the spectrum.

TABLE 6.1   Detection Parameters[a]

Parameter	Symbol	Definition	Unit	Comments
Spectral responsivity	$R(\lambda)$	i_λ/Φ_λ	A W^{-1}	i_λ = Output current (spectral) Φ_λ = Spectral radiant power input
Total responsivity[b]	R	i/Φ	A W^{-1}	$\Phi = \int \Phi_\lambda \, d\lambda$ $i = \int i_\lambda \, d\lambda$
Signal-to-noise ratio	SNR	$i_s/(\overline{i_n^2})^{1/2}$	Dimensionless	$\dfrac{i_s}{(\overline{i_n^2})^{1/2}}$ rms signal current rms noise current
Noise equivalent power (spectral)	NEP(λ)	Radiant power required to produce SNR = 1	W	$\mathrm{NEP}(\lambda) = (\overline{i_n^2})^{1/2}/R(\lambda)$
Detectivity (spectral)	$D(\lambda)$	$1/\mathrm{NEP}(\lambda)$	W^{-1}	$D(\lambda)$ increases as system improves
Normalized detectivity[c] (spectral)	$D^*(\lambda)$ (D-star)	$(AB)^{1/2}/\mathrm{NEP}(\lambda)$	cm Hz$^{1/2}$ W^{-1}	A = Detector active area (cm^2) B = Noise bandwidth (Hz)

[a] Whenever the current i appears it can be replaced by the voltage v if the corresponding voltage quantity is desired.

[b] The *blackbody responsivity* R_{bb} can be determined by using Eqs. (6.5) and (6.6) with the relative spectral power distribution ϕ_λ of a blackbody at the appropriate temperature.

[c] The *blackbody D-Star* is $D_{bb}^* = [(AB)^{1/2}/(\overline{i_n^2})^{1/2}]R_{bb} = D^*(\lambda \text{ peak})\alpha$, with α given by Eq. (6.6).

6.3 DETECTION PARAMETERS

The purpose of this section is to introduce the most fundamental detection parameters as a preliminary to the description of detector properties in Sections 6.5–6.6. Although we introduce them as detector parameters, these parameters can also be used to describe total system performance if determined and interpreted appropriately.

Table 6.1 shows the detection parameters and their defining relationships. Whenever the current i appears, it can be replaced by the voltage v if the corresponding voltage quantity is desired.

6.3.1 Responsivity

Consider a beam of radiant energy incident on a photon detector. The spectral radiant power in the beam is Φ_λ. Since each photon has energy $Q = h\nu$, the number of photons per unit time arriving at the detector is $\Phi_\lambda/h\nu$. Each of these photons produces an electron with quantum efficiency η. The number of signal electrons produced per unit time is therefore $\dot{n} = \eta\Phi_\lambda/h\nu$. If each signal electron contributes to the output current, then the electronic current is $\dot{n}q$, where q is the electronic charge. The current is then $i_\lambda = (\eta q/h\nu)\Phi_\lambda$ so that, by the definition of *spectral responsivity* in Table 6.1,

$$R(\lambda) = i_\lambda/\Phi_\lambda = \eta q/h\nu. \tag{6.3}$$

This simple relationship for spectral responsivity ignores any internal gain mechanism or finite electron–hole lifetime effects as discussed in Section 6.5.

Figure 6.3a shows a typical spectral responsivity curve for a photon detector. The slope for $\lambda < \lambda_c$ is a consequence of the wavelength dependence

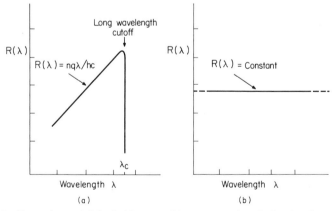

Fig. 6.3 Spectral responsivity (arbitrary scale) versus wavelength showing the consequence of photon detection mechanism (a) on shape of $R(\lambda)$ in comparison with essentially flat $R(\lambda)$ for thermal detector (b).

in (6.3) which can be written as $R(\lambda) = \eta q \lambda / hc$, where $v = c/\lambda$. The thermal detector responsivity, in Fig. 6.3b is essentially spectrally neutral since the detection mechanism depends only on total absorbed power or energy. If the detector contains an internal gain mechanism, then $R(\lambda) = (\eta q / hv)G$, where G is the current gain. To determine the *total responsivity* as defined in Table 6.1 it is necessary to consider the dependence of these parameters on wavelength. This is most conveniently done by introducing the spectral matching factor.

6.3.2 Spectral Matching Factor (SMF)

The *spectral matching factor* is a measure of the spectral overlap between the spectral power distribution incident on a detector and the detector spectral responsivity distribution (Eberhardt, 1968). Consider the spectral power distribution Φ_λ to be peak normalized so that $\Phi_\lambda = s_p \phi_\lambda$, where s_p is the radiant power at the maximum value of Φ_λ and ϕ_λ is the norUalized relative spectral power distribution. Also consider the detector spectral responsivity to be peak normalized so that $R(\lambda) = d_p r(\lambda)$, where d_p is the spectral responsivity at the maximum of $R(\lambda)$ and $r(\lambda)$ the normalized relative spectral responsivity distribution. Then the detector output current is found from $i_\lambda = R(\lambda)\Phi_\lambda$ to be

$$i = \int_0^\infty i_\lambda \, d\lambda = s_p d_p \int_0^\infty r(\lambda)\phi_\lambda \, d\lambda. \tag{6.4}$$

From the definition of total responsivity in Table 6.1,

$$R = i/\Phi = s_p d_p \int_0^\infty r(\lambda)\phi_\lambda \, d\lambda \bigg/ s_p \int_0^\infty \phi_\lambda \, d\lambda = d_p \alpha, \tag{6.5}$$

where the dimensionless integral ratio

$$\alpha = \int_0^\infty r(\lambda)\phi_\lambda \, d\lambda \bigg/ \int_0^\infty \phi_\lambda \, d\lambda \tag{6.6}$$

is the *spectral matching factor*. The spectral matching factor (SMF) can be seen from (6.6) to be a weighted average of the normalized spectral responsivity of the detector where the weighting factor is the relative spectral power distribution incident on the detector.

Figure 6.4 illustrates the calculation and interpretation of the SMF. The denominator of (6.6) is the area under the ϕ_λ curve and the numerator is the area under the product curve $\phi_\lambda r(\lambda)$. If the two spectral distributions do not overlap at all, then $\alpha = 0$. On the other hand if the relative spectral responsivity of the detector $r(\lambda)$ is one across the extent of the relative radiant

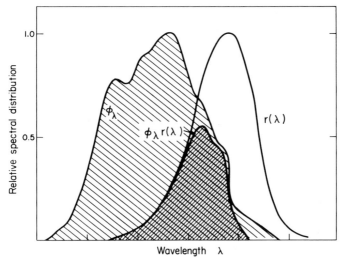

Fig. 6.4 Calculation and interpretation of the spectral matching factor (SMF).

power spectrum ϕ_λ, then $\alpha = 1$. Thus $0 \leq \alpha \leq 1$. From (6.5) we see that if the peak spectral responsivity of the detector can be determined, then the only additional information needed to determine the total responsivity R is the SMF. The SMF can in many cases be calculated approximately from standardized relative spectral distributions of sources and detectors (Eberhardt, 1968).

6.3.3 Signal-to-Noise Ratio

The dimensionless *signal-to-noise ratio* defined in Table 6.1 can be expressed equally well in terms of voltages as $v_s/(\overline{v_n^2})^{1/2}$, where v_s and $(\overline{v_n^2})^{1/2}$ are the rms signal and noise voltages, respectively. Sources of noise are described in Section 6.4. The SNR is the basis for the definition of the remaining parameters in Table 6.1 and for the statistical problem of calculating the probability of detection of a signal in noise. In some applications it may be more useful to calculate an electrical power SNR defined as $\mathrm{SNR}_p = i_s^2/\overline{i_n^2} = v_s^2/\overline{v_n^2}$.

6.3.4 Noise Equivalent Power (NEP)

Noise equivalent power (NEP) is the incident radiant signal power which will yield an electrical current or voltage with SNR = 1. From the definitions of spectral responsivity and of SNR in Table 6.1 we find that the spectral noise equivalent power is

$$\mathrm{NEP}(\lambda) = (\overline{i_n^2})^{1/2}/R(\lambda), \qquad (6.7)$$

where the result clearly depends on the noise bandwidth. For convenience the noise bandwidth effect is often removed by using the normalized quantity $\mathrm{NEP}(\lambda)/\sqrt{B}$ which has units of watts per Hertz$^{1/2}$. The total NEP is not the integral of $\mathrm{NEP}(\lambda)$ over wavelength; rather it is $\mathrm{NEP} = (\overline{i_n^2})^{1/2}/R$, where R is the total responsivity which depends on the radiant spectral power distribution as shown by (6.5) and (6.6).

6.3.5 Detectivity and Normalized Detectivity (D, D^*)

Unfortunately, as shown by (6.7), $\mathrm{NEP}(\lambda)$ decreases as performance improves, that is as $(\overline{i_n^2})^{1/2}$ decreases. In the interest of having a measure of performance (Jones, 1959a, b) which increases with improved performance, the *spectral detectivity* $D(\lambda)$ is defined as $1/\mathrm{NEP}(\lambda)$.

Analyses have shown that the detectivity is in many situations inversely proportional to the quantity $(AB)^{1/2}$, where A is the active area of the detector and B the noise bandwidth. In order to be able to compare detectors of unequal area or bandwidth (Jones, 1959a) on an equal basis, the *normalized spectral detectivity* $D^*(\lambda)$ (pronounced D-star) is used and defined as shown in Table 6.1. The detection parameter $D^*(\lambda)$ is widely used, particularly in the infrared region of the spectrum, to describe detector performance capability. However, it should be noted that the dependence on area is assumed to be $\sqrt{A}$. This assumption is valid when the dominant noise mechanism is background photon shot noise as described in Section 6.4.1, but may not be representative of the dependence on detector area when other forms of noise are dominant.

In specifying detection parameters the measurement conditions should be stated. For example $D^*(\lambda)$ depends on the signal modulation frequency and, when photon shot noise from the background is dominant, also depends on the background temperature and the solid angle subtended at the detector by the background. As a practical matter it should be noted that a nonideal performance of detectors can arise from the nonuniformity of a detection parameter. For example, the spectral responsivity of a photomultiplier or a semiconductor photodiode may vary from point to point across the photosensitive surface (Potter *et al.*, 1959; Budde and Kelly, 1971; Budde and Dodd, 1971) or it may be sensitive to the polarization characteristics of the incident beam (Budde and Dodd, 1971). Similarly a thermal detector may exhibit small departures from uniform spectral responsivity at different wavelengths due to the absorption characteristics of the black surface material (Blevin and Geist, 1974a, b). The dependence on position, angle, polarization, and other parameters must be considered in each application.

6.4 NOISE

Noise is a random variation in the results of a measurement (van der Ziel, 1956; Bennett, 1956; Schwartz, 1970). It can arise at various points in the detection process and can depend on a variety of measurement conditions such as detector temperature, radiant power from the background, detector bias voltage, or signal modulation frequency. In this section we shall look at the most important sources of noise and their effect on the current or voltage at the output of the detector. Figure 6.5 shows a generalized measurement system and shows the point at which these noise currents and voltages shall be determined.

When all other sources of noise have been removed, the final limiting noise in the detection of radiant energy is found to be the *photon shot noise* which arises from the quantum nature of the signal itself. The *signal* is the output which is due to the object of interest. The statistical properties of shot noise were described in Section 3.8.1. In most applications it is not practical to plan to achieve this signal shot noise limit. The dominant noise source may then be background photon shot noise, dark current shot noise, Johnson noise (also called *thermal noise*), amplifier noise, $1/f$ noise, temperature noise, or some combination of these. Each shall be discussed in the following section.

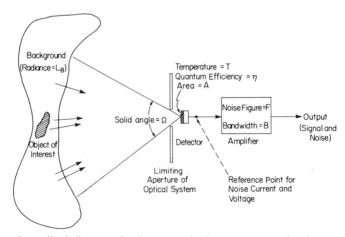

Fig. 6.5 Generalized diagram of a detector and other components showing some of the sources of noise in the detection process.

6.4.1 Shot Noise

Shot noise, with statistical properties described in Section 3.8, can arise from three different sources. These are (a) the signal, (b) the background, (c) and the dark current. "Signal" refers here to radiant energy coming from the object of interest; "background" refers to the radiant energy arriving at the detector from all other sources within the field of view (discrete objects, atmosphere, optical elements, etc.); and "dark" refers to the condition of the detector when it is shielded from all incident signal or background radiant energy. In all three cases the statistical analysis of shot noise (Rice, 1945) shows that the mean square noise current at the detector output is

$$\overline{i_n^2} = 2q\bar{i}B, \tag{6.8}$$

where q is the electronic charge, $\bar{i}$ the mean current, and B the noise bandwidth. For dark current shot noise $\bar{i}$ is simply $\bar{i}_d$, the mean dark current. For signal or background photon shot noise the mean current $\bar{i}$ is found from Table 6.1 to be $\bar{i} = R\Phi_s$ or $\bar{i} = R\Phi_B$, where R is the total responsivity and Φ_s and Φ_B are the signal or background radiant power, respectively. Then from (6.8)

$$\overline{i_n^2} = 2qR\Phi_s B \qquad \text{(signal photon shot noise)}, \tag{6.9}$$

$$\overline{i_n^2} = 2qR\Phi_B B \qquad \text{(background photon shot noise)}. \tag{6.10}$$

To determine the NEP(λ), the minimum power which can be detected, for signal or background photon shot noise limited detection we use (6.3), (6.7) and either (6.9) or (6.10) with the definition in Table 6.1 to find

$$\text{NEP}(\lambda) = 2hvB/\eta \qquad \text{(signal photon shot noise)}, \tag{6.11}$$

$$\text{NEP}(\lambda) = [(2hvB/\eta)\Phi_B]^{1/2} \qquad \text{(background photon shot noise)}, \tag{6.12}$$

where in finding (6.11) we used $\Phi_s = \text{NEP}(\lambda)$ at SNR $= 1$. Equations (6.11) and (6.12) show the limit due to generation of electron–hole pairs by photons. If recombination of electrons and holes is also important, as it is in a photoconductor, then (6.8) is increased by a factor of 2 with the result that (6.11) and (6.12) are increased by a factor of 2 and $\sqrt{2}$, respectively. When the background radiance is uniform, then $\Phi_B = L_B A\Omega$, where L_B is the background radiance, A the detector area, and Ω the solid angle seen by the detector as indicated in Fig. 6.5. The product $A\Omega$ is the *geometrical extent* discussed in Section 3.2.3.3. Figure 6.6 shows the NEP(λ) at several wavelengths from (6.12) for unit bandwidth and $\eta = 1$ as a function of the field of view $\Omega^{1/2}$. The field of view is the linear angle within which background energy is collected. The dashed lines in Fig. 6.6 are the fundamental signal photon shot noise limits obtained from (6.11).

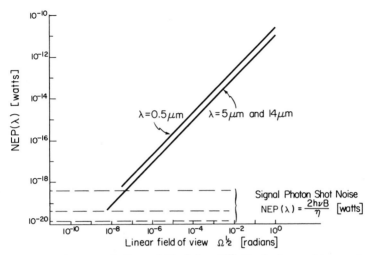

Fig. 6.6   NEP(λ) versus the linear field of view $\Omega^{1/2}$. The background temperature $T = 300$ K. $B = 1$ Hz, and $\eta = 1$.

From the bandwidth dependence in (6.12) and the area dependence of $\Phi_B = L_B A \Omega$ we see the basis for the use of $(AB)^{1/2}$ in the definition of $D^*(\lambda)$ in Table 6.1: $D^*(\lambda)$ is a relevant detection parameter when the detector is limited by background photon shot noise. Using (6.12) and the $D^*(\lambda)$ definition in Table 6.1 we find

$$D^*(\lambda) = [\eta/(2hvL_B\Omega)]^{1/2}. \qquad (6.13)$$

This $D^*(\lambda)$ is shown as a function of wavelength in Fig. 6.16 as the upper dashed line. Again we note that for a photoconductor the additional recombination noise reduces (6.13) by $\sqrt{2}$.

Having examined background and signal photon shot noise in some detail, we simply point out that the mechanisms of dark current shot noise are not as easily analyzed. However, once the mean dark current $\bar{i}_d$ is known, then NEP(λ) is found directly from (6.8) and Table 6.1. This concludes our general discussion of shot noise and we now turn our attention to the next important type of noise, Johnson noise.

6.4.2 Johnson Noise

Johnson noise (or thermal noise) is a random variation in the current or voltage at the electrical contacts of a resistor due to the random thermal motion of electrons within the resistor (Oliver, 1965). It is present in all resistive materials at temperatures above absolute zero, and can be interpreted as being due to the excitation of electrons by the blackbody radiant energy which is present within any material. Since the total noise current or voltage is the sum of a large number of independent current or voltage pulses, the total noise current or voltage is a Gaussian random process.

From the application of statistical mechanics to an electronic circuit we learn that the average free energy which can be delivered by this resistor to a matched load is kT, where k is Boltzmann's constant and T the absolute temperature. The power is then kTB, where B is the noise bandwidth. The open circuit noise power of the resistor is four times as great as the kTB determined for matched conditions, or $4kTB$. Since the relation between power and current for a resistor is power $= i^2 R_L$, where R_L is the effective load resistance, then

$$\overline{i_n^2} = 4kTB/R_L \qquad \text{(Johnson noise)}. \qquad (6.14)$$

Similarly, the noise voltage is found from power $= v^2/R_L$ to be

$$\overline{v_n^2} = 4kTBR_L \qquad \text{(Johnson noise)}. \qquad (6.15)$$

Johnson noise is reduced by reducing the temperature T through detector or preamplifier cooling, or by reducing the noise bandwidth B. Note that in (6.14) and (6.15) the temperature T is the effective *noise temperature* of the detector load resistance R_L.

6.4.3 Amplifier Noise

The next important type of noise which we consider is *amplifier noise*. The position of the amplifier after the detector is shown in Fig. 6.5. Consider the output which we would obtain from the amplifier if the amplification process were noiseless and the only noise were due to the Johnson noise at the output of the detector. Under these conditions the mean square noise voltage at the amplifier output is found from (6.15) to be

$$\overline{v_n^2} = G^2 4kT_D BR_L, \qquad (6.16)$$

where G is the voltage gain of the amplifier and T_D the noise temperature of the detector load resistance. A convenient way to describe the amount of additional noise introduced by the amplifier is then to write the actual output noise as

$$\overline{v_n^2} = G^2 4k(T_A + T_D)BR_L, \qquad (6.17)$$

where T_A is called the *amplifier noise temperature*. This is equivalent to assuming that the detector is at a temperature $T_A + T_D$ and that the amplifier is ideal. In this way we can relate the amplifier noise effect back to the reference point shown in Fig. 6.5 at the detector output.

Another useful measure of amplifier noise is the *noise figure F*. This is the ratio of the signal-to-noise ratio at the input of the amplifier to the signal-to-noise ratio at the output (Davenport and Root, 1958). Assuming that the signal voltage is simply multiplied by G in passing through the amplifier, then

$$F = 1 + (T_A/T_D). \tag{6.18}$$

In practice the noise figure is often expressed in logarithmic form as

$$NF = 10\log_{10} F \quad (dB), \tag{6.19}$$

where dB is the abbreviation for decibels. A 3-dB noise figure then corresponds to $F = 2$ or $T_A = T_D$. From (6.17) and (6.18) it can then be seen that the mean square noise voltage at the detector output due to amplifier noise is

$$\overline{v_n^2} = 4k(F\text{-}1)T_D BR_L. \tag{6.20}$$

By convention, amplifier noise figures are specified by assuming that the detector temperature is $T_D = 290$ K.

6.4.4   1/f Noise

The type of detection noise known as $1/f$ *noise* has a dependence on signal modulation frequency of the form $1/f^n$, where f is the modulation frequency and n some number. The origins of this noise are not well understood. In some cases it can be reduced by careful manufacturing methods. A method that has been used to reduce $1/f$ noise is to insert a chopper into the beam path in front of the detector so that all radiant energy incident on the detector is modulated at some frequency. If this frequency is high enough, typically in the several kilohertz range, then the signal-to-noise ratio can be significantly improved since $1/f$ noise decreases rapidly with increasing frequency f.

6.4.5   Temperature Noise

A thermal detector which is in contact with its surroundings by both conduction and radiation exhibits random fluctuations in temperature. This fluctuation is simply due to the statistical nature of the energy exchange or heat transfer between the detector and its environment. To determine the noise limitations of this thermal detector we must therefore consider some results from thermodynamics.

Before the radiant energy from a source is applied to the detector, the heat transfer equation is

$$d/dt(\Delta Q) = K\,\Delta T, \tag{6.21}$$

where ΔQ is a small energy increment, ΔT a small temperature increment between the detector and its surroundings, and K the total heat transfer coefficient for conduction and/or radiation. We know that ΔQ and ΔT are also related by the definition $\Delta Q = -C_T \Delta T$, where C_T is the heat capacity. Since ΔQ is negative when heat flows from the detector to its surroundings, the negative sign appears in this relation.

If the external source of radiant energy is now applied to the detector, represented by adding the incremental radiant power $\Delta\Phi(t)$ to (6.21), the

$$C_T(d/dt)(\Delta T) + K\,\Delta T = \Delta\Phi(t). \tag{6.22}$$

To solve (6.22) we use $\Delta\Phi(t) = \Delta\Phi_f \exp(2\pi i f t)$, where f is the modulation frequency. The solution of (6.22) for the temperature increment at frequency f is

$$|\Delta T_f| = \Delta\Phi_f/[K^2 + (2\pi f C_T)^2]^{1/2}. \tag{6.23}$$

We now assume that the applied radiant power is independent of modulation frequency so that $\Delta\Phi_f^2 = P_0\,\Delta f$ and integrate the mean square of (6.23) over all frequencies to find that

$$\overline{\Delta T^2} = \int_0^\infty \{P_0/[K^2 + (2\pi f C_T)^2]\}\,df = P_0/4KC_T. \tag{6.24}$$

From thermodynamics it is known that $\overline{\Delta T^2} = (kT^2/C_T)$, where k is Boltzmann's constant and T is the detector temperature. Equating this with (6.24), we find that $P_0 = 4kT^2K$. Then $\overline{\Delta\Phi_f^2} = 4kT^2K\,\Delta f$ and the mean square radiant power fluctuation over a total bandwidth B is

$$\overline{\Delta\Phi^2} = 4kT^2KB, \tag{6.25}$$

so the power fluctuations are independent of the heat capacity C_T.

The result, shown in (6.23) shall be used in Section 6.6 to define the performance of thermal detectors. The power fluctuations described by (6.25) set an ultimate limit to the NEP or D^* of a thermal detector. From the definition in Table 6.1, $\text{NEP}(\lambda) = (\overline{\Delta\Phi^2})^{1/2}$ when temperature noise is dominant. In the specific case where the temperature noise is due to a background which is in thermal equilibrium with the detector we find from (4.33) and (6.21) that the value of K in (6.25) is $K = 4\varepsilon\sigma T^3 A$ where ε is the detector emissivity and A the detector area.

6.5   PHOTON DETECTORS

Photon detectors are based on the photoelectric effect shown in Fig. 6.1a. The materials used in most photon detectors are semiconductors with the properties described in Section 6.2. The external photoelectric effect described in Section 6.2 is employed in the photomultiplier while the other photon detectors examined here (the semiconductor photodiode, the avalanche photodiode, and the photoconductor) use the internal photoelectric effect.

6.5.1   Photomultipliers

Photomultipliers are widely used in the near ultraviolet, visible, and near infrared regions of the spectrum. They provide high internal current gain by combining the external photoelectric effect with low noise *secondary electron multiplication* and are approximately linear over a wide dynamic range (Sommer, 1968).

Figure 6.7 shows three typical photomultiplier configurations. In Fig. 6.7a a beam of radiant energy is incident on an opaque photocathode (point 0), where the absorbed photons produce photoemission. Electrostatic fields guide and accelerate the electrons from the photocathode to the first dynode

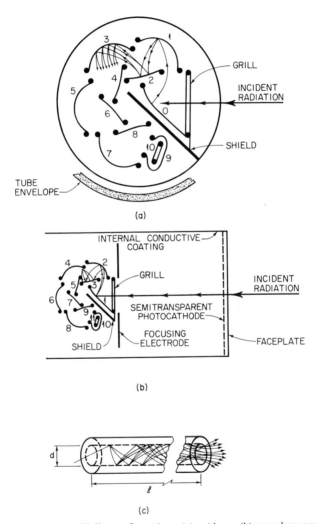

Fig. 6.7 Typical photomultiplier configurations: (a) a side-on, (b) an end-on, and (c) a continuous channel structure. (a) Point 0 is an opaque photocathode, points 1–9 are electron multiplier dynodes, and point 10 is an anode. (b) Points 1–10 are electron multiplier dynodes, and point 11 is an anode. (Courtesy RCA.)

(point 1) and to each succeeding dynode. The incident electrons at this and each succeeding dynode (points 2–9) release secondary electrons. If the average secondary multiplication at each dynode is m and there are N dynodes, the current gain is

$$G = m^N, \qquad (6.26)$$

which for $m = 6$ and $N = 9$ gives $G = 10^7$, a considerable current gain.

An important property of the secondary multiplication process is that it is low noise. The total shot noise is due to shot noise at the cathode and at each dynode. However, the total dynode shot noise is a factor m^{-1} smaller than the photocathode shot noise. Since $m = 6$ typically, the dynode shot noise can usually be neglected. Furthermore, the internal gain G does not amplify the Johnson noise due to the detector load resistance so that the signal is greatly increased relative to the Johnson noise. The other source of noise in the photomultiplier is dark current shot noise.

The relationship of all these noise sources is seen in the power signal-to-noise ratio. From (6.3), (6.8), (6.14) and Table 6.1 this is

$$\mathrm{SNR_p} = i^2/\overline{i_n^2} = 2G^2(\eta q/h\nu)^2 \Phi^2/[2G^2 q(\overline{i_s} + \overline{i_d})B + (4kTB/R_L)], \quad (6.27)$$

where $\overline{i_s} = R\Phi$ is the mean signal current, $\overline{i_d}$ the mean dark current, and R the responsivity given by $\eta q/h\nu$. When the current gain G is large enough so that the first term in the denominator is much larger than the Johnson noise term $4kTB/R_L$, then the $\mathrm{SNR_p}$ is independent of the gain. Thus the purpose of introducing internal gain is to overcome Johnson noise. Of course, without internal gain it might be necessary to use an external amplifier and this could introduce additional noise as shown by (6.20).

When the dominant noise is dark current, as it frequently is, then (6.27) and the definition of $\mathrm{NEP}(\lambda)$ in Table 6.1 give

$$\mathrm{NEP}(\lambda) = (h\nu/\eta)(\overline{i_d}B/q)^{1/2} \qquad (6.28)$$

as the minimum detectable power. This result can be improved by cooling the photomultiplier to reduce the dark current. The $\mathrm{NEP}(\lambda)/\mathrm{Hz}^{1/2}$ is also referred to as the ENI, the equivalent noise input. By choosing $\overline{i_d}$ and η from Table 6.2 the $\mathrm{NEP}(\lambda)/\mathrm{Hz}^{1/2}$ can be calculated.

The spectral responsivity of some typical photocathodes is shown in Fig. 6.8. The wavelength range of photomultipliers is limited by the work function of the photocathode at long wavelengths as shown in Section 6.2. At short wavelengths the limit is usually set by the transmittance characteristics of the window. Materials most commonly used in photocathodes are silver–oxygen–cesium (Ag–O–Cs), cesium antimony (Cs_3Sb), multialkali or trialkali [(Cs)Na$_2$KS$_b$] and bialkali ($K_2C_sS_b$). The first of these corresponds to the S-1 curve shown in Fig. 6.8. The curve designated ERMA in Fig. 6.8 is an

TABLE 6.2 Characteristics of Various Photocathodes[a]

Nominal Composition	Response Designation	Type of Photocathode	Envelope Material[b]	Conversion Factor[c] (lumen/watt at λmax)	Luminous Sensitivity (μA/lumen)	Wavelength of Maximum Response, λmax (nm)	Sensitivity at λmax (mA/watt)	Quantum Efficiency at λmax (percent)	Dark Emission at 25°C A x 10^{-15}/cm²
Ag-O-Cs	S-1	O	0080	92.7	25	800	2.3	0.36	900
Ag-O-Rb	S-3	O	0080	285	6.5	420	1.8	0.55	—
Cs_3Sb	S-19	O	SiO_2	1603	40	330	64	24	0.3
Cs_3Sb	S-4	O	0080	1044	40	400	42	13	0.2
Ca_3Sb	S-5	O	9741	1262	40	340	50	18	0.3
Cs_3Bi	S-8	O	0080	757	3	365	2.3	0.77	0.13
Ag-Bi-O-Cs	S-10	S	0080	509	40	450	20	5.6	70
Cs_3Sb	S-13	S	SiO_2	799	60	440	48	14	4
Cs_3Sb	S-9	S	0080	683	30	480	20	5.3	—
Cs_3Sb	S-11	S	0080	808	60	440	48	14	3
Cs_3Sb	S-21	S	9741	783	30	440	23	6.7	—
Cs_3Sb	S-17	O[h]	0080	667	125	490	83	21	1.2
Na_2KSb	S-24	S	7056	1505	32	380	64	23	0.0003
K-Cs-Sb	—	S	7740	1117	80	400	89	28	0.02
$(Ca)Na_2KSb$	—	S	SiO_2	429	150	420	64	18.9	0.4
$(Cs)Na_2KSb$	S-20	S	0080	428	150	420	64	19	0.3
$(Cs)Na_2KSb$	S-25	S	0080	276	160	420	44	13	—
$(Cs)Na_2KSb$	ERMA[d]	S	7056	169	265	575	45	10	1
Ga-As	—	O[i]	9741	148	250	450	37	10	0.1
Ga-As-P	—	O[j]	Sapphire	310	200	450	61	17	0.01
InGaAs-CsO[e]	—	O[i]	0080	266	260	400	71	22	1[d]
Cs_2Te	—	S	LiF	f	f	120	12.6	13	g
CsI	—	S	LiF	f	f	120	24	20	g
CuI	—	S	LiF	f	f	150	13	10.7	g

[a] By permission of RCA.

[b] Numbers refer to the following glasses: 0080, Corning lime; 9741, Corning ultraviolet transmitting; 7056, Corning borosilicate; 7749, Corning pyrex; SiO_2, Fused silica (Suprasil, trademark of Englehard Industries, Inc., Hillside, New Jersey.

[c] These conversion factors are the ratio of the radiant responsivity at the peak of the spectral response characteristic in amperes per watt to the luminous responsivity in amperes per lumen for a tungsten test lamp operated at a color temperature of 2854 K.

[d] An RCA designation for Extended Red Multialkali.

[e] An experimental photocathode (Private communication from B. F. Williams, RCA Laboratories, Princeton, New Jersey 1969). The dark emission indicated is a calculated value based on a bandgap of 1.1 eV for the particular composition studied. See also Williams (1969).

[f] Not relevant.

[g] Data unavailable; expected to be very low.

[h] Reflecting substrate.

[i] Single crystal.

[j] Polycrystalline.

O = opaque and S = semitransparent.

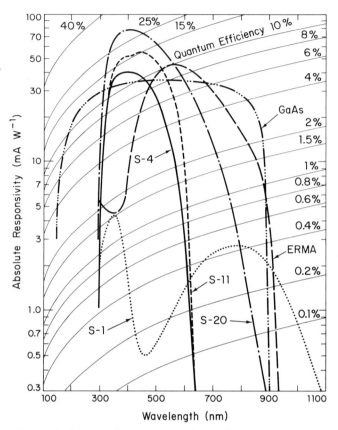

Fig. 6.8 Photocathode spectral responsivity characteristics of commercially available photomultipliers. (Courtesy RCA.)

Extended Red Multi-Alkali photocathodes in which improved red response is obtained at the expense of short wavelength response. That designated GaAs is the recently developed negative electron affinity gallium arsenide material. Table 6.2 shows the characteristics of a number of commercially available photocathode materials.

The constant quantum efficiency curves indicated in Fig. 6.8 are based on relation (6.3) between spectral responsivity $R(\lambda)$, quantum efficiency η, and wavelength λ, which is

$$R(\lambda) = \eta q \lambda / hc.$$

Since Fig. 6.8 shows log $R(\lambda)$ versus λ, the curves are logarithmic in shape.

Voltage gradients between dynodes in the photomultiplier can be provided by individual voltage sources but usually a resistive voltage divider is used with a high voltage source. This is shown in Fig. 6.9.

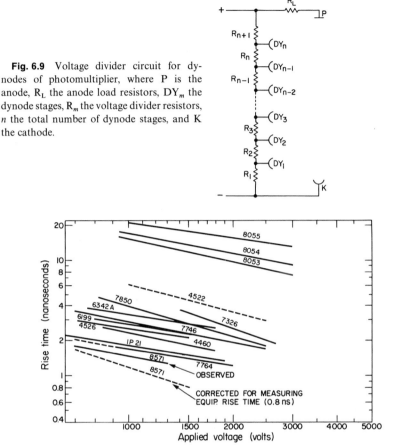

Fig. 6.9 Voltage divider circuit for dynodes of photomultiplier, where P is the anode, R_L the anode load resistors, DY_m the dynode stages, R_m the voltage divider resistors, n the total number of dynode stages, and K the cathode.

Fig. 6.10 Time response of photomultiplier tubes at various total dynode voltages. (Courtesy RCA.)

Time response in photomultipliers is limited from the photocathode to the anode. This is related ultimately to the initial velocity distribution of the photoelectrons leaving the cathode, but also to the electron path geometries and electrostatic field strengths. The latter degradation is reduced by the crossed field multiplier which uses both magnetic and electrostatic focussing fields. Typical response time values at dynode voltages of approximately 100 V/stage are between 1 and 5 ns. Increasing the voltage per stage reduces the response time. Figure 6.10 shows measured *response times* for typical photomultipliers as a function of total applied dynode voltage.

The *linearity* of a typical photomultiplier is shown in Fig. 6.11. Notice that for convenience the input flux is measured in photometric units as defined in Chapter 2. With the bias voltage per stage fixed at 100, this photomultiplier

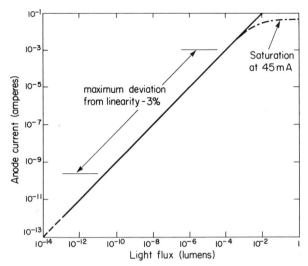

Fig. 6.11 Range of linearity of a typical photomultiplier (type 931A). The input flux is measured in photometric units (see Chapter 2) and the volts per stage is 100. (Courtesy, RCA.)

is found to be approximately linear over an anode current range of 10^6, or 6 decades. However, if the nominal input flux is near the upper or lower end of the indicated range, then the linearity of the detector for flux changes about this nominal point can be more limited.

6.5.2 Semiconductor Photodiodes

Semiconductor photodiodes are solid-state devices which are used to detect radiant energy from the near ultraviolet, through the visible, and into the far infrared regions of the spectrum (Melchior, 1973; Emmons et al., 1975; Melchior et al., 1970; Putley, 1973; Levinstein, 1965). They are described as diodes because they use a semiconductor *p–n junction*. Because the electrons or holes which constitute the signal current are generated by the internal photoelectric effect, as described in Section 6.2, operation at the longer infrared wavelengths is made possible.

The essential feature of the semiconductor photodiode is the *p–n* junction, an interface between *p* type and *n* type semiconductor materials.[†] A junction of this type can be produced by controlled diffusion of a donor impurity into a *p* type semiconductor or of an acceptor impurity into an *n* type material. The energy levels near the junction of the *n* and *p* regions are then as shown in Fig. 6.12. In the figure we see that electron–hole pairs can be produced by photon absorption in three regions. In the *p* region the positively charged hole is immobile but the electron can diffuse toward the junction region, called the *depletion layer*. Once in the depletion layer it rapidly drifts across

---

[†] A junction formed between metals and semiconductors is known as a *Schottky barrier*.

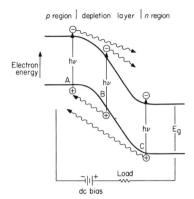

Fig. 6.12 The p–n junction in a photo-diode. The upper and lower curves indicate the edges of the conduction and valence bands as shown in Fig. 6.2.

the junction under the influence of the junction field and contributes to the signal current. Similarly, in the n region the electron is immobile but the hole can move toward the junction by diffusion and then rapidly drift across it. In the depletion layer itself both the electron and the hole are mobile and can rapidly drift in opposite directions, both contributing to the signal current. Since drift in the junction field is faster than the diffusion process, the best time response (and highest current) is produced when the photons are absorbed in the depletion layer.

Electrons and holes swept across the junction make up the current which generates a voltage across the load resistor shown in Fig. 6.12. This signal voltage is the output of the semiconductor photodiode.

The electrical characteristics of the photodiode are illustrated in Fig. 6.13. In this figure the uppermost curve shows the relationship between current

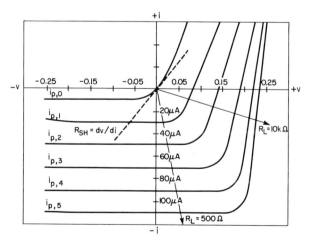

Fig. 6.13 Current–voltage relationships in a semiconductor photodiode. Photocurrents $i_{p,n}$ are at increasing levels of irradiance. Linearity is achieved with small R_L values.

and voltage when no radiant energy is incident on the photodiode detector. The bend in the curve occurs at $v = 0$ in this case.

Figure 6.12 showed a dc bias voltage. If a positive bias voltage is introduced across the junction, the detector is said to be *forward biased* and if the bias voltage is negative, then the detector is said to be *reverse biased*. As shown in Fig. 6.13 the semiconductor photodiode when operated as a radiometric detector is operated with reverse bias or with no external bias. When silicon p–n junctions are used in solar energy conversion applications, maximum power transfer occurs when $R_L = R_{SH}$, the shunt resistance.

Referring again to Fig. 6.13, the other curves represent the v–i relationship when increasing levels of radiant power are incident on the detector. The radiant power level increases in equal steps. Although it might at first appear to be desirable to operate with forward bias, i.e. at a point on the v–i curve where the slope dv/di is maximum, this would not yield a linear relationship between radiant power and signal voltage. To achieve linearity reverse bias is used. For the silicon photodiode example shown in Fig. 6.13 this corresponds to using a load resistor of 500 Ω rather than one of 10 kΩ.

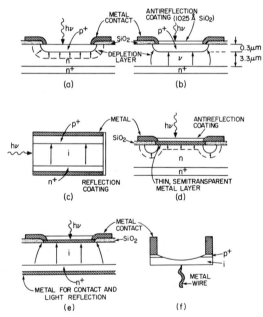

Fig. 6.14   Configurations of semiconductor photodiodes. Construction of different high-speed photoiodides: (a) p–n diode; (b) p–i–n diode (Si optimized for 0.63 μm); (c) p–i–n diode with irradiation parallel to junction; (d) metal–semiconductor diode; (e) metal–i–n diode; (f) semiconductor–point contact diode. (After Melchior, Fisher, and Arams, 1970.)

In Fig. 6.14 some configurations of semiconductor photodiodes are shown. The SiO$_2$ indicated in some devices is a surface passivation designed to prevent leakage currents from moving along the surface. The silicon photodiodes indicated in this figure are used from the near ultraviolet to the near infrared. In the middle and far infrared other materials, including doped silicon, are used. As indicated in this figure, some photodiodes use a high resistivity insulating layer i. Since most of the potential drop occurs across this layer, the layer thickness typically can be adjusted to ensure that most of the photons are absorbed within the junction region.

The spectral responsivities of some silicon photodiodes are shown in Fig. 6.15. Notice that the characteristic shape for this photon detector is essentially that shown in Fig. 6.3a and described in Section 6.3.1. The long wavelength cutoff in the region of $\lambda = 1.1\ \mu\text{m} = 1100$ nm is due to the bandgap energy for pure Si, an intrinsic semiconductor. At the short wavelengths the rapid decrease is due to increased absorption at the short wavelengths by the Si material before the photons can reach the p–n junction. In the intermediate wavelength region the shape is essentially as shown in Fig. 6.3a.

Photodiodes operating in the infrared are of two types:

(1) intrinsic semiconductors which have bandgap energy $e_G < hc/\lambda_c$, where λ_c is the maximum desired wavelength of operation (for example, mercury–cadmium telluride HgCdTe or indium antimonide InSb), or

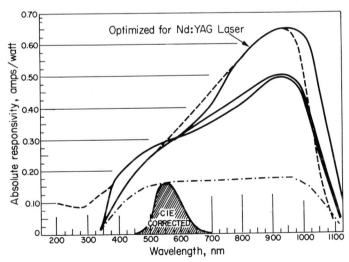

Fig. 6.15 Spectral responsivities for some typical silicon photodiode detectors. (Courtesy EG & G.)

(2)   extrinsic semiconductors in which the donor or acceptor levels are chosen to yield e_d or $e_a < hc/\lambda_c$ (for example, silicon doped with various impurities Si:XX).

In the infrared it is common practice to show the spectral characteristics of a photon detector in terms of $D^*(\lambda)$ rather than responsivity. When the detection process is background shot noise limited, the power signal-to-noise ratio is found from (6.13) and $v = c/\lambda$ to be

$$D^*(\lambda) = [\eta\lambda/2hcL_B\Omega]^{1/2}. \tag{6.29}$$

In addition to the explicit dependence on wavelength λ, the background radiance L_B for a blackbody depends on λ. Figure 6.16 shows this theoretical $D^*(\lambda)$ limit for a photodiode as well as for a photoconductor (see Section 6.5.4). Also known are the measured $D^*(\lambda)$ for some commercially available infrared

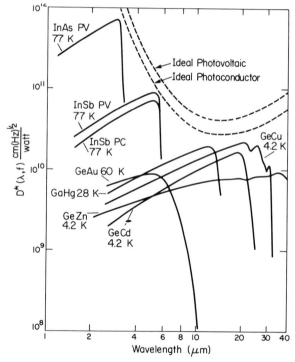

Fig. 6.16 Normalized spectral detectivity of infrared semiconductor detectors with a field of view of 2π sr and background temperatures of 295 K. PV (photovoltaic) denotes a photodiode and PC a photoconductor. (Courtesy Santa Barbara Research Center.)

photodiodes. These detectors approach the background photon shot noise limit at their spectral peaks under the measurement conditions shown (2π s and 295 K background temperature). As the field of view narrows and Ω becomes smaller, the detectivity is theoretically improved by the factor

$$\text{improvement factor} = 1/\sin(\theta/2), \tag{6.30}$$

where θ is the total linear angle seen by the detector. At small angles this improvement is limited by the increasing importance of other noise mechanisms relative to the background photon shot noise.

Time response or *frequency response* is another important property of photon detectors. The principal mechanisms which limit the frequency response of a photodiode are

(1) the RC time constant of the effective load resistance R and the parallel junction capacitance C [The RC time constant is approximately $\tau = RC$ and the resulting upper frequency limit is approximately $f_{max} = 1/2\pi RC$ (Since the device capacitance is proportional to area, the detector area is kept small for high-frequency operation.)],

(2) the diffusion time for carriers generated in the p and n regions shown in Fig. 6.12 and discussed earlier in this section [In order to minimize this effect the photodiode is designed to absorb as many photons as possible in the depletion layer by proper choice of material or layer thickness.], and

(3) the transit time of the carriers drifting across the depletion layer [This is minimized by increasing the reverse bias voltage until dark current becomes a limiting factor or the drift velocity saturates.].

Photodiodes have been developed which provide gigahertz frequency response. Operational amplifier–photodiode combinations provide linear operation with high gain over extended frequency ranges.

6.5.3 Avalanche Photodiodes

Avalanche photodiodes are semiconductor photodiodes which provide internal gain through the use of *avalanche multiplication*. Avalanche multiplication occurs when the electric field within the semiconductor accelerates the electrons to kinetic energy levels which are large enough to produce new electron–hole pairs by collision (McIntyre, 1970; Webb *et al.*, 1974; McKay, 1954; Lindley *et al.*, 1969). The new carriers are also accelerated and contribute to the multiplication process. Figure 6.17 shows the current multiplication factor M in an avalanche photodiode as a function of the electric field.

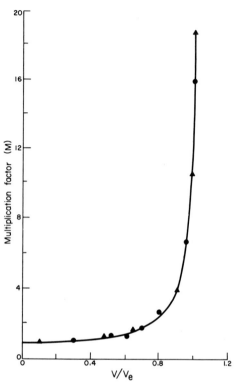

Fig. 6.17 Current multiplication factor in avalanche photodiode for $T = 25°C$ (▲) and $-132°C$ (●). The quantity V/V_e is the ratio of the applied voltage to the avalanche breakdown voltage. (After McKay, 1954.)

As in the case of the photomultiplier, the purpose of internal gain is to overcome the Johnson noise limit expressed by (6.14) or (6.15). From Table 6.1, (6.8), (6.14), and (6.20) the total power SNR can be written as

$$\text{SNR}_p = \frac{i_s^2}{(\overline{i_n^2})^{1/2}}$$

$$= \frac{M^2 R^2 \Phi_s^2}{2qM^2F'(R\Phi_s + R\Phi_B + \overline{i_d}) + (4kTB/R_L) + \left[4(F-1)kT_{290}B/R_L\right]},$$

$$(6.31)$$

where M is the avalanche current gain and F' represents an increase in the shot noise due to the avalanche process. Examination of (6.31) shows that if M becomes large enough, the shot noise is larger than both the Johnson noise $4kTB/R_L$ and the amplifier noise. Also, in this case the SNR_p would become independent of the avalanche current gain M.

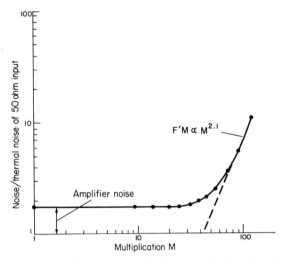

Fig. 6.18 Ratio of total noise to thermal noise for an avalanche photodiode showing the additional shot noise due to the avalanche multiplication process. (After Lindley *et al.*, 1969.)

However, F' is dependent on the gain M. Figure 6.18 shows the ratio of total noise to thermal noise for a typical avalanche photodiode. At avalanche gains larger than approximately $M = 100$ or $M = 200$ the factor F' is proportional to M^n, where typically $1 < n < 2$. The optimum performance of an avalanche photodiode is obtained by increasing the gain M only until the shot noise is approximately equal to the Johnson noise.

In other respects the avalanche photodiode is essentially the same as the semiconductor photodiodes described in Section 6.5.2. The principal materials in which low noise avalanche gain has so far been achieved are Si and Ge.

6.5.4 Photoconductors

The *photoconductor* or photoconductive detector is also a solid-state semiconductor device. At low modulation frequencies it provides *photoconductive gain* to overcome Johnson noise. Because it uses the internal photoelectric effect, the photoconductor can be used at wavelengths as long as those in the far infrared. However, at these wavelengths the detector requires cooling to avoid thermal excitation of electrons into the conduction band or holes into the valence band.

Figure 6.19 shows the use of the photoconductive detector to produce an electronic signal. Radiant energy incident on the detector produces an electron–hole pair which lowers the detector resistance by producing more carriers. The change in the photoconductor resistance ΔR_p produces a change

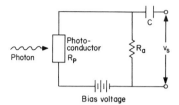

Bias voltage

Fig. 6.19   Electronic circuit for photo-conductor detector.

in the voltage drop across R_a. For $\Delta R_p / R_p \ll 1$ the signal voltage is then

$$\Delta v_s = [R_a/(R_a + R_p)] i \, \Delta R_p, \qquad (6.31)$$

where i is the current due to the bias voltage before irradiance.

To find Δv_s and the detector responsivity we examine the carrier generation mechanism. The conductivity before irradiance is

$$\sigma = nq\mu_n + pq\mu_p, \qquad (6.32)$$

where n and p are the electron and hole concentrations (number per volume), q the electronic charge, and μ_n and μ_p the electron and hole mobilities. The mobility is $\mu = v/\mathscr{E}$, where v is the average carrier velocity and $\mathscr{E}$ the electric field due to the bias voltage. When radiant energy is incident, the number of photons per unit area per unit time is E_e/hv, where E_e is the irradiance. The number of electrons per unit area per unit time is then $\eta E_e/hv$, where η is the quantum efficiency. If each carrier has a lifetime τ, then the change in carrier concentration is

$$\Delta n = \Delta p = (\eta E_e/hv)(\tau/d), \qquad (6.33)$$

where d is the thickness between electrodes.

From (6.32) and (6.33) the conductivity change is

$$\Delta\sigma = q(\mu_n \, \Delta n + \mu_p \, \Delta p) \qquad (6.34)$$
$$\Delta\sigma = q(\mu_n + \mu_p)(\eta E_e/hv)(\tau/d). \qquad (6.35)$$

The current change through the detector is $\Delta i_p = \Delta\sigma \mathscr{E} A$, where A is the cross-sectional area. From (6.35), with the definition $b = \mu_n/\mu_p$, and using $E_e A = \Phi$, the current change is therefore

$$\Delta i_p = (\eta q/hv)(\Phi)[\mu_p(1+b)\tau\mathscr{E}/d], \qquad (6.36)$$

and the voltage responsivity is simply $R = \Delta v_s/\Phi = R_a \, \Delta i_p/\Phi$ with Δi_p given by (6.36).

The quantity in square brackets in (6.36) is the *photoconductive gain* G. Since $\mu_p \mathscr{E} = v$, the carrier drift velocity, and $d/v = T$, the transit time between electrodes, the photoconductive gain can be written as

$$G = (1+b)\tau/T. \qquad (6.37)$$

Equation (6.37) shows that if $b \gg 1$ or if $\tau \gg T$, the photoconductive gain can be much greater than unity.

If the applied voltage is increased to decrease T, the transit time, a point will be reached where space charge effects will limit the transit time to $T = T_{min}$. Since the bandwidth, or maximum modulation frequency, is $B_{max} = 1/2\pi\tau$, then, from (6.37),

$$GB_{max} = (1 + b)(1/2\pi T_{min}). \qquad (6.38)$$

Equation (6.38) is the *gain–bandwidth product* for a photoconductive detector. It shows that in this detector there is a maximum gain which can be realized for a given bandwidth, and vice versa. Photoconductive gains up to $G = 10^3$ have been achieved at low bandwidths.

Figure 6.16 shows $D^*(\lambda)$ for a number of commercially available photoconductors. The temperature indicated on each curve is the cooled detector temperature. The lower of the two dashed lines at the top is the theoretical limit of detectivity for background photon shot noise limited detection as shown in (6.13), but reduced by $\sqrt{2}$ due to the additional recombination noise of the photoconductor. The random recombination of electron–hole pairs generates shot noise which must be considered because the lifetime of the carriers is so long in the photoconductor.

6.6 THERMAL DETECTORS

Thermal detectors convert radiant energy into an electronic signal by a primary detection mechanism in which the absorbed radiant energy raises the detector temperature. Depending on which type of thermal detector is considered, this temperature increase may then

(1) change the detector resistance,
(2) produce a thermoelectric voltage,
(3) alter the detector capacitance, or
(4) heat a medium, such as water, in which a conventional contact thermocouple is used to generate a voltage.

These four mechanisms are the basis for the thermal detectors considered in this section.

In general, thermal detectors are essentially spectrally neutral. The temperature increase which is registered does not depend directly on the photoelectric effect, and so does not exhibit the fundamental wavelength-dependent responsivity of the photon detector as shown in Fig. 6.3. With the exception of the pyroelectric detector, the response time of thermal detectors is usually much longer than that of photon detectors.

Usually thermal detectors are used in situations where spectrally flat responsivity is important. However, the pyroelectric detector in addition to being spectrally neutral has fast time response and does not require cooling. As a result it has recently received a good deal of attention for applications where this combination of characteristics is important.

In the following sections we examine the properties of the bolometer, the thermocouple and thermopile, the pyroelectric detector, and the laser calorimeter. Our primary emphasis is on time or frequency response, noise limitations, and responsivity (Smith *et al.*, 1968; Kruse *et al.*, 1962).

6.6.1 Bolometers

A *bolometer* detector consists of a thin blackened metal or semiconductor strip arranged as shown in Fig. 6.20. Radiant energy incident on the blackened strip raises its temperature, and this increase produces a change in the strip resistance. The resistance change leads to a change of current through the strip due to the bias voltage V. Since this is the same current which flows through the resistor R_L, there is a change in the voltage v_s across R_L. The voltage change Δv_s is the electronic output signal from the bolometer. For measuring a steady beam of radiant energy, the chopper is used to modulate the signal thus avoiding problems inherent in dc measurements.

With no radiant energy falling on the strip, the heating of the strip by the bias voltage is balanced by the heat loss due to both conduction through the strip supports and thermal radiation from the strip. This is expressed by

$$d/dt(\Delta Q) = K\,\Delta T - i^2 R_s, \qquad (6.39)$$

where $d(\Delta Q)/dt$ is the total energy loss per unit time, ΔT the temperature difference between the strip and its environment, i the current due to the bias voltage V, and R_s the strip resistance in the absence of radiant energy. The resistance depends on temperature according to

$$R_s = R_0(1 + \alpha\,\Delta T), \qquad (6.40)$$

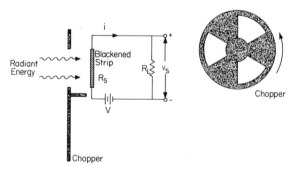

Fig. 6.20   Simplified diagram of bolometer for measurement of chopped radiant energy.

where R_0 is the strip resistance when $\Delta T = 0$ and α the *temperature coefficient of resistance*. In metals α is positive but in semiconductors it is negative.

If radiant energy is now directed onto the bolometer strip, the strip resistance changes by an amount which we find from (6.40) to be

$$\Delta R_s = R_0 \alpha \, \Delta T. \tag{6.41}$$

The ΔT in (6.40) is the incremental temperature difference. To find ΔT we return to Section 6.4.5 where it was found in (6.23) that

$$|\Delta T| = \Delta \Phi / [K^2 + (2\pi f C_T)^2]^{1/2}, \tag{6.42}$$

where $\Delta \Phi$ is the absorbed radiant power and C_T the heat capacity of the strip.

A relation between the resistance change ΔR_s and the voltage change Δv_s can be found from Fig. 6.20. From this figure $\Delta v_s = R_L \, \Delta i$, and from analysis of the current loop, $\Delta i = (-i/R_L) \Delta R_s$ when $R_L \gg R_s$. Then $\Delta v_s = -i \Delta R_s$. Using this with (6.41) and (6.42) the output signal voltage from the bolometer is

$$|\Delta v_s| = i R_0 \alpha \varepsilon \, \Delta \Phi' / [K^2 + (2\pi f C_T)^2]^{1/2}, \tag{6.43}$$

where we used $\Delta \Phi = \varepsilon \, \Delta \Phi'$ with ε the emissivity of the blackening material and $\Delta \Phi'$ the incident (rather than the absorbed) radiant power.

The voltage responsivity of the bolometer is found from (6.43) to be

$$R = |\Delta v_s| / \Delta \Phi' = i R_0 \alpha \varepsilon / [K^2 + (2\pi f C_T)^2]^{1/2}. \tag{6.44}$$

For low frequency operation where $f \ll 1/(2\pi(C_T/K))$, (6.44) is $R = i R_0 \alpha \varepsilon R_e$, where $R_e = 1/K$ is the effective thermal resistance. Thus high responsivity is obtained by operating with a high temperature coefficient of resistance α, high emissivity ε, high strip resistance R_0, and high effective thermal resistance R_e. Although it also seems that a very high current i should also be used, there is an optimum current at which to operate.

As a numerical example we consider a metal bolometer in which $K = 10^{-4}$ W/C, $i = 10$ mA, $R_0 = 50\ \Omega$, $\alpha = 0.01/C$, and $\varepsilon = 1$. For these typical conditions the low frequency responsivity is $R = 50$ V/W.

Notice that the low frequency responsivity does not depend directly on the heat capacity of the blackened strip. However, from (6.44) the maximum frequency at which the bolometer can be operated is approximately

$$f_{\max} = 1/2\pi(C_T/K), \tag{6.45}$$

which depends directly on C_T. To maximize $f_{\max}$ the heat capacity C_T should be a minimum. This implies that the bolometer strip should have a small area and thickness.

Bolometers which use semiconductor strips (*thermistor* bolometers) have a much higher absolute value of α which leads to much higher responsivity than for the metal bolometer. Typically thousands of volts per watt can be obtained. The superconductive bolometer operating at cryogenic temperatures is an important development (Zwerdling *et al.*, 1968) also, but shall not be discussed here.

6.6.2 Thermocouples and Thermopiles

A *thermocouple* generates a voltage at the junction of two dissimilar metals in response to a temperature change at the junction. This is the *Seebeck effect*. When an absorber of radiant energy is placed in thermal contact with the junction, the device becomes a detector of radiant energy known as the *radiation thermocouple*. A number of junctions connected in series so that the voltages add form the *thermopile* (Stevens, 1970; Putley, 1973).

Figure 6.21 shows two arrangements of the junction. The basis for the voltage generation in these arrangements is as follows. In Fig. 6.22 two dissimilar metals are joined to form an electrical circuit. The junction at temperature T is thermally isolated from the region at temperature T_0, the ambient temperature. When radiant energy absorption at the black absorber

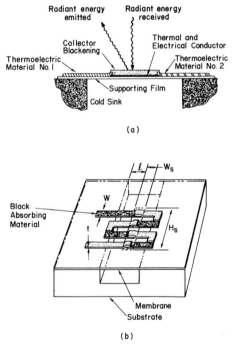

(a)

(b)

Fig. 6.21 Configurations of thermocouple and thermopile detectors. (a) Thermocouple structure on thin supporting film for high thermal resistance; (b) series connection of thermocouples to form thermopile. (After Stevens, 1970.)

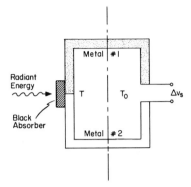

Fig. 6.22 Generation of signal voltage in radiation thermocouple based on Seebeck effect.

changes the junction temperature by an amount $\Delta T = T - T_0$, then a voltage Δv_s is generated at the open contact. The ratio, which is characteristic of metal 1 and metal 2, is the *Seebeck coefficient* given by

$$\alpha_{12} = \Delta v_s / \Delta T. \tag{6.46}$$

Generation of a signal voltage in response to a temperature change is a common characteristic of both the thermopile and the bolometer (Section 6.6.1), although the mechanism is different. As might be expected, the responsivity is determined in a similar way. From (6.21) or (6.39) with the current $i = 0$, and using the definition $\Delta Q = C_T \Delta T$ we have the energy balance equation

$$C_T [d/dt](\Delta T) + K \Delta T = \Delta \Phi(t), \tag{6.47}$$

where C_T is the heat capacity of the absorber, K the heat transfer coefficient, and $\Delta \Phi(t)$ the radiant power increment.

The solution of (6.47) for the magnitude of ΔT is, again, as in Section 6.6.1 and 6.4.5, given by

$$|\Delta T| = \Delta \Phi_0 / [K^2 + (2\pi f C_T)^2]^{1/2} \tag{6.48}$$

with $\Delta \Phi(t) = \Delta \Phi_0 \exp(2\pi i f t)$. However, the relation between ΔT and the signal voltage for the thermocouple is (6.46) so that the responsivity of the thermocouple is

$$R = \Delta v_s / \Delta \Phi_0' = \alpha_{12} \varepsilon / [K^2 + (2\pi f C_T)^2]^{1/2}, \tag{6.49}$$

where ε is the emissivity of the absorber defined by $\Delta \Phi_0 = \varepsilon \Delta \Phi_0'$ and $\Delta \Phi_0'$ the incident radiant power (after transmission through the device window).

At low frequencies where $f \ll 1/2\pi(C_T/K)$, the responsivity for a *thermopile* of n junctions is then

$$R = n\alpha_{12}\varepsilon R_T, \tag{6.50}$$

where $R_T = 1/K$ is the effective thermal resistance. High responsivity then requires a large Seebeck coefficient α_{12}, high emissivity absorber, a large thermal resistance (minimum heat loss due to conduction down supports or electrical leads), and a large number of junctions. Although increasing n improves responsivity it does not necessarily improve signal-to-noise ratio.

Since the heat capacity of the thermopile C_T and the heat transfer coefficient $K = 1/R_T$ are not radically different from those of the bolometer, it can be expected that the maximum frequency response $[f_{\max} = 1/2\pi(C_T/K)]$ devices are unsuited for high frequency measurements.

There are two principal types of thermopiles in use: (a) thin wire windings, and (b) thin film. The small connections and thermal elements of the thin film device provide a range of impedance and time constants. However, a good wire-wound thermopile is capable of Johnson noise limited performance (see Section 6.4.2).

Black absorber materials used with thermopiles are most often assumed to be spectrally flat. However, this is not exactly true but is difficult to measure quantitatively. Some recent work has, however, contributed to the determination of spectral properties of nominally black absorbers (Blevin and Geist, 1974a,b).

Also, it should be noted that like other thermal detectors, the thermopile can be calibrated by electrical techniques which can produce the same heating effect produced by radiant energy.

6.6.3 Pyroelectric Detectors

Pyroelectric detectors are a relatively recent addition to the family of thermal detectors. The *pyroelectric effect* is a change in the surface charge of a material in response to a temperature change ΔT. This is shown in Fig. 6.23.

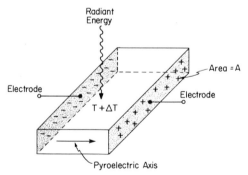

Fig. 6.23   Orientation of pyroelectric axis and electrode geometry in a pyroelectric detector with edge electrodes.

TABLE 6.3 Properties of Some Pyroelectric Detector Materials at 300 K[a]

Material	Pyroelectric coefficient p (C cm^{-2} K^{-1})	Dielectric constant	Specific heat (J gm^{-1} K^{-1})	Thermal conductivity K (W cm^{-1} K^{-1})	Density (gm cm^{-3})
Tourmaline	4 × 10^{-10}				
BaTiO$_3$	2 × 10^{-8}	160 (∥ polar axis) 4100 (⊥ polar axis)	0.5	9 × 10^{-3}	6.0
Triglycine sulphate (TGS) (NH$_2$CH$_2$COOH)$_3$·H$_2$SO$_4$	(2–3.5) × 10^{-8}	25–50	0.97	6.8 × 10^{-3}	1.69
Li$_2$SO$_4$·H$_2$O	1.0 × 10^{-8}	10	~0.4	17 × 10^{-3}	2.05
LiNbO$_3$	4 × 10^{-9}	30 (∥ polar axis) 75 (⊥ polar axis)			4.64
LiTaO$_3$	6 × 10^{-9}	58			
SbSI	2.6 × 10^{-7}	10^4	0.29		8.2
NaNO$_2$	1.2 × 10^{-8}	8.0	0.96		2.1

[a] After Putley (1970).

The surface charge variations produce a current in an external circuit given by

$$i_p = p_T A \, dT/dt, \qquad (6.51)$$

where p_T is the *pyroelectric coefficient* at temperature T, A the electrode surface area, and dT/dt the rate of temperature change. The properties of some pyroelectric detector materials (Putley, 1970) are shown in Table 6.3.

Because the current i is proportional to dT/dt, the pyroelectric detector does not respond to constant radiant energy. If the signal energy is unmodulated, it must be chopped. However, the responsivity for high modulation frequencies is much higher than in other thermal detectors. Like other thermal detectors the responsivity of the pyroelectric device is essentially independent of wavelength. Also, it is not necessary to cool the detector to achieve its optimum performance. Finally, the pyroelectric detector can be electrically calibrated by using the heating effect of a known electrical current.

To determine the responsivity, frequency response, and NEP of the pyroelectric detector we must examine both its thermal properties and its electrical properties. Its thermal properties are qualitatively similar to those of the bolometer and thermopile (Sections 6.6.1 and 6.6.2). The energy balance equation is

$$C_T [d/dt](\Delta T) + K \, \Delta T = \Delta\Phi(t), \qquad (6.52)$$

where $\Delta\Phi(t)$ is the absorbed radiant power, C_T the heat capacity, K the heat transfer coefficient, and $\Delta\Phi(t)$ assumed to be $\Delta\Phi(t) = \Delta\Phi_0 \exp(2\pi i f t)$. Then the solution of (6.52) is

$$|\Delta T| = \varepsilon \Delta\Phi_0'/[K^2 + (2\pi f C_T)^2]^{1/2}, \qquad (6.53)$$

where ε is the emissivity. This result assumes that the radiant power is absorbed uniformly throughout the material. The *thermal time constant* from (6.53) is defined to be $\tau_T = C_T/K$.

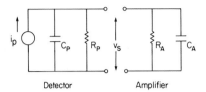

Detector   Amplifier

Fig. 6.24   Equivalent circuit for pyroelectric detector and amplifier combination. In the
amplifier R_A = amplifier equivalent resistance
and C_A = amplifier equivalent capacitance.

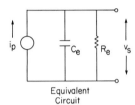

Equivalent
Circuit

The electrical characteristics of the pyroelectric detector are found from
the equivalent circuit shown in Fig. 6.24. The pyroelectric element is represented as a capacitor C_p in parallel with a resistor R_p. The current source
i_p is that given by (6.51). If $\Delta T = |\Delta T| \exp(2\pi i f t)$, then $i_p = 2\pi f p_T A \Delta T$. The
voltage at the output of the equivalent circuit is

$$v_s = i_p R_e / [1 + (2\pi f \tau_E)^2]^{1/2}, \tag{6.54}$$

where $\tau_E = R_e C_e$ is the *electrical time constant*.
From (6.54) and $i_p = 2\pi f p_T A \Delta T$ we find that

$$v_s = 2\pi f p_T A R_e \Delta T / [1 + (2\pi f \tau_E)^2]^{1/2}, \tag{6.55}$$

and by using (6.53) and the definition of the voltage responsivity as $R = v_s/\Delta\Phi_0'$ we find that

$$R = [2\pi f \varepsilon p_T A R_e / K] / [1 + (2\pi f \tau_E)^2]^{1/2} [1 + (2\pi f \tau_T)^2]^{1/2}. \tag{6.56}$$

It is obvious from (6.56) that the pyroelectric detector has a responsivity
which involves two time constants, the thermal time constant τ_T and the
electrical time constant τ_E.
Figure 6.25 shows the voltage responsivity of a typical commercially
available pyroelectric detector. The vertical dashed line is the modulation
frequency corresponding to the thermal time constant, approximately $f_T = 1/2\pi\tau_T$. Below this frequency the responsivity drops rapidly because, as
shown by (6.53), the temperature at these low frequencies is essentially un-
changing.

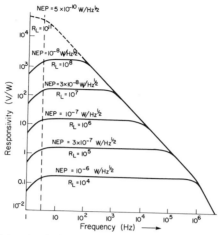

Fig. 6.25 Responsivity of typical commercially available pyroelectric detector as a function of signal modulation frequency, where R_L is the load resistance $(=R_c)$. (Courtesy Molectron Corp.)

On the other hand, at frequencies above $f_T = 1/2\pi\tau_T$ but below $f_E = 1/2\pi\tau_E$ the responsivity is constant as shown by (6.56). This is the useful working frequency range of the device. At frequencies above the electrical cutoff frequency, the responsivity drops as shown. Notice that since this cutoff frequency is determined by $\tau_E = R_e C_e$, the cutoff frequency varies with load resistance $R_e = R_L$. However, since the signal voltage across the load resistor (within the working frequency range below $f_E = 1/2\pi\tau_E$) is $v_s = i_p R_e$, the maximum responsivity is obtained at the largest values of load resistance. In the pyroelectric detector there is therefore an important tradeoff between responsivity and frequency response. Maximum frequency response and maximum responsivity are not obtained simultaneously.[†]

6.6.4 Laser Calorimeters

A *calorimeter* is any instrument in which the incident radiant energy is converted to heat and measured as such. In principle then, all thermal detectors can be classified as calorimeters. In practice, however, the term finds most use in describing detectors for measuring laser power or energy (Gunn, 1973, 1974; West and Churney, 1970; Watt, 1973; Franzen and Schmidt, 1976; West *et al.*, 1972; Thacher, 1976).

Figure 6.26 shows two calorimeter configurations which are used. Incident radiant energy, in Figure 6.26a, raises the temperature of the disk. The temperature change in the disk changes the temperature of the conducting

[†] Recent developments are surveyed by Liu and Long (1978).

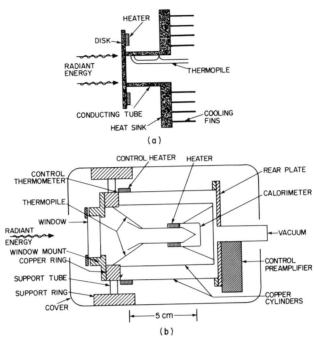

Fig. 6.26 Configurations of laser calorimeters. (a) Disk calorimeter; (b) cone calorimeter. (After West *et al.*, 1972.)

tube to which a conventional contact thermocouple or thermopile is connected. The voltage produced by the thermopile is the output signal of the calorimeter. Figure 6.27b shows a more detailed diagram of a calorimeter for laser measurements. In both cases the heater is used to calibrate the calorimeter.

Calorimeters for laser measurements can be classified as *isoperibol* or *conduction* devices. The isoperibol calorimeter is thermally isolated from its surroundings by minimizing all heat transfer by conduction, convection, or radiation. A conduction calorimeter, on the other hand, uses a thermally conducting path between the radiant energy absorber and heat sink. When constant radiant power is received, a thermal gradient is established along the conducting path. By measuring this gradient, which is proportional to power, the radiant power is determined.

Although Fig. 6.26 shows solid absorber calorimeters, it is also common to use water flow designs. The temperature rise of a constant water flow which is in thermal contact with the absorber provides a measurement of the radiant power absorbed.

It is desired to convert as much of the incident radiant energy to heat as possible. However, the high peak power of a laser pulse may damage a blackened surface. For this reason calorimeters are often designed in a cavity configuration with high reflectivity walls so that the beam must undergo multiple reflections. This also makes the effective absorptance of the calorimeter less sensitive to small changes of single surface reflectivity.

If a square pulse of radiant energy is incident on an absorbing surface, the peak temperature increase is

$$\Delta T = 2(\Phi/A)(t_{\text{p}}/\pi p c_{\text{T}} K)^{1/2}, \tag{6.57}$$

where Φ is the power absorbed, t_{p} the pulse width, ρ the density of the absorbing material, c_{T} the specific heat, and K the heat transfer coefficient.

The categories of laser calorimeter configuration are

(1) disk (as in Fig. 6.27a),
(2) cone,
(3) hollow sphere,
(4) bolometric, and
(5) volume absorption.

Types (1) and (5) are usually designed for partial absorption. Type (4) refers to any calorimeter which relies on a resistance change in the heated element and so is essentially identical in its detection mechanism to the bolometer examined in Section 6.6.1.

Many calorimeters require a window as shown in Fig. 6.26b. To avoid this problem it is possible to use a long entrance tube as shown in Fig. 6.27. This is the Q calorimeter used by the National Bureau of Standards for standard measurements of high power laser pulses. In this calorimeter, which is a volume absorption calorimeter [type (5)] the absorbing material is a neutral density glass (Schott NG-10) at 1.06 μm, the wavelength of the Nd:YAG laser and $M_{\text{G}}F_2$ single crystal material at 10.6 μm, the wavelength of the CO_2 laser.

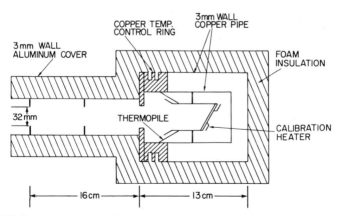

Fig. 6.27 Laser calorimeter using long entrance tube to avoid the use of a window. (After Franzen and Schmidt, 1976.)

TABLE 6.4 Specification of NBS Q Calorimeter
at 1.06 μm[a]

Specification	Valve
Aperture size	32 mm $\times$ 32 mm
Cooling constant	0.00265 sec^{-1}
Maximum energy injection time	40 sec
Random error (one standard deviation)	0.2%
Systematic error	2.3%
Energy range	0.4–15 J
Maximum energy density per pulse	3 J/cm^2
Maximum pulse repetition rate, where E is the energy density per pulse in J/cm^2	10/E pps
Minimum pulse width	20 nsec documented 0.2 nsec probable
Wavelength range	1.06 μm documented 1.2–0.5 μm probable

[a] After Franzen and Schmidt (1976).

As an example of the measurement characteristics of a pulsed laser calorimeter, Table 6.4 shows the specifications for the NBS device shown in Fig. 6.27.

REFERENCES

Bennett, W. (1956). *Proc. IRE* **44**, 609.
Blevin, W., and Geist, J. (1974a). *Appl. Opt.* **13**, 1171.
Blevin, W., and Geist, J. (1974b). *Appl. Opt.* **13**, 2212.
Budde, W., and Dodd, C. (1971). *Appl. Opt.* **10**, 2607.
Budde, W., and Kelly, P. (1971). *Appl. Opt.* **10**, 2612.
Davenport, W., and Root W. (1958). "Random Signals and Noise." McGraw–Hill, New York.
Eberhardt, E. H. (1968). *Appl. Opt.* **7**, 2037.
Emmons, R., Hawkins, S., and Cuff, K. (1975). *Opt. Eng.* **14**, 21.
Franzen, D., and Schmidt, L. (1976). *Appl. Opt.* **15**, 3115.
Gunn, S. (1973). *J. Phys. E* **6**, 105.
Gunn, S. (1974). *Rev. Sci. Instrum.* **45**, 936.
Jones, R. C. (1959a). *Proc. IRE* **47**, 1495.
Jones, R. C. (1959b). *Proc. IRE* **47**, 1481.
Kruse, P., McGlauchlin, L., and McQuistan, R. (1962). "Elements of Infrared Technology." Wiley, New York.
Levinstein, H. (1965). *Appl. Opt.* **4**, 639.
Liu, S. I., and Long, D. (1978). *Proc. IEEE* **66**, 14.
Lindley, W., Phelan, R., Wolfe, C., and Foyt, A. (1969). *Appl. Phys. Lett.* **14**, 197.
McKay, K. (1954). *Phys. Rev.* **94**, 877.
Melchior, H. (1973). *J. Lumin* **7**, 390.
Melchior, H., Fisher, M., and Arams, F. (1970). *Proc. IEEE* **58**, 1466.
McIntyre, R. (1970). *IEEE Trans. Electron. Dev.* **ED-17**, 347.
Oliver, B. (1965). *Proc. IEEE* **53**, 436.
Potter, R., Bennett, J., and Naugle, A. (1959). *Proc. IRE* **47**, 1503.

Putley, E. H. (1970). "Semiconductors and Semimetals" R. Willardson and A. Beer, editors New York. Academic Press, Vol. 5.

Putley, E. H. (1973). *Phys. Tech.* **4**, 202.

Richtmeyer, F., Kennard, E., and Cooper, R. (1968). "Introduction to Modern Physics," 5th ed. McGraw-Hill, New York.

Schwartz, M. (1970). "Information Transmission, Modulation, and Noise," 2nd ed. McGraw–Hill, New York.

Senturia, S., and Wedlock, B. (1975). "Electronic Circuits and Applications." Wiley, New York.

Smith, R. A. (1961). "Semiconductors." Cambridge Univ. Press, London and New York.

Smith, R., Jones, F., and Chasmar, R. (1968). "The Detection and Measurement of Infrared Radiation." Oxford Univ. Press, London and New York.

Sommer, A. H. (1968). "Photoemissive Materials." Wiley, New York.

Stevens, N. (1970). *In* "Semiconductors and Semimetals" (R. Willardson and A. Beer, eds.), Vol. 5. Academic Press, New York.

Thacher, P. (1976). *Appl. Opt.* **15**, 1815.

van der Ziel, A. (1956). "Noise." Prentice Hall, Englewood Cliffs, New Jersey.

Watt, B. (1973). *Appl. Opt.* **12**, 2373.

Webb, P., McIntyre, R., and Conradi, J. (1974). *RCA Rev.* **35**, 234.

West, E. D., and Churney, K. L. (1970). *J. Appl. Phys.* **41**, 2705.

West, E., Case, W., Rasmussen, A., and Schmidt, L. (1972). *J. Res. Nat. Bur. Std.* **76A**, 13.

Williams, B. F. (1969). *Appl. Phys. Letters* **14**(9), 273.

Zwerdling, S., Smith, R., and Theriault, J. (1968). *Infrared Phys.* **8**, 271.

7

Spectral Analyzers

7.1 INTRODUCTION

In the two previous chapters we have examined sources and detectors used in radiometry. Most radiometric measurement systems also employ filters, monochromators, or other components between the source and detector, with the choice of components depending on the application. In this chapter we examine the properties of absorption filters, interference filters, prism monochromators, grating monochromators, Fourier spectrometers, and multichannel spectrometers. The combination of these components with sources and detectors to form complete radiometric systems is described in Chapters 8 and 9.

7.2 FILTERS

In the most general sense a filter is any material or device which changes the spectral distribution of a beam of radiant energy. In practice, however, the term "filter" is used in a narrower sense to refer to glass filters, interference filters, and similar components, and excludes the atmosphere, monochromators, and other devices which are treated separately. Spectral filtering techniques include absorption (Scharf, 1965; Wyszecki and Stiles, 1967), interference (Baumeister, 1965; Heavens, 1955; Vasicek, 1960; Walter, 1956; Holland, 1960), reflection, refraction, scattering (Christiansen, 1884, 1885), and polarization (Lyot, 1933; Evans, 1949a,b), but the most widely used filters are those involving absorption or interference.

7.2.1 Absorption Filters

Common absorption filters are made of glass, dyed gelatin, semiconducting materials such as germanium, or dyes in liquid cells. Figure 7.1 shows a

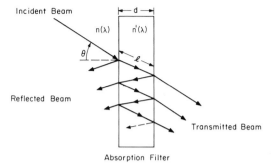

Fig. 7.1 Diagram of absorption filter showing the multiple reflections which occur at the surfaces.

diagram of a beam of radiant energy incident on an absorption filter at an angle θ from the normal. The incident and transmitted radiant power are related by

$$\Phi_{\lambda\tau} = \tau(\lambda)\Phi_{\lambda}, \qquad (7.1)$$

where Φ_{λ} is the spectral radiant power of the incident beam, $\Phi_{\lambda\tau}$ the spectral radiant power of the transmitted beam, and $\tau(\lambda)$ the spectral transmittance of the filter.

To determine the spectral transmittance of the absorption filter we follow the beam through the filter as shown in Fig. 7.1. At the first surface a part of the beam is transmitted and a part reflected. The transmitted part is attenuated in passing from the first to the second surface. At the second surface a part is transmitted and a part reflected, and so on. A calculation of the filter transmittance must therefore include both the surface effects and the internal absorption effects.

Consider the incident beam. The part which is transmitted through the first surface is $[1 - \bar{\rho}(\lambda)]\Phi_{\lambda}$, where $\bar{\rho}(\lambda)$ is the spectral reflectance of the surface. The part of this transmitted beam which reaches the second surface is $[1 - \bar{\rho}(\lambda)]\tau_i(\lambda)\Phi_{\lambda}$, where $\tau_i(\lambda)$ is the internal spectral transmittance. On reaching the second surface, the part that is transmitted is then

$$[1 - \bar{\rho}(\lambda)]^2\tau_i(\lambda)\Phi_{\lambda}.$$

If we add all of the transmitted beams to find the total transmitted beam to the right of the filter, we find

$$\Phi_{\lambda\tau} = [1 - \bar{\rho}(\lambda)]^2\tau_i(\lambda)\Phi_{\lambda} + [1 - \bar{\rho}(\lambda)]^2\tau_i(\lambda)[\bar{\rho}(\lambda)\tau_i(\lambda)]^2\Phi_{\lambda} + \cdots. \qquad (7.2)$$

From (7.2), (7.1), and the series relation $1 + x + x^2 + \cdots = 1/(1 - x)$ we find that the overall spectral transmittance of the absorption filter is

$$\tau(\lambda) = \Phi_{\lambda\tau}/\Phi_{\lambda} = [1 - \bar{\rho}(\lambda)]^2\tau_i(\lambda)/[1 - \bar{\rho}^2(\lambda)\tau_i^2(\lambda)]. \qquad (7.3)$$

Equation (7.3) assumes that the transmitted beams are incoherent. This will be true when $d \gg \lambda^2/\Delta\lambda$, where d is the filter thickness and $\Delta\lambda$ the spectral bandwidths. The thickness of an absorption filter and the spectral bandwidth of nonlaser sources will usually meet this requirement. However, the bandwidth $\Delta\lambda$ for a laser source can be small enough so that interference effects can take place between the transmitted beams. In that case (7.3) is invalid.

For a homogeneous isotropic filter material the internal spectral transmittance can be expressed in alternative ways as

$$\tau_i(\lambda) = e^{-a(\lambda)l} = 10^{-a'(\lambda)l} = 10^{-a'_c(\lambda)cl}, \tag{7.4}$$

where $\tau_i(\lambda)$ is the internal spectral transmittance and $a(\lambda) = 2.3a'(\lambda)$ the *spectral absorption coefficient* defined in Section 2.7.3 and in Section 3.5.1. The quantity $a'_c(\lambda) = a'(\lambda)/c$ is the *molar spectral absorption coefficient* and is appropriate for describing the internal transmittance of a liquid absorber of concentration c. The relation $\tau_i(\lambda) = e^{-a(\lambda)l}$ is *Bouguer's law* and the relation $\tau_i(\lambda) = 10^{-a'_c(\lambda)cl}$ is *Beer's law*. Beer's law is found to be experimentally valid except at high concentrations. Notice that at normal incidence ($\theta = 0$) the path length l is equal to the filter thickness d.

At normal incidence ($\theta = 0$) the spectral reflectivity of each surface is found from Section 3.4 to be

$$\bar{\rho}(\lambda) = [n'(\lambda) - n(\lambda)]^2/[n'(\lambda) + n(\lambda)]^2, \tag{7.5}$$

where $n'(\lambda)$ and $n(\lambda)$ are the indices of refraction shown in Fig. 7.1. The overall spectral transmittance of the absorption filter, at normal incidence, is then found by using (7.4) and (7.5) in (7.3) with $l = d$. For angles other than normal, the angle dependence of $\bar{\rho}(\lambda)$ must be considered.

In many cases $\bar{\rho}(\lambda)\tau_i(\lambda) \ll 1$ so that (7.3) simplifies to

$$\tau(\lambda) \cong [1 - \bar{\rho}(\lambda)]^2 \tau_i(\lambda) \tag{7.6}$$

which is a simple product of the internal and surface transmittances. Also, it is frequently true that the surface transmittance $[1 - \bar{\rho}(\lambda)]$ is not strongly dependent on angle so that (7.6) is approximately correct for beams of finite but small cone angle. In practice it is best, whenever possible, to measure the transmittance of a filter in the same geometry in which it will be used rather than rely entirely on calculations.

Data on the index of refraction and absorption coefficient or attenuation coefficient of filter materials are available in the literature (Wyszecki and Stiles, 1967). Manufacturers data for glass and IR semiconductor filter materials are frequently presented in the form of spectral transmittance $\tau(\lambda)$ for a standard filter thickness d_s. The spectral transmittance for other thicknesses can be found from (7.6) and (7.4) with $l = d$ to be $\tau(\lambda) =$

$\tau_s(\lambda)\exp[-(d-d_s)a(\lambda)]$, where the subscript s represents data for the standard thickness. Calculation of $\tau(\lambda)$ from $\tau_s(\lambda)$ therefore requires information on $a(\lambda)$, the spectral absorption coefficient.

Some examples of filter transmittance data are shown in Fig. 7.2. In practice the choice of a filter for a given application will depend on a number of properties such as (1) strength, (2) hardness, (3) available sizes, (4) thermal

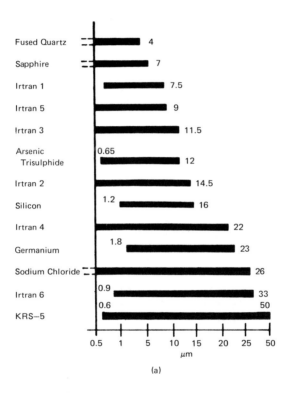

(a)

Fig. 7.2 Typical spectral transmittance properties of absorption filters. (a) The transmittance ranges (between short wavelength cuton and long wavelength cutoff) of 2 mm thicknesses of IRTRAN materials and other infrared optical materials. Limits are set at approximately 10% transmittance. (Courtesy, Eastman Kodak Company.) (b) Absorption filter materials used in the near UV, visible, and near IR regions of window glass of 300 mm (---), crown glass (—·—·), dense flint glass of 486 mm (· · ·), fused quartz of 2.69 mm (————), and distilled water in a quartz cell with 1-mm walls and 10-mm water column (——). (After Wyszecki and Stiles, 1967.) (c) Filters which transmit visible energy but absorb UV energy. These filters are 2.5 mm thicknesses of silicate glasses. (Courtesy, Hoya Corporation.) (Figures 7.2b and 7.2c on following page.)

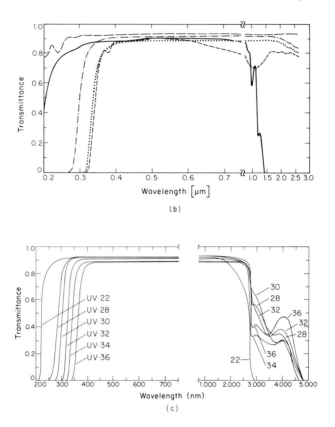

(b)

(c)

coefficients of expansion, (5) solubility, (6) density, (7) emissivity, (8) scattering, and (9) thermal shock resistance.

7.2.2 Interference Filters

In an interference filter, the transmittance at each wavelength is determined by constructive and destructive interference between waves (Baumeister, 1965; Heavens, 1955; Vasíček, 1960; Wolter, 1956; Holland, 1960). Interference filters typically provide very narrow passbands or very sharp transition between high and low spectral transmittance with high transmittance at the desired wavelengths and low transmittance at undesired wavelengths. Since interference filters are thin film devices, they must be deposited on a substrate and the spectral transmittance of the supporting substrate will contribute to the overall filter transmittance.

There are four basic types of construction for interference filters. These are the

(1)   Fabry–Perot filter (metal–dielectric–metal),
(2)   all-dielectric filter,
(3)   all-dielectric multicavity filter, and
(4)   induced transmission filter.

In addition, the spectral characteristics of interference filters can be classified as (a) narrowband, (b) wideband, (c) short wave pass, and (d) long wave pass. Supporting substrates in the near UV and visible are usually quartz or glass while in the infrared they include germanium, silicon, Irtran II, Irtran IV, Irtran V, sapphire, indium arsenide, indium antimonide, and arsenic trisulfide.

The basic principle of operation of the interference filter (Heavens, 1955) can be illustrated by considering the simple *Fabry–Perot filter* in Fig. 7.3. A beam of radiant energy is incident on a thin film structure composed of two thin metallic partially transmitting layers separated by a dielectric spacer layer of optical thickness $\lambda/2$. The wavelength λ is the wavelength at which maximum filter transmittance is desired. The transmitted beam components shown in Fig. 7.3 will be in phase if the optical path length difference between them is an integral number of wavelengths or when

$$2n'd \cos \theta' = m\lambda, \tag{7.7}$$

where n' is the index of refraction of the dielectric, d the dielectric layer thickness, θ' the angle in the dielectric, and m an integer representing the order of interference. From (7.7), when $\theta = 0$ the peak transmittance occurs when $n'd = m(\lambda/2)$, that is, when the optical thickness is an integral number of half wavelengths. The longest wavelength at which a peak occurs is $\lambda = 2n'd$ but harmonic peaks also occur at $\lambda = 2n'd/m$. For wavelengths at

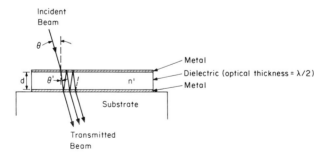

Fig. 7.3   Fabry–Perot interference filter with dielectric spacer layer of optical thickness $n'd = \lambda/2$.

which the optical thickness is not an integral number of half wavelengths, an out-of-phase condition exists and the incident radiant energy is for the most part reflected. Notice from (7.7) that the peak wavelength depends on the angle of incidence through θ'.

Although the Fabry–Perot structure was the first interference filter, it has the disadvantages that the absorption in the metal layers limits the peak transmittance, the filter bandwidth cannot easily be controlled, and transmission occurs at the harmonic wavelengths. An *all-dielectric filter* as shown in Fig. 7.4b provides higher peak transmittance and control of bandwidth. The alternating high and low index dielectric layers of optical thickness $n'd = \lambda/4$ on each side of the $\lambda/2$ spacer layer perform the function of the reflecting metal layers which were used in the Fabry–Perot structure. The thickness of the spacer layer still controls the peak wavelength while the number of $\lambda/4$ layers controls the bandwidth. Since only dielectrics are used the peak transmittance can be high. A further modification which makes the filter transmittance more uniform over the passband and provides a sharper cutoff is the *all-dielectric multicavity filter* shown in Fig. 7.4c. Here a coupling layer is used between two cavities each having the structure shown in Fig. 7.4b. The advantage of using more than one cavity is shown in Fig. 7.5. As the number of cavities increases the passband becomes more rectangular and the cutoff is sharper. The fourth filter structure is the *induced transmission filter* shown in Fig. 7.4d. This filter is a hybrid which combines the high transmittance and bandwidth control of the all dielectric filter with

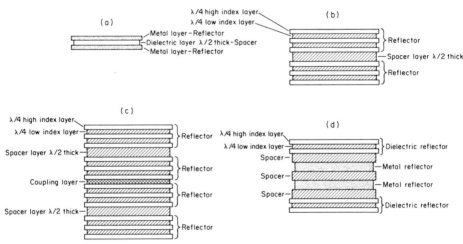

Fig. 7.4 Construction of interference filters; (a) the basic Fabry–Perot metal dielectric metal (MDM) filter; (b) One-cavity all-dielectric filter; (c) two-cavity all-dielectric filter; (d) induced transmission filter.

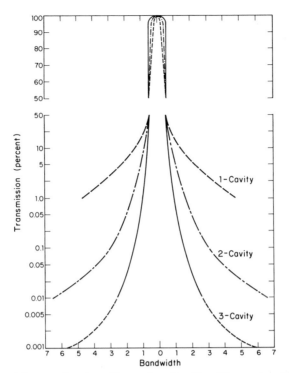

Fig. 7.5 Band shape as a function of the number of cavities. All curves have been normalized so that the peak transmittance is 100% (Courtesy Corion Corp.)

the long wavelength absorbing characteristics of the metals. The dielectric stacks and spacer layers act as an antireflection system for the metal layers at the peak wavelength of the filter. All of these structures can be considered to be extensions of the simple Fabry–Perot structure shown in Fig. 7.3.

Due to the complex structure of these multilayer filters, an analytical treatment would be lengthy and complex. For further details, we refer the reader to several excellent books on this subject (Heavens, 1955; Vasicek, 1960; Baumeister, 1965).

As indicated by (7.7), the wavelength of peak transmittance shifts to shorter wavelengths when θ' increases. The shift can be calculated from (7.7) for the Fabry–Perot structure by using Snell's law of refraction $n \sin \theta = n' \sin \theta'$. The value of $\cos \theta'$ is then found from

$$\cos \theta' = (1 - \sin^2 \theta')^{1/2} = [1 - (n/n')^2 \sin^2 \theta]^{1/2}. \tag{7.8}$$

From (7.7) and (7.8) the peak wavelength for a filter in air ($n = 1$) is

$$\lambda_\theta = \lambda_0 [1 - (\sin \theta/n')^2]^{1/2}, \tag{7.9}$$

where λ_θ and λ_0 are the peak wavelengths at incident angle θ and at normal incidence. When the other forms of filters shown in Fig. 7.4 are used, the wavelength shift is primarily a function of the index of refraction of the spacer layer. For a spacer layer with higher index than the adjacent dielectric layers, the effective index of the total filter is

$$n^* = (n_H n_L)^{1/2}, \tag{7.10}$$

where n_H and n_L are the indices of refraction of the high and low index layers. On the other hand, for a spacer layer with lower index

$$n^* = n_L / [1 - (n_L/n_H) + (n_L/n_H)^2]^{1/2}. \tag{7.11}$$

For these filters n^* replaces n' in (7.9). When a convergent or divergent beam is used, then a weighted average of (7.9) over the range of angles gives the effective peak wavelength shift.

Interference filters are used in combination with absorption filters to reduce out-of-band energy and to reduce the transmission at harmonic wavelengths. When possible the substrate is chosen to perform this absorption function. Other practical considerations in the choice of interference filter are (1) temperature effects on peak wavelength, (2) mechanical durability, (3) temperature durability, and (4) humidity effects.

7.3 MONOCHROMATORS

A *monochromator* is a tunable device which is designed to isolate a narrow wavelength band of radiant energy. It is composed of (1) a spectral dispersing element (prism or grating), (2) focusing lenses or mirrors, and (3) entrance and exit slits. The principal distinction between monochromator types is in the choice of a prism or a grating as the dispersing element.

7.3.1 Prism Monochromators

A simple prism monochromator is shown in Fig. 7.6. Radiant energy passing through the entrance slit is collimated by the first lens (or mirror)

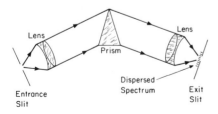

Fig. 7.6 Configuration of a simple prism monochromator.

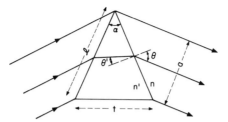

Fig. 7.7 Dispersion by a prism at minimum deviation. The width of the emergent beam is a and the length of the prism base is t.

and dispersed by the prism into its wavelength components. The dispersed energy is focused by the second lens (or mirror) onto the plane of the exit slit. Since the dispersed spectrum appears in the plane of the exit slit, the desired narrow wavelength band can, in principle, be isolated by sequentially adjusting either the position of the exit slit or the prism angle.

Figure 7.7 shows the prism in more detail. The collimated beam is refracted by the prism and emerges at an angle θ from the normal to the second surface. For each wavelength the angle θ is different. To find the rate of change of angle with wavelength, we calculate the *angular dispersion* $d\theta/d\lambda$ which can be expressed as

$$d\theta/d\lambda = (d\theta/dn)(dn/d\lambda). \tag{7.12}$$

In (7.12), $d\theta/dn$ depends only on geometry and $dn/d\lambda$, which is the *dispersion*, is characteristic of the prism material. The quantity $d\theta/dn$ is obtained from Snell's law of refraction $n \sin\theta = n' \sin\theta'$. In air ($n = 1$) and at minimum deviation ($\theta' = $ const.), we find $d\theta/dn = \sin\theta'/\cos\theta$ at each face. The total for both faces is then $2\sin\theta'/\cos\theta$. From Fig. 7.7 we see that $\alpha = 2\theta'$ and so

$$d\theta/dn = 2l\sin(\alpha/2)/l\cos\theta = t/a, \tag{7.13}$$

where t is the prism base length and a the width of the emerging beam. Using (7.13) in (7.12) we conclude therefore that

$$d\theta/d\lambda = (t/a)(dn/d\lambda). \tag{7.14}$$

This expression for the angular dispersion of a prism permits us to find an important fundamental parameter of a monochromator, the resolving power. The *resolving power* R of a monochromator is defined as

$$R = \lambda/d\lambda, \tag{7.15}$$

where λ is the mean wavelength and $d\lambda$ the wavelength difference of two spectral lines which are just resolved. The theoretical limit on resolving power is diffraction as described in Section 3.3.2. For a rectangular aperture

TABLE 7.1 Refractive Indices and Dispersions for Several Common Types of Glass in the Visible Region of the Optical Spectrum[a,b]

Wavelength λ (nm)	Telescope crown		Borosilicate crown		Barium flint		Vitreous quartz	
	n	$-dn/d\lambda$	n	$-dn/d\lambda$	n	$-dn/d\lambda$	n	$-dn/d\lambda$
C 656.3	1.52441	0.35×10^{-4}	1.50883	0.31×10^{-4}	1.58848	0.38×10^{-4}	1.45640	0.27×10^{-4}
643.9	1.52490	0.36×10^{-4}	1.50917	0.32×10^{-4}	1.58896	0.39×10^{-4}	1.45674	0.28×10^{-4}
D 589.0	1.52704	0.43×10^{-4}	1.51124	0.41×10^{-4}	1.59144	0.50×10^{-4}	1.45845	0.35×10^{-4}
533.8	1.52989	0.58×10^{-4}	1.51386	0.55×10^{-4}	1.59463	0.68×10^{-4}	1.46067	0.45×10^{-4}
508.6	1.53146	0.66×10^{-4}	1.51534	0.63×10^{-4}	1.59644	0.78×10^{-4}	1.46191	0.52×10^{-4}
F 486.1	1.53303	0.78×10^{-4}	1.51690	0.72×10^{-4}	1.59825	0.89×10^{-4}	1.46318	0.60×10^{-4}
G' 434.0	1.53790	1.12×10^{-4}	1.52136	1.00×10^{-4}	1.60367	1.23×10^{-4}	1.46690	0.84×10^{-4}
H 398.8	1.54245	1.39×10^{-4}	1.52456	1.26×10^{-4}	1.60870	1.72×10^{-4}	1.47030	1.12×10^{-4}

[a] After Jenkins and White (1975).
[b] Unit of dispersion, nm^{-1}.

the diffraction-limited angular separation of two lines which are just resolved is approximately $d\theta = \lambda/a$, where a is the beam width shown in Fig. 7.7. Using this in (7.14) we find that the maximum resolving power of a prism, which we represent by R_0, is

$$R_0 = a\, d\theta/d\lambda = t\, dn/d\lambda, \tag{7.16}$$

and the actual resolving power R will be in the range $0 < R < R_0$. From relation (7.16) we see that in principle a prism of larger base length t and of higher material dispersion will yield higher resolving power. Typical values (Jenkins and White, 1957) of dispersion for prism materials used in the near UV and visible are shown in Table 7.1. Similar data for IR prism materials (Smith *et al.*, 1968) is shown in Fig. 7.8. The theoretical *wavelength resolution* of a prism monochromator is

$$\Delta\lambda = \lambda/R_0 = \lambda/(t\, dn/d\lambda), \tag{7.17}$$

and so is a function of wavelength. It should be noted that the theoretical maximum resolving power R_0 in Eq. (7.16) assumes that the entrance and exit slit have zero width. As the slit widths are increased, the resolving power R drops below the theoretical maximum R_0. However, the slit widths must have some finite size to transmit adequate radiant energy as discussed in Section 8.3.1.

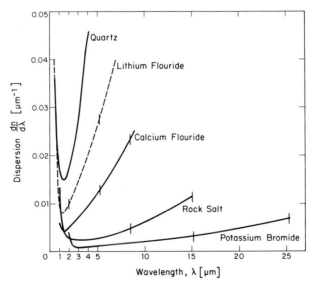

Fig. 7.8 Dispersion of IR prism materials used in prism monochromators. The vertical lines indicate the limits of typical spectral regions within which the prism would be used.

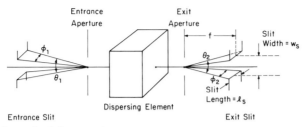

Fig. 7.9 Generalized diagram of a monochromator. The dispersing element may be a prism or a grating.

Transmission of radiant energy is a second fundamental property of a monochromator which involves both geometrical and spectral considerations. Consider the generalized diagram of a monochromator shown in Fig. 7.9. The dispersing element shown may be either a prism or a grating. In either case the definition of radiance and its conservation along a ray enable us to express the spectral radiant power passing through the exit slit as

$$\Phi_\lambda = L_\lambda G\tau(\lambda) = L_\lambda A \Omega \tau(\lambda) \tag{7.18}$$

as shown in Section 3.2.4.7. Here L_λ is the spectral radiance at the entrance slit, G the geometrical extent of the monochromator, A the exit aperture area, Ω the solid angle subtended by the exit slit at the exit aperture, and $\tau(\lambda)$ the spectral transmittance of the monochromator between entrance and exit slit. As shown in Fig. 7.9 the solid angle is $\Omega = \theta_2 \phi_2$, where ϕ_2 is the angular length of the exit slit. Also, from the definition of resolving power R given in (7.15), which applies to both prisms and gratings, we find that the slit width angle θ_2 is $\Delta\lambda \, (d\theta/d\lambda) = (\lambda/R)(d\theta/d\lambda)$. Using these in (7.18) leads to the relation

$$\Phi_\lambda = L_\lambda A \phi_2 (\lambda/R)(d\theta/d\lambda)\, \tau(\lambda) \quad (\text{W } \mu\text{m}^{-1}) \tag{7.19}$$

for the spectral radiant power passing through the exit slit of a monochromator. The total power transmitted in a wavelength interval is $\Phi = \Phi_\lambda \, d\lambda$ and this is found from (7.15) and (7.19) to be

$$\Phi = \Phi_\lambda \, d\lambda = L_\lambda A \phi_2 (\lambda/R)^2 \, (d\theta/d\lambda)\, \tau(\lambda). \quad (\text{W}) \tag{7.20}$$

As shown by (7.19) and (7.20) the radiant power (spectral or total) transmitted by a monochromator is inversely related to the resolving power R. These simple relations are based on geometrical optics which is valid when the resolving power R is much less than R_0. Otherwise the right-hand sides of (7.19) and (7.20) must be multiplied by the diffraction correction factor K_d shown in Fig. 7.10 (Jacquinot, 1954). After considering the grating monochromator in Section 7.3.2, we shall compare the radiant energy transmission of these two monochromators and shall find that, for equal resolving power

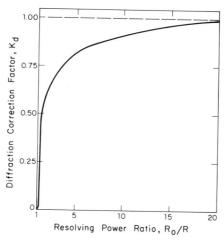

Fig. 7.10 Diffraction correction factor K_d for the radiant power transmitted by a monochromator (prism or grating). (After Jacquinot, 1954.)

R and with certain other assumptions, the grating monochromator is superior to the prism monochromator (Jacquinot, 1954).

For a prism monochromator the spectral radiant power transmitted through the exit slit can now be found. Notice that the area of the exit aperture is $A = ha$, where h is the prism height and a the beam width shown in Fig. 7.7. Also, the area of the base of the prism is $A_p = th$ and the angular length of the exit slit is $\phi_2 = l_s/f$ as shown in Fig. 7.9. With these relations and the angular dispersion (7.14) we see that for the prism monochromator (7.19) is

$$\Phi_\lambda = L_\lambda A_p (l_s/f)(\lambda/R)(dn/d\lambda)\,\tau(\lambda), \qquad (7.21)$$

where L_λ is the spectral radiance at the entrance slit, A_p the area of the base of the prism ($= th$), l_s the exit slit length, f the focal length on the output side of the monochromator, λ the wavelength, R the resolving power ($R \leq R_0$), $dn/d\lambda$ the dispersion of prism material, and $\tau(\lambda)$ the spectral transmittance of the monochromator.

A practical consideration in using a monochromator is the presence of stray energy at the exit slit. This unwanted energy arises from surface reflections and from scattering by dust and imperfections in the optical components within the monochromator. To reduce this effect and improve the spectral purity it is common to use a *double monochromator* in which the exit slit of the first becomes the entrance slit of the second. Two prism monochromators can be combined to give subtractive or additive dispersion, the former giving more uniform attenuation of all rays within the beam.

7.3.2   Grating Monochromators

A grating monochromator is composed of a spectral dispersing element, focusing lenses or mirrors, and entrance and exit slits. The spectral dispersing element is a diffraction grating. A grating monochromator has the general configuration shown in Fig. 7.9 with the focusing lenses or mirrors located in or near the entrance and exit apertures. When reflecting elements are used, the optical path between the entrance and exit slits will be folded.

The fundamental principle in operating a diffraction grating (Born and Wolf, 1964) is illustrated in Fig. 7.11. A collimated beam of radiant energy is incident on the grating surface. Each groove in the grating diffracts the incident energy. The angle θ of the diffracted beam from the grating is determined by the condition that the components from the different grooves add up in phase with each other. This will occur when the optical path difference between the diffracted components from adjacent grooves is an integral multiple of a wavelength, that is when $P_1 - P_2 = m\lambda$, where P_1 and P_2 are shown in Fig. 7.11 and m is an integer. Examination of the geometry in Fig. 7.11 will show that $P_1 = b \sin i$ and $P_2 = \pm b \sin \theta$, where the minus sign is used when θ and i are on opposite sides of the normal (as shown) and the plus sign when they are on the same side. With this understanding,

$$P_1 - P_2 = b(\sin i \pm \sin \theta) = m\lambda \qquad (7.22)$$

where b is the groove spacing, i the angle of incidence, θ the angle of diffraction, m the order of diffraction (an integer), and λ the wavelength. Equation (7.22) is the fundamental *grating equation*. The integer m is called the *order*. When $m = 0$, the grating simply acts as a mirror and all wavelengths are superimposed at the same angle. However, when $m \neq 0$, (7.22) shows that a beam incident at a given angle i will be diffracted into an angle θ which depends on the wavelength. This diffraction phenomenon provides the angular

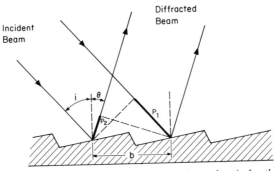

Fig. 7.11   Enlarged section of grating surface showing the total optical path length difference $P_1 - P_2$ between the diffracted components coming from adjacent grooves.

separation of wavelengths in a grating monochromator, and by sequentially varying the angle θ, the desired narrow wavelength band can be isolated.

Angular separation is described by the *angular dispersion* $d\theta/d\lambda$. This can be obtained by differentiating (7.22) with the incident angle i fixed. Then

$$d\theta/d\lambda = m/(b\cos\theta). \tag{7.23}$$

The *linear dispersion* is the product of (7.23) with the focal length f of the focusing element. Since m and b are not independent quantities we can solve (7.22) for m/b and insert this in (7.23) to find

$$\frac{d\theta}{d\lambda} = \frac{1}{\lambda}\frac{\sin i \pm \sin\theta}{\cos\theta}, \tag{7.24}$$

which shows the important conclusion that, for a fixed length, the angular dispersion of a grating depends only on the angles of incidence and diffraction. In the important and common case of the Littrow mounting ($i = \theta$ on the same side of the normal), (7.24) reduces to $d\theta/d\lambda = (2/\lambda)\tan\theta$ thus showing the increase in angular dispersion which can be obtained when the grating is used at large angles θ.

The *resolving power* of a monochromator was defined in (7.15) to be $R = \lambda/d\lambda$. To find the resolving power of the grating monochromator, we consider another view of the grating shown in Fig. 7.12a. The diffracted beam at angle θ has width a. Diffraction will spread the emerging beam into an angle $d\theta = \lambda/a$. We can then write the theoretical resolving power for the diffraction limited beam as

$$R_0 = \lambda/d\lambda = (\lambda/d\theta)(d\theta/d\lambda) = a\,d\theta/d\lambda, \tag{7.25}$$

which is actually the same relation as was found for the prism monochromator of emerging beam width a. Notice that the beam width a is related to the

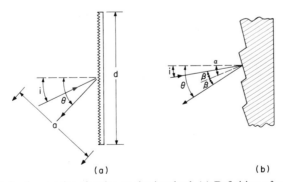

(a)   (b)

Fig. 7.12 Reflection grating showing angles involved. (a) Definition of angles; (b) corresponding angles for a magnified portion of a blazed grating. The blaze angle is α.

total grating width by $a = d \cos \theta$ so that when (7.24) (with the plus sign) is used in (7.25) the diffraction limited resolving power is

$$R_0 = (d/\lambda)(\sin i + \sin \theta), \tag{7.26}$$

where d is the total grating width. Again, for the Littrow mounting $(i = \theta)$ this is $R_0 = (2d/\lambda) \sin \theta$. The maximum possible resolving power of a grating monochromator $(\sin \theta = 1)$ is then $R_0 = 2d/\lambda$.

If we now consider the *wavelength resolution* $\Delta\lambda$ for the diffraction limit, we find from (7.15) and (7.26) that

$$\Delta\lambda = \lambda/R_0 = \lambda^2/d(\sin i + \sin \theta) \tag{7.27}$$

with the minimum possible value $(\sin i + \sin \theta = 2)$ being $\Delta\lambda = \lambda^2/2d$. To indicate the magnitude of this minimum value of $\Delta\lambda$, when $d = 10$ cm we have $\Delta\lambda = 10^{-3}$ nm at $\lambda = 500$ nm and $\Delta\lambda = 5 \times 10^{-8}$ μm at $\lambda = 10$ μm. In practice, the wavelength resolution will be much worse than indicated by these numbers due to the

(1)   use of the grating at smaller angles i and θ,
(2)   finite slit widths required to pass adequate energy,
(3)   optical quality and aberrations of focusing and auxiliary optical components, and
(4)   vibrations and other mechanical effects.

Spectral data from grating monochromators are frequently presented in terms of *wavenumber* $\sigma = 1/\lambda$. The reason can be seen from the relation $\Delta\sigma = (-1/\lambda^2)\Delta\lambda$, which is found by differentiation. Then (7.27) leads to the simple form for the magnitude of the *wavenumber resolution*

$$|\Delta\sigma| = 1/a(\sin i + \sin \theta). \tag{7.28}$$

Notice that unlike (7.27), (7.28) is independent of wavelength and so the wavenumber resolution is a constant throughout the spectrum. The usual units of wavenumber σ are reciprocal centimeters.

In addition to the resolving power, a second fundamental property of a monochromator is the transmission of radiant energy. As in the case of the prism monochromator, this transmission depends on both geometrical and spectral effects. We showed in Section 7.3.1 that for any monochromator the spectral radiant power passing through the exit slit is

$$\Phi_\lambda = L_\lambda A \phi_2(\lambda/R)(d\theta/d\lambda)\tau(\lambda). \tag{7.29}$$

Since the diffracted beam shown in Fig. 7.12a emerges at an angle θ, the exit aperture which is matched to the beam size has area $A = A_g \cos \theta$, where A_g is the grating area. With this relation and the angular dispersion (7.24) the form of (7.29) for a grating is

$$\Phi_\lambda = L_\lambda A_g \phi_2(1/R)(\sin i \pm \sin \theta)\tau(\lambda). \tag{7.30}$$

Most modern radiometric measurements using grating monochromators employ a blazed grating, a section of which is shown in Fig. 7.12b. The angle α is the *blaze angle*. From this figure $i = \alpha - \beta$ and $\theta = \alpha + \beta$ so that $\sin i + \sin \theta = 2 \sin \alpha \cos \beta$. This has a maximum value when $\beta = 0$ or when $i = \theta = \alpha$, showing that the maximum radiant power Φ_λ is transmitted through the grating monochromator when the Littrow arrangement ($i = \theta$) is used and when the incident and diffracted beams are both at the blaze angle α. For this optimum arrangement the spectral radiant power (7.30) passing through the exit slit, using $\phi_2 = l_s/f$, can then be calculated from

$$\Phi_\lambda = L_\lambda A_g(l_s/f)(1/R)(2 \sin \alpha)\tau(\lambda), \tag{7.31}$$

where L_λ is the spectral radiance at the enrance slit, A_g the grating area, l_s the exit slit length, f the focal length on the output side of the monochromator, R the resolving power ($R \leq R_0$), α the blaze angle of the grating, and $\tau(\lambda)$ the spectral transmittance of the monochromator.

While the results (7.31) and (7.21) are separately important, it is also interesting to compare the grating and prism monochromators in radiometric transmission performance (Jacquinot, 1954). From (7.21) and (7.31) we obtain the ratio of transmitted spectral radiant powers as

$$P = \frac{\Phi_{\lambda p}}{\Phi_{\lambda g}} = \frac{A_p}{A_g} \frac{\tau_p(\lambda)}{\tau_g(\lambda)} \frac{\lambda (dn/d\lambda)}{2 \sin \alpha} \tag{7.32}$$

for the condition where the resolving powers R are equal. If we assume that the prism base area and grating area are equal ($A_g = A_p$) and that $\tau(\lambda)$ is the same for both instruments, then $P = \lambda (dn/d\lambda)/2 \sin \alpha$. At a typical blaze angle ($\alpha = 30°$) this ratio is $P = \lambda (dn/d\lambda)$. Figure 7.13 shows this performance ratio P for a variety of common prism materials. It can be seen that $P < 1$ under all

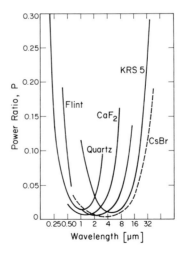

Fig. 7.13 Ratio P of transmitted spectral radiant powers for prism and grating monochromator (for blaze angle $\alpha = 30°$) for a variety of common prism materials. (After Jacquinot, 1954.)

conditions. This shows that the prism monochromator always transmits less radiant power than the grating monochromator when both have the same resolving power. Notice that $\lambda(dn/d\lambda)$ is approximately constant only when it is very small, further indicating the advantage of the grating monochromator.

We should point out that although the dependence of monochromator resolving power and transmission on the relevant parameters has been shown, it is desirable to measure these quantities whenever possible rather than rely entirely on calculations.

As mentioned at the beginning of this section, a grating produces a number of orders as shown by (7.22). At each angle θ the diffracted beam will contain the wavelength λ in the first order ($m = 1$), the wavelength $\lambda/2$ in the second order ($m = 2$), and so on. Multiple wavelengths therefore appear at the same angle θ and will all pass through the exit slit set for this angle. The range of wavelengths for which this overlapping does not occur is the *free spectral range* $\Delta\lambda_F$. The order $m + 1$ will just overlap the order m when $m(\lambda + \Delta\lambda_F) = (m + 1)\lambda$ so that

$$\Delta\lambda_F = \lambda/m, \tag{7.33}$$

showing that the free spectral range is proportional to wavelength and inversely proportional to the order number m. The problem of overlapping orders is reduced in the visible region by using absorption filters at the monochromator output to absorb energy at wavelengths outside the region of measurement interest. In the infrared an auxiliary prism or grating monochromator may be required. These devices are called *order sorters*. The free spectral range in terms of wavenumber corresponding to (7.33) is $\Delta\sigma_F = \sigma/m$. For further information regarding the selection and use of a grating monochromator and the design of grating instruments for specific applications, we refer the reader to the literature.

7.4 SPECTROMETERS

The prism and grating monochromators which we have examined in previous sections of this chapter are designed to isolate a narrow wavelength band of radiant energy. To measure the spectral distribution of radiant energy it is necessary to scan the monochromator sequentially through the wavelength range of interest. Obviously this procedure is inefficient since at any point in time the part of the total radiant energy which is not passing through the exit slit of the monochromator is lost. A more efficient device is the spectrometer.

A *spectrometer* is a device which permits the measurement of the spectral distribution of radiant energy by collecting radiant energy at all wavelengths simultaneously. The first type of spectrometer which we shall examine is the *Fourier spectrometer*, which requires that the collected energy data be mathematically Fourier transformed to produce the spectrum. Another type of spectrometer is the *multichannel spectrometer* which uses a spatial array of detectors to collect the dispersed energy from a spectral dispersing element such as a grating.

7.4.1 Fourier Spectrometer

The *Fourier spectrometer* derives its name from the mathematical Fourier transformation which is applied to the measurement data to determine the spectrum (Vanasse and Sakai, 1967; Vanasse, 1977; Bell, 1973). It is also referred to as a *multiplex spectrometer* since a frequency multiplexing operation is involved as shall be seen. The principle application of the Fourier spectrometer has been in the infrared for situations in which detector noise limits the signal-to-noise ratio, although interest in applications in the visible region has recently increased.

By considering Fig. 7.14 we can see the basic operating principle of the Fourier spectrometer. A beam of radiant energy of unknown spectral distribution is incident on a beamsplitter where it is divided into transmitted and reflected components. If the incident beam has wave amplitude A, then the transmitted and reflected amplitudes are $U_1 = (A/\sqrt{2}) \exp(i\alpha_1)$ and $U_2 = (A/\sqrt{2}) \exp(i\alpha_2)$, where α_1 and α_2 are the phase changes introduced on transmission and reflection. The two components pass along the optical paths to mirrors M_1 and M_2 where they are reflected back to the beamsplitter. On returning to the beamsplitter the components have amplitudes $U_1' = (A/\sqrt{2}) \exp(i\alpha_1 + i2\pi d_1/\lambda)$ and $U_2' = (A/\sqrt{2}) \exp(i\alpha_2 + i2\pi d_2/\lambda)$, where d_1 and d_2 are the total optical path lengths. The effect of the beamsplitter is again to divide the two beams. We can see that the emerging beam has two

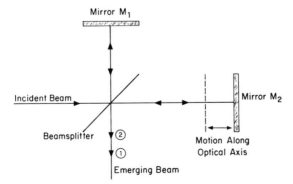

Fig. 7.14 Basic configuration of Michelson–Fourier spectrometer. The emerging beam contains two components which interfere.

components, each of which has undergone a transmission and a reflection at the beamsplitter, giving a total amplitude

$$U_T = U_1'(1/\sqrt{2})\exp(i\alpha_2) + U_2'(1/\sqrt{2})\exp(i\alpha_1). \qquad (7.34)$$

In the emerging beam the spectral radiant power is proportional to $|U_T|^2$ so that

$$\Phi_\lambda = \text{const.} \times |U_T|^2 = \text{const.} \times (A^2/2)[1 + \cos(2\pi\Delta/\lambda)], \qquad (7.35)$$

where $\Delta = d_1 - d_2$ is the optical path length difference. This last equation shows that the output of the Fourier spectrometer contains two parts, one steady and one modulated by the term $\cos(2\pi\Delta/\lambda)$. Notice that if one of the mirrors in Fig. 7.14 is moved at a constant velocity v the path length difference becomes $\Delta = vt$, where t is the time, so that a modulation frequency $v_m = v/\lambda$ is imposed. Also notice that with only a single constant velocity v the frequency v_m is different for each wavelength. This spectral modulation or encoding process is the fundamental basis for the operation of the Fourier spectrometer.

To put (7.35) in radiometric terms we recognize that the spectral radiant power Φ_λ depends on Δ and write, as in Eq. (7.18),

$$\Phi_\lambda(\Delta) = GL_\lambda\tau(\lambda,\Delta) = G(L_\lambda/2)\tau(\lambda,0)[1 + \cos(2\pi\Delta/\lambda)], \qquad (7.36)$$

with L_λ the input radiance, G the geometrical extent of the spectrometer, and $\tau(\lambda,0)$ the spectral transmittance at $\Delta = 0$. Suppose now that a single detector collects the total emerging radiant power at all wavelengths simultaneously. It is convenient to use the wavenumber $\sigma = 1/\lambda$ rather than λ and express the output current from this detector using (7.36) as

$$i(\Delta) = \int_0^\infty R(\sigma)\Phi_\sigma(\Delta)\,d\sigma = \overline{i} + \int_0^\infty S(\sigma)\cos 2\pi\sigma\Delta\,d\sigma, \qquad (7.37)$$

where $\overline{i}$ is the average output current, $S(\sigma) = \frac{1}{2}GR(\sigma)\tau(\sigma,0)L_\sigma$, G the spectrometer geometrical extent, $R(\sigma)$ the spectral responsivity of the detector, $\tau(\sigma,0)$ the spectral transmittance of the spectrometer ($\Delta = 0$), L_σ the input spectral radiance, and Δ the optical path difference. Equation (7.37) expresses the fundamental Fourier transformation relation between the quantity $S(\sigma)$ and the measurement data $i(\Delta) - \overline{i}$. To find the input spectral radiance L_σ we must obviously perform two steps:

(1)   Calculate the inverse Fourier transform of $i(\Delta) - \overline{i}$ to obtain $S(\sigma)$.
(2)   Use the calibration data for $GR(\sigma)\tau(\sigma,0)$ to obtain

$$L_\sigma = 2S(\sigma)/[GR(\sigma)\tau(\sigma,0)]$$

As we have just shown, the essence of the Fourier spectrometer is that a single detector collects the total energy from all parts of the spectrum simultaneously, and that the energy at each wavelength or wavenumber is encoded at the modulation frequency $v_m = v/\lambda = \sigma v$ due to the mirror motion at velocity v. It is this encoding process which is referred to as *multiplexing*.

Under certain conditions a Fourier spectrometer has two basic advantages over a monochromator. A monochromator must use a slit which is narrow in the direction of dispersion to obtain spectral resolution. The Fourier spectrometer, on the other hand, does not have this restriction. It can accept radiant energy over a circular field of view with diameter as large as the long dimension of the slit. The gain in solid angle over which energy is collected is known as the *etendue advantage* of the Fourier spectrometer. Also, a spectral measurement using a monochromator to measure m spectral elements must divide the total measurement time t among the spectral elements so that each element is observed for only t/m seconds. On the other hand, the Fourier spectrometer observes each spectral element for the total time t by using the multiplex encoding technique. When the signal-to-noise ratio is limited by detector noise the resulting gain in signal-to-noise is a factor $m^{1/2}$. This is known as the *multiplex advantage* of the Fourier spectrometer.

Interferometers other than the Michelson arrangement shown in Fig. 7.14 have been used with the multiplex encoding method. For further details, we refer the reader to the literature (Bell, 1973; Vanasse, 1977).

7.4.2 Multichannel Spectrometers

A multichannel spectrometer is similar to a monochromator in the sense that it employs a grating or prism as a dispersing element but, unlike the monochromator, collects the dispersed energy in a number of wavelength channels simultaneously. The energy in each channel may be collected by a discrete detector element of a detector array or may be detected by using the spatial resolution capability of a vidicon or similar sensor. Figure 7.15 shows

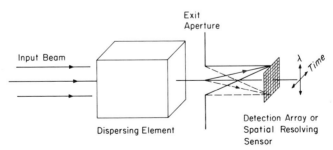

Fig. 7.15 Basic components of a multichannel spectrometer. Focusing optical elements are used at the exit aperture so that the spectral power at each wavelength appears along the vertical direction.

the arrangement of the essential components in a multichannel spectrometer. The dispersed spectrum which is produced by the dispersing element is focused by optical elements located in or near the exit aperture. The focused beam is incident on a detection plane in which the various wavelengths are spread vertically. A scanning optical element may also be used so that the spectrum at various times appears at various horizontal locations. The resultant two-dimensional array of data shows Φ_λ versus time. Variations on this basic arrangement may employ no scanner and only a one-dimensional vertical array of detectors which are read out at discrete time intervals. As mentioned, the array may also be replaced by a sensor which provides spatial resolution, such as a vidicon tube. Spectrometers which provide time resolution by scanning as described or by periodic readout of the detector array are often referred to as *rapid-scan spectrometers.*

The fundamental principles of operation of the dispersing elements in multichannel spectrometers are as described in Section 7.3 for monochromators. Optical elements may, however, require further aberration correction since they now work over a larger field of view. A principal reason for employing multichannel techniques is to obtain improved signal-to-noise by collecting energy at all wavelengths of interest simultaneously. The signal-to-noise advantage is identical to that of the Fourier multiplex spectrometer since if m spectral resolution elements are detected simultaneously, the signal-to-noise advantage is $m^{1/2}$ when the measurement is limited by detector noise. However, if time-resolved measurements are made, the integration time T between readout times will limit the signal-to-noise also. It can be shown that

$$\text{SNR} = \text{const.} \times T^{1/2}(\Delta\lambda)^2, \tag{7.38}$$

where SNR is the signal-to-noise ratio when limited by detector noise, T the integration time, and $\Delta\lambda$ the wavelength resolution. From (7.38) we can see that for a fixed SNR requirement, the spectral resolution $\Delta\lambda$ is inversely proportional to the fourth root of the scanning time. This conclusion is based on geometrical optics and applies when the detector spatial resolution is the limiting factor in determining wavelength resolution $\Delta\lambda$. Multichannel spectrometers and rapid scan spectrometers are commercially available and have been described in the literature (Vanasse, 1977).

REFERENCES

Baumeister, P. (1965). Interference and optical interference coatings, *In* "Applied Optics and Optical Engineering" (R. Kingslake, ed.), Vol. 1. Academic Press, New York.
Bell, R. (1973). "Introduction to Fourier Transform Spectroscopy." Academic Press, New York.
Born, M., and Wolf, E. (1964). "Principles of Optics," 2nd revised ed. Pergamon, Oxford.
Christiansen, C. (1884). *Wied. Ann. Phys. Chem.* **23**, 298.
Christiansen, C. (1885). *Wied. Ann. Phys. Chem.* **24**, 439.

Evans, J. (1949a). *J. Opt. Soc. Am.* **39**, 229.

Evans, J. (1949b). *J. Opt. Soc. Am.* **39**, 412.

Heavens, O. S. (1955). "Optical Properties of Thin Solid Films." Butterworths, London.

Holland, L. (1960). "Vacuum Deposition of Thin Films." Chapman and Hall, London.

Jacquinot, P. (1954). *J. Opt. Soc. Am.* **44**, 761.

Jenkins, F., and White, H. (1957). "Fundamentals of Optics." McGraw–Hill, New York.

Lyot, B. (1933). *C. R. Acad. Sci. Paris* **197**, 1593.

Scharf, P. T. (1965). Filters, *In* "Applied Optics and Optical Engineering" (R. Kingslake, ed.), Vol. 1. Academic Press, New York.

Smith, R., Jones, F., and Chasmar, R. (1968). "The Detection and Measurement of Infrared Radiation," 2nd ed. Oxford Univ. Press, London and New York.

Vasicek, A. (1960). "Optics of Thin Films." North–Holland Publ., Amsterdam.

Vanasse, G. A. (ed.) (1977). "Spectrometric Techniques," Vol. I. Academic Press, New York.

Vanasse, G. A., and Sakai, H. (1967). Fourier Spectroscopy, *In* "Progress in Optics" (E. Wolf, ed.), Vol. VI. North–Holland Publ., Amsterdam.

Wolter, H. (1956). *In* "Encyclopedia of Physics" (S. Flügge, Ed.), Vol. 24. Springer–Verlag, Berlin and New York.

Wyszecki, G., and Stiles, W. S. (1967). "Color Science." Wiley, New York.

8

Measurements of Radiant Power and Radiant Energy

8.1 INTRODUCTION

In earlier chapters we considered radiometric concepts and components. This chapter and the two following chapters are concerned with *techniques*— the methods by which concepts and components are integrated to form an effective measurement system. In the present chapter we consider techniques for the measurement of radiant power and energy while the measurement of ratios of power or energy quantities such as transmittance, reflectance, and absorptance are treated in Chapter 9. We reserve for Chapter 10 the discussion of the available standards and calibration techniques used in radiometry.

It is useful to begin a discussion of radiant power and energy measurement techniques by examining the general configuration of system components shown in Fig. 8.1. The system components are shown in Part (a) and the ray paths from source to detector are shown in (b). In 8.1b the instrument axis is inclined at an angle θ to the source normal. As shown in Fig. 8.1b the instrument entrance aperture subtends a half-angle α_0 at the source. Notice that the surface of the detector acts as a field stop which limits the instantaneous field of view of the instrument. The geometrical projection of this field stop onto the source is shown by the dashed lines in Fig. 8.1b. As indicated, the half-angle field of view is then β_0. These two angles, α_0 and β_0, are directly related to the energy collecting properties of the instrument for incoherent sources but are also important when measuring laser sources. Further discussion of these topics is found in Chapter 3.

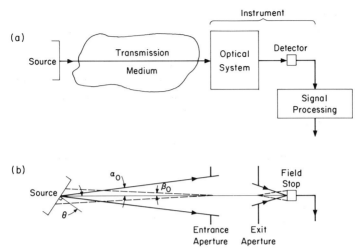

Fig. 8.1   System components for measurement of radiant power or radiant energy: (a) system components; (b) ray paths.

In an *ideal instrument* of the type shown in Fig. 8.1 we would find that

(a)   the instrument responds uniformly to radiant power or energy at all wavelengths in a desired band $\lambda_1 \leq \lambda \leq \lambda_2$ but has zero response outside the band;

(b)   the system is linear so that the output voltage v (or current i) is directly proportional to the incident radiant power or energy throughout the required dynamic range;

(c)   the optical system exhibits no vignetting or aberrations so that the field and aperture are both defined sharply and independently of each other at all wavelengths;

(d)   the instrument response is the same for all polarizations; and

(e)   the temporal frequency response of the instrument covers all temporal frequencies present in the incident radiant power or energy.

The output voltage of the ideal instrument is then related directly to the incident radiant power by the instrument responsivity. Depending on the radiometric quantity of interest, the *responsivity* is defined as

$$\text{power responsivity} = R_\Phi = v/\Phi_m \quad (\text{VW}^{-1}), \qquad (8.1)$$

$$\text{irradiance responsivity} = R_E = v/E_m \quad (\text{VW}^{-1}\ \text{cm}^2), \qquad (8.2)$$

$$\text{radiance responsivity} = R_L = v/L_m \quad (\text{VW}^{-1}\ \text{cm}^2\ \text{sr}), \qquad (8.3)$$

where v is the instrument output voltage, Φ_m the incident radiant power at the instrument entrance aperture, E_m the incident irradiance at the entrance aperture, and L_m the incident radiance at the entrance aperture. When the quantity of interest is clear, the subscript may be omitted. Also, the current responsivities can be similarly defined. Note that in the absence of radiant energy from the background and attenuation by the medium the radiance at the instrument entrance aperture is the same as the radiance at the source as shown in Chapter 3.

The purpose of a radiant power or energy measurement is to find a *field quantity* which is independent of the instrument being used. The field quantity of interest may be the radiance, the irradiance, or any of the other radiometric quantities. It is usually convenient to find this field quantity in the instrument entrance aperture shown in Fig. 8.1b. We refer to the field quantity in this aperture as the *measured* field quantity. Independence of the instrument being used is obtained by a prior calibration of the instrument under environmental and geometrical conditions which closely approximate the actual conditions in which it will be used. When the desired field quantity is found in the entrance aperture, it is then possible to use information about the transmission medium or background radiant energy to infer the desired source properties. Alternatively, with a known source and a calibrated instrument the field quantities measured at the instrument aperture can be used to determine the transmission properties of the medium. As a third possibility, when the source and transmission medium are known so that the field quantities in the entrance aperture are determined, the voltage measured at the detector output can be used to find the instrument responsivity.

8.2 TOTAL RADIOMETRIC MEASUREMENTS

According to the definition of responsivity the spectral output voltage from an instrument at wavelength λ is

$$v_\lambda(\lambda) = R_L(\lambda)L_{\lambda m}(\lambda), \qquad (8.4)$$

where $R_L(\lambda)$ is the spectral radiance responsivity of the instrument and $L_{\lambda m}(\lambda)$ is the measured spectral radiance at the entrance aperture shown in Fig. 8.1b. When the instrument responds to radiant energy over a broad wavelength range extending from λ_1 to λ_2, the output voltage is

$$v = \int_{\lambda_1}^{\lambda_2} v_\lambda(\lambda)\,d\lambda = \int_{\lambda_1}^{\lambda_2} R_L(\lambda)L_{\lambda m}(\lambda)\,d\lambda. \qquad (8.5)$$

Relations similar to (8.4) and (8.5) can also be written for the spectral radiant power or spectral irradiance at the entrance aperture.

Some important conclusions arise from an examination of (8.4) and (8.5). The *total radiance* between λ_1 and λ_2 is defined as

$$L_m = \int_{\lambda_1}^{\lambda_2} L_{\lambda m}(\lambda)\, d\lambda. \tag{8.6}$$

We can express the spectral responsivity as $R_L(\lambda) = d_p r_L(\lambda)$, where d_p is the peak value and $r_L(\lambda)$ the peak normalized relative spectral responsivity. Also we can express the spectral radiance as $L_{\lambda m}(\lambda) = f_p l_{\lambda m}(\lambda)$, where f_p is the peak value and $l_{\lambda m}(\lambda)$ the peak normalized relative spectral radiance. With these relations we can write the *total responsivity* as

$$R_L = v/L_m = \int_{\lambda_1}^{\lambda_2} R_L(\lambda)L_{\lambda m}(\lambda)\, d\lambda \Bigg/ \int_{\lambda_1}^{\lambda_2} L_{\lambda m}(\lambda)\, d\lambda, \tag{8.7}$$

$$R_L = v/L_m = d_p \int_{\lambda_1}^{\lambda_2} r_L(\lambda)l_{\lambda m}(\lambda)\, d\lambda \Bigg/ \int_{\lambda_1}^{\lambda_2} l_{\lambda m}(\lambda)\, d\lambda. \tag{8.8}$$

The ratio of integrals in (8.8) is the spectral matching factor defined in Chapter 6 in connection with detectors, although in this discussion it applies to the entire instrument.

8.2.1 Detector-Based Measurements

Notice from (8.8) that

(a) when the instrument responsivity is spectrally flat, that is when $r_L(\lambda)$ is unity, the total responsivity R_L is independent of the measured relative spectral radiance $l_{\lambda m}(\lambda)$.

For this reason thermal detectors, which have approximately flat spectral responsivity over a wide wavelength range, are conveniently used in measurements of total field quantities such as the total radiance L_m. When the relative spectral responsivity $r_L(\lambda)$ is unity, then (8.7) and (8.8) show that the total radiance is simply $L_m = v/R_L = v/d_p$. In practice the spectral properties of windows, filters, and detector-absorbing materials in the instrument must also be considered to determine whether the instrument responsivity is sufficiently flat over the wavelength band.
On the other hand (8.7) and (8.8) show that

(b) when the instrument responsivity is not spectrally flat, that is when $r_L(\lambda)$ is not unity, the total responsivity R_L depends on the measured relative spectral radiance $l_{\lambda m}(\lambda)$.

In this case we again have $L_m = v/R_L$, but R_L is given by (8.8).

Measurements of total radiance with a spectrally selective instrument therefore require prior knowledge of the measured relative spectral radiance as well as the relative spectral radiance responsivity. Similar conclusions are obtained for other total field quantities.

8.2.2   Source-Based Measurements

It is possible to perform measurements of total radiance, or of other total field quantities, which are based on the use of a known source rather than a known detector. If we represent the quantities associated with the known source by a prime to distinguish them from the unprimed quantities associated with the unknown source, then using $R_L(\lambda) = d_p r_L(\lambda)$ and (8.5) we see that the measured voltage with the known source is

$$v' = d_p \int_{\lambda_1'}^{\lambda_2'} L'_{\lambda m}(\lambda) r_L(\lambda) \, d\lambda, \tag{8.9}$$

where $L'_{\lambda m}(\lambda)$ is the measured spectral radiance due to the known source. From (8.7) the voltage for the unknown source is $v = L_m R_L$, so that the total radiance is $L_m = v/R_L$. When an instrument with flat spectral response is used so that $r_L(\lambda) = 1$, the total radiance is found from (8.9) and (8.8) to be

$$L_m = v/R_L = (v/v') \int_{\lambda_1'}^{\lambda_2'} L'_{\lambda m}(\lambda) \, d\lambda = (v/v') L_m' \quad [\text{W cm}^{-2} \text{ sr}^{-1}], \tag{8.10}$$

where v is the instrument output voltage with the unknown source, v' the instrument output voltage with the known source, and L_m' the measured total radiance at the entrance aperture due to the known source. Notice that relation (8.10) does not involve the instrument responsivity. Also, it does not require a knowledge of the relative spectral radiance. Only the total radiance and voltage due to the known source are required. However, it should be noted also that the instrument spectral response is assumed to be flat over the ranges $\lambda_1 - \lambda_2$ and $\lambda_1' - \lambda_2'$. As a result, thermal detectors are used for this type of measurement. Depending on the accuracy required, it may be necessary to examine closely the actual spectral responsivity of the total instrument even when thermal detectors are used.

Finally we note again that the relations found here involving radiance and radiance responsivity can also be expressed in terms of irradiance or radiant power and the appropriate responsivity. Similarly, the output voltage can be replaced by the output current with a corresponding change in the responsivity quantity being used. With these concluding comments regarding total radiometric measurements we now turn our attention to spectral measurements.

8.3 SPECTRAL MEASUREMENTS

Spectral measurements, as opposed to total measurements, require the use of a spectral component to distinguish between various wavelengths. The spectral component may be a monochromator, a filter, a Fourier spectrometer, or other device. In this section we examine the radiometric properties of spectral measurement systems emphasizing the relative advantages of each technique. Most of the comments in Chapter 9 regarding the origins of spectra and the differences between continuous sources and line sources are also applicable here.

8.3.1 Monochromators

As described in Chapter 7 a monochromator employs a prism or grating to disperse a beam of radiant energy into its wavelength components. Figure 8.2 shows schematically the monochromator and other components which are present in a spectroradiometer. The components include (a) a collecting optical system which focuses a source image onto the entrance slit, (b) entrance and exit slits, (c) collimating and refocusing optics within the monochromator, (d) the spectral dispersing element (prism or grating), (e) a detector which collects the radiant energy emerging from the exit slit, and (f) signal processing. When a grating is used, the spectroradiometer components may also include absorption filters or other order-sorting spectral components

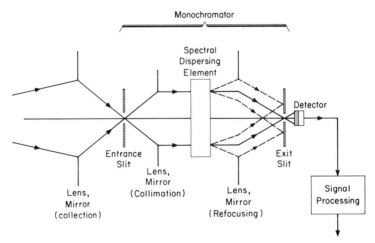

Fig. 8.2 Schematic arrangement of monochromator with collection optics, detector, and signal processing.

to suppress multiple spectral orders which can emerge from a grating along the same direction. Also, a double monochromator may replace the single monochromator shown in Fig. 8.2 to achieve higher dispersion to improve the spectral purity by reducing stray energy. In this arrangement the exit slit of the first monochromator becomes the entrance slit of the second. In all arrangements the spectral scanning is accomplished by changing the effective angle of the dispersing element. The relation between angle and wavelength is as discussed in Chapter 7.

Figure 8.3 shows a practical arrangement of components which uses a blazed grating in a Littrow mounting. As indicated, a Littrow prism could also have been used as the dispersing element. The wavelength of measurement λ_m is selected by rotating the drive mechanism which changes the grating angle. The concave mirror M_1 focuses the source onto the entrance slit and the source image fills the entrance slit. The f-number of this mirror is arranged to be the same as that of the first collimating mirror M_2 so that the monochromator solid angle is filled without introducing stray energy. The f-number is also made as small as possible to deliver maximum radiant power to the detector consistent with keeping optical aberrations under control. The parallel beams dispersed by the grating or prism are focused by M_3 onto the exit slit. Absorption filters can be placed at A. A chopper may be placed just after the entrance slit. Also, when atmospheric absorption within the instrument could be important, it is enclosed and flushed with a low-absorption dry gas.

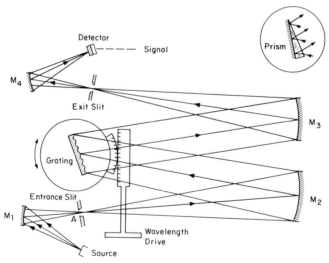

Fig. 8.3 Practical monochromator arrangement using grating or prism. The components M_1–M_4 are concave mirrors. This is the Czerny–Turner form of the Ebert–Fastie monochromator.

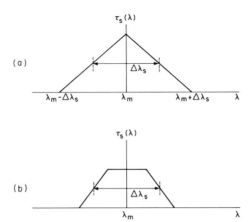

Fig. 8.4 Slit spectral transmittance factor for a monochromator in the geometrical optics limit. The entrance and exit slit widths in wavelength units are S_e and S_x. (a) $\Delta\lambda_s = S_e = S_x$; (b) $\Delta\lambda_s = S_e$ or S_x, whichever is larger.

The wavelength λ_m at which the measurement is being made corresponds to the ray from the center of the entrance slit to the center of the exit slit. Obviously, with slits of finite size which are necessary to pass adequate energy, a finite range of wavelengths about λ_m is actually transmitted. Figure 8.4 shows the relative spectral transmittance factor $\tau_s(\lambda)$ of a monochromator due to finite slit widths. In Fig. 8.4a the entrance and exit slits are assumed to be of equal width and large enough so that diffraction effects are negligible. In Fig. 8.4b diffraction is again assumed negligible but the slit widths are unequal. What Fig. 8.4a shows is that the peak transmittance occurs when the wavelength of the radiant energy is equal to λ_m, the central wavelength at which the measurement is being made. Other wavelengths are also transmitted but with decreased energy relative to λ_m. Wavelengths beyond the limits $\lambda_m \pm \Delta\lambda_s$ are completely blocked, neglecting scattered energy effects within the monochromator. In the opposite situation when diffraction completely determines the slit spectral transmittance factor $\tau_s(\lambda)$, then

$$\tau_s(\lambda) = [(\sin \gamma)/\gamma]^2, \tag{8.11}$$

where $\gamma = (\pi w_s/\lambda_m) \sin \delta$, w_s is the slit width of the monochromator, and δ the angle between the ray of wavelength λ and the ray of wavelength λ_m.

These slit spectral transmittance factors $\tau_s(\lambda)$ are combined with the spectral transmittance $\tau_a(\lambda)$ due to absorption and other effects within the spectroradiometer so that the combined instrument spectral transmittance is $\tau(\lambda) = \tau_s(\lambda)\tau_a(\lambda)$. The measured voltage at the detector for each wavelength setting λ_m is then

$$v(\lambda_m) = \int_{\lambda_m - \Delta\lambda_s}^{\lambda_m + \Delta\lambda_s} R_L(\lambda) L_{\lambda m}(\lambda) \, d\lambda, \tag{8.12}$$

where $R_L(\lambda)$ is the spectral radiance responsivity of the spectroradiometer and $L_{\lambda m}(\lambda)$ the measured spectral radiance at the entrance aperture of the spectroradiometer. By separating the overall responsivity into detector, slit width, and instrument absorption factors we have, from (8.12),

$$v(\lambda_m) = d_p \int_{\lambda_m - \Delta\lambda_s}^{\lambda_m + \Delta\lambda_s} \tau_s(\lambda) \tau_a(\lambda) r_d(\lambda) L_{\lambda m}(\lambda) \, d\lambda, \tag{8.13}$$

where $v(\lambda_m)$ is the detector output voltage at wavelength setting λ_m, λ_m the measurement wavelength determined by the monochromator setting, $\Delta\lambda_s$ the spectral slit width, d_p the peak spectral radiance responsivity of the spectroradiometer, $\tau_s(\lambda)$ the relative spectral transmittance due to the slits, $\tau_a(\lambda)$ the relative spectral transmittance due to instrument absorption and other effects, $r_d(\lambda)$ the relative spectral responsivity of the detector (V W^{-1} cm^2 sr μm), and $L_{\lambda m}(\lambda)$ the spectral radiance in the entrance aperture of the spectroradiometer (W cm^{-2} sr^{-1} μm^{-1}).

In the simplest approximation to (8.13) for continuous sources all factors are regarded as being approximately constant over the range $\lambda_m - \Delta\lambda_s \leq \lambda \leq \lambda + \Delta\lambda_s$ so that

$$v(\lambda_m) = d_p \tau_s(\lambda_m) \tau_a(\lambda_m) r_d(\lambda_m) L_{\lambda m}(\lambda_m) \Delta\lambda_s, \tag{8.14}$$

where $\Delta\lambda_s$ is the effective bandwidth corresponding to the 50% points shown in Fig. 8.4. In some cases the variations within the bandwidth are significant, however, and in this situation (8.13) must be used. Also, in some cases the image of the entrance slit which falls on the exit slit may be curved or other optical aberrations may be present and these effects must be considered.

When line sources are used, then the output voltage is the sum of the voltages for each line as found by evaluating (8.13) at the discrete wavelength of each line. Equation (8.14) is not applicable to line sources since $L_m(\lambda)$ is not constant over the bandwidth $\Delta\lambda_s$. For line sources, $v(\lambda_m) = \sum_{i=1}^{n} v(\lambda_i)$, where λ_i is the wavelength of the ith line and n the number of lines within the range $\lambda_m - \Delta\lambda_s \leq \lambda \leq \lambda_m + \Delta\lambda_s$. A spectral line which occurs at λ_m will produce an image of the entrance slit centered on the exit slit. In the geometrical optics limit no radiant energy in the line will be blocked by the exit slit of the monochromator when the two slits are of equal width. However, for other lines to pass with no loss due to the exit slit it is necessary to increase the exit slit width.

For the detector following the monochromator the exit slit acts as a source. Table 8.1 shows the irradiance or radiant power at the detector for continuous sources and line sources. As shown by the table the irradiance or radiant power depends on whether an auxiliary image formation system is

TABLE 8.1 Irradiance or Radiant Power at Detector in a Spectroradiometer

	Continuous source[a]	Line source[a]
Without image formation	$E = \dfrac{I}{b^2} = \dfrac{L w_x l_x}{b^2}$	$E = \dfrac{w_e l_x}{b^2} \sum\limits_{i=1}^{n} L(\lambda_i)$
With image formation	$E = \dfrac{L A_i}{S'^2}$	$\Phi = \dfrac{w_e l_x A_i}{S^2} \sum\limits_{i=1}^{n} L(\lambda_i)$

[a] The following notation is used in the table: L is the radiance at the exit slit $(= L_m \tau_a)$, τ_a the spectroradiometer transmittance from entrance aperture to exit aperture, L_m the radiance at the entrance aperture, w_x the width of the exit slit, l_x the length of the exit slit, b the distance from the exit slit to the detector, A_i the area of the exit aperture of the auxiliary imaging optics, S the distance from the exit slit to the entrance aperture of the auxiliary optics, S' the distance from the exit aperture of the auxiliary optics to the detector, w_e the width of the entrance slit, n the number of lines in a group, and λ_i the wavelength of the ith line.

used between the exit slit and the detector. When an imaging system is used with a source consisting of discrete spectral lines, it is necessary to specify the incident radiant power Φ at the detector since the irradiance is nonuniform. The formula for the line source with image formation assumes that the exit slit has been widened to include the n lines in a group so that the entrance slit width w_e is the limiting slit. The radiance L in the exit slit which appears in Table 8.1 is related to the radiant power Φ found in Chapter 7 by the appropriate area and solid angle.

Spectral measurements on sources can be based on the use of a known detector or a known source. Table 8.2 shows the relationships for determining

TABLE 8.2 Relationships for Spectral Measurement of an Unknown Source with Monochromator

Reference	Continuous source[a]	Line source[a]
Known detector	$l_\lambda(\lambda_m) = \dfrac{v(\lambda_m) r(\lambda_r) \tau(\lambda_r) d_m}{v(\lambda_r) r(\lambda_m) \tau(\lambda_m) d_r}$	$l(\lambda_m) = \dfrac{v(\lambda_m) r(\lambda_r) \tau(\lambda_r)}{v(\lambda_r) r(\lambda_m) \tau(\lambda_m)}$
Known source	$L_\lambda(\lambda_m) = \dfrac{v}{v'} L_\lambda'(\lambda_m)$	$L(\lambda_m) = \dfrac{v}{v'} L'(\lambda_m)$

[a] The following notation are used in the table: l_λ is the relative spectral radiance $(= dl/d\lambda)$, $v(\lambda_m)$ the measured voltage at wavelength λ_m, $r(\lambda_r)$ the relative spectral responsivity, $\tau(\lambda_r)$ the spectral transmittance of the spectroradiometer from entrance aperture to detector, l the relative radiance, d the linear dispersion of the monochromator $(= f d\theta/d\lambda)$, λ_m the measured wavelength, λ_r the reference wavelength, v' the voltage for the reference source, and L_λ the spectral radiance $(= dL/d\lambda)$.

the spectral radiance of an unknown continuous source or line source. Notice that the relative spectral radiance is determined when a known detector is used. To determine absolute values it is easier to use a known source as a reference. It should also be noted that when a reference source is used, it is not necessary to know the transmittance of the system, the dispersion, or the detector spectral responsivity. In this case it is assumed that the geometrical conditions remain the same for the reference and unknown sources.

In some cases it is necessary to perform measurements of a line source and to use a continuous source as reference. For this measurement the desired radiance of the line source is $L(\lambda_m) = L_\lambda(v/v')(w_x/d)$, where L_λ is the spectral radiance of the continuous source, v and v' the voltages for the unknown and reference sources, w_x the exit slit width, and d the linear dispersion ($= f \, d\theta/d\lambda$) of the monochromator.

8.3.2   Filters

Absorption filters and interference filters were described in Chapter 7. While these filters do not provide the flexibility of a monochromator which is continuously tunable in wavelength, they do have several advantages over monochromators in spectral measurements. The principal advantages are that (a) the spectral transmittance of a filter is more readily measured than that of a monochromator, (b) the path of the rays in a beam of radiant energy can remain essentially unchanged when filters are used, and (c) the absolute transmittance of a filter is generally higher than that of a monochromator thus providing higher energy levels at the detector. In comparing the spectral properties of filters and monochromators it is possible to think of the mono-chromator at each wavelength setting as a filter with relative spectral trans-mittance as shown in Fig. 8.4 or Eq. (8.11).

In using a filter to perform spectral measurements the filter should be used in a parallel beam whenever possible. In any case the spectral transmittance of the filter should be determined under geometrical and environmental conditions which closely approximate the conditions in which it will be used.

For spectral measurements with filters the measured voltage is

$$v = \int_{\lambda_1}^{\lambda_2} E_\lambda(\lambda)\tau(\lambda)R(\lambda) \, d\lambda, \qquad (8.15)$$

where $E_\lambda(\lambda)$ is the spectral irradiance measured, $\tau(\lambda)$ the filter spectral trans-mittance, $R(\lambda)$ the spectral irradiance responsivity, and λ_1, λ_2 the limiting wavelengths, usually set by filter transmittance. Table 8.3 shows special cases derived from (8.15) when the unknown source is a continuous source or a line source and when either a known detector or a known source is used as a reference. For the line source with the known detector it is assumed that

TABLE 8.3 Relationships for Spectral Measurements
of an Unknown Source with Filters

Reference	Continuous source[a]	Line source[a]
Known detector	$E_\lambda(\lambda_{\mathrm{m}})^b = \dfrac{v(\lambda_{\mathrm{m}})}{R(\lambda_{\mathrm{m}})\tau(\lambda_{\mathrm{m}})\,d\lambda}$	$E(\lambda_{\mathrm{m}}) = \dfrac{v(\lambda_{\mathrm{m}})}{R(\lambda_{\mathrm{m}})\tau(\lambda_{\mathrm{m}})}$
	$E_\lambda{}^c = \dfrac{v e_\lambda}{e_\lambda(\lambda)\tau(\lambda)R(\lambda)\,d\lambda}$	
Known source	$E_\lambda(\lambda_{\mathrm{m}}) = \dfrac{v}{v'}\,E_\lambda{'}(\lambda_{\mathrm{m}})$	$E(\lambda_{\mathrm{m}}) = \dfrac{v}{v'}\,E'(\lambda_{\mathrm{m}})$

[a] The following notation is used in the table: λ_{m} is the measurement wavelength, $E(\lambda)$ the measured irradiance, $E_\lambda(\lambda)$ the spectral irradiance at the detector, $e_\lambda(\lambda)$ the relative spectral irradiance (peak normalized), v the voltage measured with the unknown source, v' the voltage measured with the known source, $\tau(\lambda)$ the spectral transmittance of the filter, and $R(\lambda)$ the spectral irradiance responsivity.
[b] Narrow filter.
[c] Wide filter.

only one line is present within the filter bandpass and this line is at wavelength λ_{m}. For continuous sources measurements can be made with narrow bandpass filters for which $E_\lambda(\lambda)$ and $R(\lambda)$ are essentially constant from λ_1 to λ_2. Alternatively, a wide bandpass filter can be used. However, in this situation it is necessary to have prior knowledge of the relative spectral irradiance distribution e_λ. Finally, when a known source is used, Table 8.3 shows that it is not necessary to know any detector responsivity properties. In all cases it is of course assumed that the detector is linear over the required dynamic range. When performing spectral measurements with filters it is convenient to measure the irradiance and determine other quantities from it. This removes, or at least postpones, any questions of nonuniformity when performing measurements at short distances from sources which may be spatially nonuniform.

Finally it should be noted that filters used in performing spectral measurements should have planar surfaces and should be free of both surface and internal detects. Since heat may affect the filter properties it is also important to locate the filter at a position which is sufficiently far from the source.

8.3.3 Fourier Spectrometer

We now consider spectral measurement techniques for the Fourier spectrometer which was discussed in Chapter 7. With the development of improved translation mechanisms and the increased availability of digital signal processing capabilities this instrument is finding increasing use for spectral measurements (Fellgett, 1951, 1967; Steel, 1967; Bell, 1973). This discussion shall be brief. A thorough treatment of Fourier spectrometric techniques is now available in the literature (Vanesse, 1977).

As discussed in Chapter 7, Fourier spectrometers provide two potential advantages over spectroradiometers which use filters or monochromators. These two advantages are the etendue advantage and the multiplex advantage. Whether these advantages are realized in practice depends on practical considerations such as the required resolving power. They are important considerations when it is necessary to collect high-resolution spectral data over a large spectral range in a short time.

Unlike the spectral measurement techniques which employ filters or monochromators, the recovery of the spectral field quantity from a Fourier spectrometer requires a Fourier transformation of the voltage signal $v(t)$ or current signal $i(t)$ from the detector. Here the time is $t = \Delta/v$, where Δ is the optical path difference introduced by the Fourier spectrometer scanning motion and v the effective scan velocity. However, similar to the situation with other spectral techniques, it is common to perform relative spectral measurements with the Fourier spectrometer and obtain the absolute radiometric level by using other supplementary instrumentation. With reference to Eqs. (7.36) and (7.37) the spectral radiance is

$$L_\sigma(\sigma) = 2 \int_0^\infty B(\Delta) \cos(2\pi \, \Delta \sigma) \, d\Delta \bigg/ GR(\sigma)\tau_0(\sigma), \qquad (8.16)$$

where Δ is the optical path difference ($= vt$) (cm), $B(\Delta) = i(\Delta) - \bar{i}$, $i(\Delta)$ the detector output current at path difference Δ, $\sigma = 1/\lambda$ the wavenumber (cm^{-1}), G the extent or etendue of the spectrometer, $R(\sigma)$ the spectral power responsivity of the detector, and $\tau_0(\sigma)$ the spectral transmittance of the spectrometer at $\Delta = 0$.

In obtaining the spectral radiance $L_\sigma(\sigma)$ a number of practical considerations arise which can be categorized as:

(a)   spectral resolution and sampling errors,
(b)   sampling interval,
(c)   path difference monitoring,
(d)   mechanical drive technique,
(e)   signal-to-noise ratio and noise source,
(f)   field of view,
(g)   recording techniques, and
(h)   signal processing.

To describe each of these considerations it is helpful to consider a practical realization of a Fourier spectrometer configuration (Haycock and Baker,

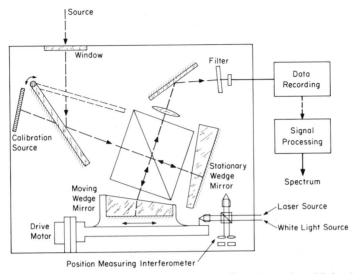

Fig. 8.5 Practical realization of a Fourier spectrometer [after Haycock and Baker (1974)].

1974) as shown in Fig. 8.5. Radiant energy from the source enters the instrument window. In the infrared this window and the entire instrument may be cryogenically cooled. The first mirror directs the energy to the beam splitter. This first mirror is movable so that the instrument can view the calibration source. At the beam splitter the beam is divided with one path leading to a stationary wedge mirror and the other to a moving wedge mirror. The wedge shape is designed to provide tilt compensation within the interferometer. As indicated in the figure, the moving wedge mirror is translated along a drive shaft to provide the path length change Δ. On recombination at the beamsplitter the combined beams pass through a spectral filter which limits the spectral band of the source energy and are incident on the detector. The recorded detector output signal is subsequently Fourier-transformed using digital Fast Fourier Transform (FFT) processing (Rabiner and Gold, 1975) to obtain the source spectra. Notice that a secondary optical interferometer provides position measurement to establish the Δ value for each $B(\Delta)$ appearing in Eq. (8.16). Use of the calibration source in this instrument is equivalent to the use of a known source in the techniques described earlier for measurements using filters and monochromators.

In a Fourier spectrometer such as that shown in Fig. 8.5 the wave number spectral resolution is essentially equal to $\sigma_R = 1/\Delta_{\max}$, where $\Delta_{\max}$ is the maximum optical path difference. Here Δ is in units of centimeters so that the resolution σ_R is in units reciprocal centimeters. Errors in the times at which the output is sampled lead to degraded spectral resolution. The time interval at which the output signal should be sampled is approximately $T = (2v\sigma_R)^{-1}$, where v is the effective velocity of translation. Notice that in Fig. 8.5 the effective velocity is less than the actual velocity due to the mirror angle.

Path difference monitoring to determine Δ is accurately done with an auxiliary interferometer. This information is recorded in synchronization with the detector output data. Notice in Fig. 8.5 that by inclining the mirror to the direction of motion large changes in position correspond to only small changes in optical path thus reducing sensitivity to position errors due to the mechanical drive technique. Both continuous and stepped motions have been used. Both techniques require accurate servo control, and the development of effective servo designs has been a key factor in Fourier spectrometer advances.

As mentioned in Chapter 7, the multiplex advantage occurs when the noise is not signal-dependent. However, signal-dependent scintillation noise which is due to (a) atmospheric turbulence, (b) unstable components in the interferometer, or (c) source fluctuations can reduce the multiplex advantage. In this case the signal-to-noise gain due to multiplexing is no longer $m^{1/2}$ where m is the number of spectral elements observed.

To collect energy from a large source area the angular field of view of the spectrometer must be adequately large. It is field widening which requires the tilt compensation shown in Fig. 8.5. One measure of the effectiveness of field-of-view widening is the time required to observe the spectrum of a line source of radiance L at fixed signal-to-noise SNR. This time is

$$t_{obs} = \frac{\pi A_d (\text{SNR})^2}{(LA\Omega\tau D^*)^2}, \tag{8.17}$$

where A_d is the detector area, SNR the signal-to-noise voltage or current ratio, L the source radiance, $A\Omega$ the instrument extent or etendue, τ the instrument transmittance, and D^* the normalized detectivity of the detector. For fixed spectral resolution and instrument size the Fourier spectrometer requires several orders of magnitude less viewing time than a grating spectrometer. By comparison a field-widened Fourier spectrometer can further shorten this viewing time several more orders of magnitude when the source subtends a large angle at the instrument (Vanasse, 1977, Chapter 2).

Recording techniques have been devised to minimize the effects of scintillation noise. These include (a) the use of a separate detector to monitor source fluctuations, (b) the use of two detectors at the two outputs which are available with any interferometer, and (c) chopping the incoming beam so that low-frequency scintillation noise can be filtered out. Frequency filtering is found to be more effective than the two-detector method.

Signal processing involves digital phase correction and interpolation of the raw data usually followed by application of the efficient Fast Fourier Transform routine. Computing efficiency is improved by using the symmetry of the interferogram data. Decimation in time is also employed for handling long sets of data. Further information regarding digital signal processing methods is available in the literature (Rabiner and Gold, 1975; Vanasse and Sakai, 1967).

8.4   RADIANCE AND RADIANT INTENSITY OF AN INCOHERENT SOURCE

When an incoherent source fills the field of view of an instrument as shown in Fig. 8.1, the general technique for finding the source radiance is as follows. Based on the conservation of radiance along rays in a lossless medium we know that the radiance measured at the entrance aperture of the instrument is $L_m = L\tau$, where L is the source radiance and τ the transmittance of the medium. When the radiance responsivity of the instrument is R_L as defined in (8.3) and the voltage output is v, the source radiance is found from

$$L = L_m/\tau = v/R_L\tau \quad (\text{W cm}^{-2}\,\text{sr}^{-1}). \tag{8.18}$$

If it is not possible to arrange for the source to fill the field of view, then it is necessary to calculate the source radiance from the radiance measured at the instrument entrance aperture and the source and instrument geometry. In this case the source radiance is

$$L = \frac{L_m}{\tau} = \frac{E}{\Omega_e'\tau} = \frac{ED^2}{\tau A_s \cos\theta} = \frac{vD^2}{R_E\tau A_s \cos\theta} \quad (\text{W cm}^{-2}\,\text{sr}^{-1}), \tag{8.19}$$

where L is the source radiance, L_m the measured radiance at the entrance aperture, E the measured irradiance at the entrance aperture, τ the atmospheric transmittance, Ω_e' the solid angle subtended by the source at the instrument entrance aperture, v the voltage output, D the distance from the source to the entrance aperture, R_E the irradiance responsivity, A_s the source area, and θ the angle between the source normal and the instrument axis. These relations obviously assume that the transmission medium is passive. If the transmission medium also emits radiant energy, then the relations found in Chapter 3 for emitting and attenuating media must be used. Also, it should be pointed out that the principal uncertainty in measurements of source radiance over long atmospheric paths is the transmittance τ.

Figure 8.6 shows the components of a simple radiometer used in the infrared region of the spectrum. A chopper modulates the incident radiant

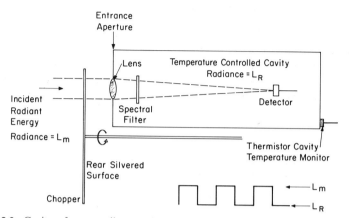

Fig. 8.6 Cavity reference radiometer for radiance measurement in the infrared region.

energy entering a temperature-controlled cavity. With the rear surface of the chopper coated with a highly reflecting material, the detector looks alternatively at the incident signal energy and at the energy reflected from inside the cavity. A waveform which is shown in idealized form in Fig. 8.6 is generated so that the peak-to-peak voltage difference Δv_m is proportional to the radiance difference $L_m - L_R$. The measured radiance L_m is then found from

$$L_m = (S\,\Delta v_m/R_L) + L_R, \qquad (8.20)$$

where L_m is the incident radiance at the entrance aperture, S the electronic attenuation factor, Δv_m the measured output voltage difference, R_L the radiance responsivity, and L_R the reference radiance from the temperature-controlled cavity. In this arrangement the chopper can be seen to provide several advantages. It provides discrimination between the incident signal energy and other energy incident on the detector; it permits the introduction of the reference radiance level; and it permits the use of ac amplifiers which are inherently more stable than dc amplifiers. A similar technique can be used for radiance measurements in the visible or ultraviolet region with the difference that the reference radiance would be provided by a different type of controlled reference source.

Measurements of radiant intensity are performed indirectly by measuring the irradiance produced by the source at a known distance where the distance is large in comparison with the source dimensions. At sufficiently large distances the source radiant intensity is

$$I = I_m/\tau = E_m D^2/\tau = v D^2/R_E \tau \quad (\text{W sr}^{-1}) \qquad (8.21)$$

where I is the source radiant intensity, I_m the measured radiant intensity, τ the radiant transmittance of the atmospheric transmission medium, E_m the measured irradiance at the instrument entrance aperture, D the distance from the source to the entrance aperture, v_m the measured voltage, and R_E the irradiance responsivity of the instrument. Since relation (8.21) depends on the inverse square law, the distance D must be large enough to justify the use of this law. Figure 3.7 shows the inverse square law for a circular source. If the maximum dimension of an arbitrary source is taken to be equal to the diameter of an equivalent circular source, then Fig. 3.7 can be used for arbitrary source shapes. A convenient result to remember is that when the distance D is more than 10 times the maximum source dimension, the deviation from the inverse square law is less than 1%.

To ensure that the irradiance at the entrance aperture shown in Fig. 8.1b is completely collected by the detector, it is essential that the source remain entirely within the field of view of the instrument. From another viewpoint, in an imaging radiometer the source image must lie entirely within the sensitive surface area of the detector. This is done by ensuring that the instrument field of view is somewhat larger than the source even at the expense of collecting increased background radiant energy. This also reduces the requirements on pointing and aligning the instrument. A second consideration in these measurements is that the detector responsivity is usually nonuniform across its surface. This problem is accommodated by using a field lens located close to the field stop to reimage the aperture of the instrument onto the detector surface. Since the irradiance in the aperture is usually uniform, this ensures that the responsivity of the instrument remains constant as long as the source remains totally within the instrument field of view.

8.5 GEOMETRICALLY TOTAL RADIANT POWER

When we integrate the radiant power from a source over all angles we obtain the geometrically total radiant power. Measurements of this quantity can be performed with the aid of an integrating sphere as shown in Fig. 8.7. When the source is placed within the sphere as shown and the inside surface of the sphere is coated with a diffuse material of high reflectance, the radiant power from the source is geometrically distributed in a uniform manner due to multiple reflections. The irradiance measured at a small aperture in the surface of the sphere is then proportional to the geotrically total radiant power from the source. In this arrangement the proportionality constant depends only on the reflectance of the coating material and the sphere radius.

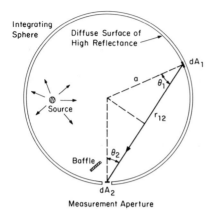

Fig. 8.7 Integrating sphere with small aperture for measurement of geometrically total radiant power from a source.

The integrating sphere technique (Sumpner, 1892; Walsh, 1958) is based on the following relations. Consider the radiant power which arrives at a small area element dA_2 in the aperture due to a small area element dA_1 on the surface of the sphere. From Chapter 3 this power is

$$d^2\Phi_{12} = L_1\, dA_1\, dA_2 (\cos\theta_1 \cos\theta_2)/r_{12}^2, \qquad (8.22)$$

where L_1 is the radiance at dA_1 and the other quantities are as illustrated in Fig. 8.7. Since $\cos\theta_1 = r_{12}/2a$ and $\theta_1 = \theta_2$, the irradiance at dA_2 due to dA_1 is

$$dE_2 = d^2\Phi_{12}/dA_2 = L_1\, dA_1\,(1/4a^2). \qquad (8.23)$$

We can express the radiance at dA_1 as $L_1 = E_1\rho/\pi$, where E_1 is the irradiance at dA_1 and ρ the coating material reflectance. Thus, from (8.23),

$$dE_2 = (\rho/4\pi a^2)E_1\, dA_1. \qquad (8.24)$$

In the ideal integrating sphere the irradiance E_1 is uniform over the surface. If we calculate the irradiance at dA_2 due to all surface area elements dA_1, we therefore find

$$E_2 = (\rho/4\pi a^2)\Phi_t, \qquad (8.25)$$

where

$$\Phi_t = \Phi(1 + \rho + \rho^2 + \rho^3 + \cdots) \qquad (8.26)$$

is the total radiant power incident on the sphere surface due to direct and reflected components and Φ the geometrically total radiant power from the source. Using the power series expansion $(1 - \rho)^{-1} = 1 + \rho + \rho^2 + \cdots$ we find from (8.25) and (8.26) that the irradiance at the measurement aperture is

$$E_2 = \rho\Phi/4\pi a^2(1 - \rho) \quad \text{(W cm}^{-2}\text{)}, \qquad (8.27)$$

where ρ is the coating material reflectance, Φ the geometrically total radiant power from the source within the sphere, and a the sphere radius. Notice that in (8.27) E_2 is proportional to Φ and that the proportionality constant depends on ρ and a.

Equation (8.27) assumes an ideal integrating sphere. Real integrating spheres have

(1) a baffle to prevent direct source energy from reaching the measurement aperture,

(2) a measurement aperture of finite size,

(3) coatings which are not perfectly diffuse or uniform and depend on wavelength, and

(4) the internal source itself.

All of these factors modify the spatial and spectral distribution and can be considered in a more detailed analysis (Walsh, 1958). In many cases a detailed analysis is unnecessary since the sphere is used in a *substitution* mode in which a standard source and the test source are successively placed in the same position and a power ratio is measured.

8.6 RADIOMETRIC TEMPERATURE MEASUREMENT

Since all materials at a temperature above absolute zero emit radiant energy and since the amount of emitted energy depends on temperature, radiometric measurements can be performed to determine the temperature of a remote material source (Plumb, 1972). The field of measurements concerned with radiometric temperature determination is known as *pyrometry*. In this section we briefly survey techniques for performing these measurements.

The basic relation for radiometric temperature measurements is

$$L(\lambda, \theta, T_R) = \varepsilon(\lambda, \theta, T)L_{bb}(\lambda, \theta, T), \tag{8.28}$$

where θ is the angle from the surface normal, T the true temperature of the material surface, T_R the radiometric (apparent) temperature of the material surface, L_{bb} the blackbody radiance, L the radiance from the material, and ε the emissivity of the material. What (8.28) shows is that angle and wavelength effects play a role in radiometric temperature measurements, and that the effects of material emissivity ε must be considered. Due to emissivity effects, the radiometric or apparent temperature T_R is not the same as the true temperature T. Some pyrometric techniques are designed to measure $L(T_R)$ and then infer T from a knowledge of ε. Other methods are designed to eliminate the need for emissivity information.

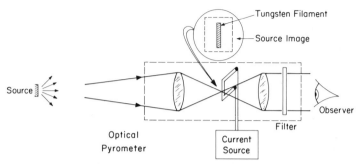

Fig. 8.8 Simple optical pyrometer for measurement of radiometric temperature T_R using a disappearing filament technique.

A relatively simple technique for measuring $L(T_R)$ is the disappearing filament method illustrated in Fig. 8.8. A source is imaged onto an intermediate image plane where a tungsten filament at a controlled current is placed. By requiring that the filter absorb radiant energy at wavelengths shorter than the red, the combination of the filter transmittance and observer's eye response produce essentially a single wavelength measurement at an effective wavelength λ_e with a bandwidth $\Delta\lambda$. The observer adjusts the filament current until the filament disappears indicating a radiance match at the wavelength λ_e. Using the Wien approximation the radiance is

$$L_\lambda \, \Delta\lambda = C_c \varepsilon(\lambda_e) \exp(-hc/\lambda_e kT) \, \Delta\lambda \quad (\mathrm{W\,cm^{-2}\,sr^{-1}}), \qquad (8.29)$$

where C_c is the calibration constant, $\varepsilon(\lambda_e)$ the source emissivity at effective wavelength λ_e, $hc/\lambda_e k$ the constant from the blackbody radiation law, and T the true temperature of the source. When the source emissivity $\varepsilon(\lambda_e)$ is known and when the radiance of the filament as a function of current has been previously calibrated, the true temperature T can be found. Most instruments would actually assume that $\varepsilon = 1$ so that the temperature reading is the radiometric or apparent temperature T_R. Typical operating ranges of optical pyrometers of this type cover 700–12,000 K source temperatures. A simple telescope may also be employed to enlarge the image of distant small sources for reliable comparison with the filament.

With a stable radiometer it is possible to eliminate the human observer and simply measure $L_\lambda \, \Delta\lambda$ directly. The relation (8.29) is then used to determine the temperature T. Direct measuring pyrometers of this type still require information concerning the source emissivity $\varepsilon(\lambda_e)$ if the true temperature T rather than the apparent or radiometric temperature T_e is to be determined.

A second method (which is attributed to M. Czerny) of measuring temperature which does not require emissivity information is illustrated in

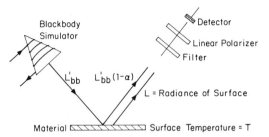

Fig. 8.9   Czerny method of determining material temperature T without emissivity data.

Fig. 8.9. A blackbody simulator which is maintained at temperature T' by a controlled heater current irradiates the surface of interest. If the radiance of the blackbody simulator is L'_{bb}, then the component reflected from the surface is $L'_{bb}(1 - \alpha)$, where α is the surface absorptivity. The radiance of the surface itself is εL_{bb}. The total radiance from the surface is then the sum

$$L_{tot} = L'_{bb}(1 - \alpha) + \varepsilon L_{bb}, \tag{8.30}$$

where $\varepsilon = \alpha$ is the surface emissivity and L_{bb} the blackbody radiance for a surface at temperature T. In Fig. 8.9 the filter and detector in combination determine the effective wavelength λ_e at which the measurement is performed. If the linear polarizer is rotated and the simulator temperature T' is varied, there will be a temperature T' at which the total radiance L_{tot} is unpolarized. When this occurs, $\alpha L'_{bb} = \varepsilon L_{bb}$ and so $L_{tot} = L'_{bb}$. The temperature T of the surface is then equal to the simulator temperature T'. Notice that in this method it is not assumed that the surface is acting as a blackbody. Instead it is the balance of absorbed and emitted energy that permits the determination of the temperature T.

8.7   MEASUREMENTS OF EMISSIVITY

To determine the radiance of a surface at a known temperature the surface emissivity must be known. Emissivity was defined in Chapter 4 as the ratio of the radiance of the actual surface to the radiance of a blackbody. There are basically two different methods available for determining emissivity and these are referred to as the *radiometric* method and the *calorimetric* method.

The radiometric method (Bedford, 1972) is based on a comparison of the radiance of the sample to that of a blackbody simulator at the same temperature. Either the spectral or the total emissivity can be measured in this way. The comparison blackbody simulator can be separate or can be constructed so that the sample is an integral part of it. The integral arrangement

TABLE 8.4 Allowed Temperature Variation (Kelvin) for 0.5% Variation
of Blackbody Radiance

T (K)	$\lambda = 0.4$ (μm)	$\lambda = 0.5$ (μm)	$\lambda = 0.75$ (μm)	$\lambda = 1.0$ (μm)	$\lambda = 5$ (μm)	$\lambda = 15$ (μm)
500	0.035	0.043	0.065	0.087	0.433	1.112
1000	0.139	0.174	0.261	0.348	1.640	3.215
1500	0.313	0.391	0.586	0.782	3.335	5.538
2000	0.556	0.695	1.042	1.389	5.302	7.949
2500	0.869	1.086	1.628	2.164	7.425	10.389
3000	1.251	1.564	2.343	3.102	9.639	12.848

is preferred at high temperatures or short wavelengths since it permits more accurate relative temperature control. To indicate the temperature control requirements, Table 8.4 shows the allowable temperature variation for 0.5% variation of blackbody radiance.

Figure 8.10 shows an arrangement for a radiometric spectral emissivity measurement using a separate blackbody (Bedford, 1972). The sample is placed in a furnace at temperature T and a blackbody simulator is operated at the same temperature. A chopper alternately views the sample and the blackbody simulator. Both beams are directed along the same optical path through a monochromator. Reflecting components are used so that wavelengths from the visible out to approximately 15 μm in the IR can be passed, within the limits of the monochromator. The IR measurements employ a thermocouple detector while the visible and near IR signal is detected with a photomultiplier. As shown in the figure a reference blackbody can be placed in the sample beam for calibration purposes.

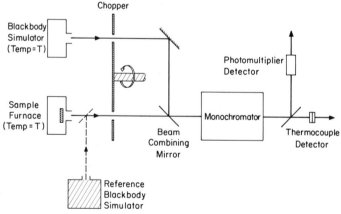

Fig. 8.10 Arrangement of components for spectral emissivity measurement of temperature-controlled sample. Focusing optical components are omitted for clarity.

The alternative radiometric method is the integral blackbody technique. In this technique the blackbody cavity walls are constructed of the material of interest. By viewing a small aperture in the wall of the cavity with an optical pyrometer and by subsequently viewing the sample material surface on the region near the hole it is possible to determine the emissivity.

The second basic method of measuring emissivity is the calorimetric method (Plumb, 1972). The emissivity or absorptivity is measured by determining the heat lost or gained by the sample through radiant heat transfer. The ratio of the observed rate of heat transfer to the theoretical rate of radiant heat transfer for a blackbody at the same temperature yields the emissivity or absorptivity. In measuring emissivity this technique is only suitable for spectrally total measurements. From the Stefan–Boltzmann law derived in Chapter 4 the hemispherical total emissivity is

$$\varepsilon(T) = \frac{\text{radiant power emitted by sample}}{\text{radiant power emitted by blackbody}} = \frac{vi}{\sigma A(T^4 - T_0{}^4)} \quad (8.31)$$

where v is the voltage applied to the sample, i the current through the sample, A the radiating sample area, σ the Stefan–Boltzmann constant, T the temperature of the sample, and T_0 the temperature of the surroundings. Relation (8.31) assumes that all electrical power delivered to the sample is converted to radiant power. This will be true when power losses due to conduction are negligible. It also assumes an equilibrium or steady-state situation.

In summary it can be said that the steady-state calorimetric method provides the most accurate measurements of emissivity. Other techniques such as reflectance measurements (Edwards *et al.*, 1961) are also used to measure emissivity using the relation $\varepsilon = 1 - \rho$, where ρ is the reflectance. However, for low emissivity materials small fractional errors in measuring ρ will produce large fractional errors in ε since the reflectivity is approximately equal to unity in this case.

8.8   LASER POWER AND ENERGY MEASUREMENTS

Laser power and energy measurements differ in several ways from measurements involving incoherent sources. All of these differences are due to the highly directional and essentially monochromatic character of the beam of radiant energy from a laser source. The principal differences are:

(1)   high energy and power densities,
(2)   use of instrumentation at essentially a single wavelength,
(3)   interference effects, and
(4)   rapid time changes.

With high energy and power densities the possibilities of linearity saturation and even component damage become important. Single wavelength operation of instrumentation means that the spectral variations of component transmittance $\tau(\lambda)$ or detector responsivity $R(\lambda)$ are avoided, thus simplifying data processing and interpretation. Interference effects can occur between the secondary beams of reflected and transmitted energy from various system components unless special precautions are taken. Finally, in the case of temporal measurements on Q-switched or mode-locked pulsed lasers which show rapid changes of power, it is necessary to employ detectors with high-frequency response. On the other hand, measurements of total pulse energy can be made with relatively slow detectors such as would be used for CW laser measurements.

Both (a) photon detectors and (b) thermal detectors as described in Chapter 6 are used for laser measurements. Photon detectors have the advantage of high-frequency response when compared with most thermal detectors, an exception being pyroelectric detectors. On the other hand, thermal detection systems can be more conveniently designed to absorb almost all incident energy and thereby minimize the complex interference effects which can occur when performing laser measurements. Thermal detection systems can also be designed to handle higher power and energy levels and are conceptually simpler than photon detection systems. In the following we emphasize thermal methods. Thermal detection systems for laser measurements are known as *calorimeters*. Laser calorimeters can be specifically classified (Gunn, 1973) as:

(a)   disk calorimeters,
(b)   cone calorimeters,
(c)   hollow sphere calorimeters,
(d)   bolometer calorimeters,
(e)   volume absorption calorimeters, and
(f)   partial absorption calorimeters.

Figure 8.11 shows an example of each of these calorimeter designs. The disk calorimeter shown in Fig. 8.11a is simple in design. A blackened metal disk absorbs incident energy and the resulting temperature rise is measured by contact thermocouples. The heater windings are used to calibrate the calorimeter through electrical standards. The cone calorimeter shown in Fig. 8.11b is widely used. If the included angle of the cone is $180°/n$ and the interior surface is specular, then an incoming ray parallel to the axis will experience n reflections before exiting. If the absorptance at each reflection is α, then the total absorptance before the ray exits is $1 - (1 - \alpha)^n$. Thus high absorptance coatings which damage easily are not required. The hollow sphere calorimeter in Fig. 8.11c achieves a similar effect through multiple

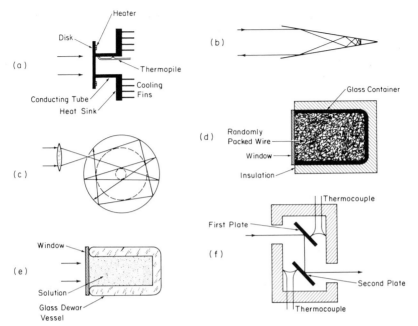

Fig. 8.11 Laser calorimeter arrangements (a) disk, (b) cone, (c) hollow sphere, (d) bolometer calorimeter, (e) volume absorption, and (f) partial absorption. (After Gunn, 1973.)

reflections as discussed later in this section. Figure 8.11d shows a rat's nest bolometer calorimeter in which the wire windings of the thermocouple serve as both the absorbing material and the temperature transducer. This calorimeter is particularly insensitive to the location of the incident beam within the absorbing area. In Fig. 8.11e is a liquid volume absorption calorimeter suitable for high-power pulsed or CW laser beams (Gunn, 1974). Solids composed of filter materials may also be used in place of the liquid. Due to the distribution of absorbed energy through the volume, the temperature rise at any point is small. The last arrangement, shown in Fig. 8.11f is a partial absorption calorimeter which is designed to absorb only a small but accurately known fraction of the beam. This form has the advantage of introducing minimum losses and also of being suitable for high-power laser measurements.

Calorimeters can be generally classified as either the *isoperibol* or *conduction* type (Gunn, 1973; West and Churney, 1970). The isoperibol calorimeter has a radiant energy absorber which is thermally isolated from the jacket of the calorimeter. The temperature rise of the absorber is measured after it has received a single pulse or a timed interval of CW laser power. By contrast, the conduction calorimeter has a coefficient of heat transfer from the absorber to the jacket which is as large as possible. The steady-state temperature gradient measured between the absorber and jacket is proportional to the absorbed radiant power. The liquid flow calorimeter is a variation on the conduction calorimeter in which the temperature rise of a liquid flowing through the absorber is measured.

Some of the design and measurement considerations (Watt, 1973; Tietz, 1977; West et al., 1972; Franzen and Schmidt, 1976; Thacher, 1976) with laser calorimeters include

(1) twin or single absorber,
(2) window effects,
(3) heater design,
(4) absorption, and
(5) thermal reradiation.

A twin design calorimeter employs two similar calorimeter absorbers, only one of which receives radiant energy. By measuring the temperature difference ΔT between the two absorbers measurement errors are reduced (Sakurai et al., 1967). The second consideration is window effects. Windows are used to contain the fluid of a liquid absorption calorimeter or to reduce thermal reradiation to the environment outside the calorimeter. A wedge-shaped window is used to minimize interference effects due to interreflections among window surfaces. Heater design, the third consideration, is important because calorimeters provide the advantage of direct electrical calibration. This electrical calibration is accomplished by measuring the electrical power dissipated in an internal heater along with the calorimeter temperature rise. Heater design is intended to ensure equivalence of electrical power with radiant power.

The next issue is absorption. With the high power levels which can be obtained from lasers it is possible to damage a blackened surface. The problem is avoided by using a cavity geometry which ensures multiple reflections. The cavity coating material can then have a high reflectivity to avoid damage and the calorimeter will still provide high absorption properties. Also, the overall absorptance becomes relatively insensitive to the coating material reflectivity.

Finally, thermal reradiation is a fundamental problem. Since the rate of energy loss between the absorber at temperature T and the environment at temperature T_e is proportional to $T^4 - T_e^4$, a large rise in absorber temperature is undesirable. The reradiation problem is minimized by using an absorber of large thermal mass, consistent with maintaining an absorber temperature change large enough to be measurable.

An analysis of the energy balance in laser calorimeters shows that

$$U_{in} = U_c + U_1, \tag{8.32}$$

where U_{in} is the input energy, U_c the energy sensed by the calorimeter, and U_1 the energy lost at the window, collector, absorber, or transducer. The loss mechanisms include reflection, absorption, scattering, conduction, and reradiation to the environment. That radiant energy which is sensed by the calorimeter is related to the temperature rise ΔT by

$$U_c = mc_p \Delta T, \tag{8.33}$$

where m is the effective mass and c_p the effective specific heat of the calorimeter. The measured energy U_m from the calorimeter is related to the temperature change ΔT by $U_m = \gamma \Delta T$, where γ is a calibration constant. Then from (8.32) and (8.33) the energy balance relation is

$$U_{in} = U_m(mc_p/\gamma) + U_1. \tag{8.34}$$

From (8.34) we can see that an important part of the accurate use of a laser calorimeter is the evaluation of all loss mechanisms and their reduction where possible.

8.9   BLACKBODY SIMULATOR EMISSIVITY

By cavity design the emissivity ε of a blackbody simulator is increased over the inner surface emissivity ε' (Bedford, 1972; Heinisch, 1972) to

$$\varepsilon = \varepsilon'(1 - k)/\{(A/s) + \varepsilon'[1 - (A/s)]\} \tag{8.35}$$

with A the aperture area, s the cavity surface area, and k a correction factor ($k \cong 0$). For a cavity of length l and aperture radius r

$$A/s = 1 - l^2/(r^2 + l^2) \qquad \text{(sphere),} \tag{8.36}$$

$$A/s = \pi r^2/[\pi r^2 + \pi r(r^2 + l^2)^{1/2}] \qquad \text{(cone),} \tag{8.37}$$

$$A/s = \pi r^2/(2\pi r^2 + 2\pi r l) \qquad \text{(cylinder).} \tag{8.38}$$

For example, for $l/r = 10$ and $\varepsilon' = 0.5$ we have $\varepsilon = 0.99$ (sphere), $\varepsilon = 0.97$ (cylinder), and $\varepsilon = 0.95$ (cone).

8.10   DETECTOR MEASUREMENTS: RESPONSIVITY, LINEARITY, NEP, AND D^*

In most radiometric systems the detector determines the basic system capability. Among the detector properties which are important in system performance are the responsivity, linearity, noise equivalent power (NEP), and normalized detectivity (D^*). Measurements of these properties are of interest for describing detectors as components and also as a preparation for performing radiant power and energy measurements.

In Section 8.1 we defined three responsivity quantities. The responsivity of most direct interest in detector measurements is the power responsivity defined as $R = v/\Phi$, where v is the detector output voltage and Φ the radiant power incident on the detector surface. The corresponding spectral power responsivity is $R(\lambda) = v_\lambda/\Phi_\lambda$. To measure the spectral power responsivity $R(\lambda)$ it is common to measure separately (a) the relative spectral power responsivity at all wavelengths, and (b) the absolute spectral power responsivity at one or a few selected wavelengths. In these measurements either a known detector or a known source can, in principle, be used as the reference. For practical reasons the following procedure is used.

(a)   Measure the relative spectral power responsivity at all wavelengths with a monochromator (or filters) and a reference thermal detector which is assumed to be spectrally flat.

(b)   Measure the absolute spectral power responsivity at one (or a few) wavelengths with a narrowband filter and a reference thermal detector of known absolute responsivity.

The monochromator provides the flexibility of wavelength selection in the relative measurements, while the narrowband filter avoids the optical characteristics of the monochromator such as beam geometry which could influence the absolute measurements.

Figure 8.12 shows a schematic arrangement of components for measuring the relative spectral power responsivity of a detector (Potter *et al.*, 1959;

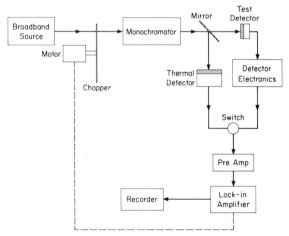

Fig. 8.12 Components for measurement of the relative spectral power responsivity of a test detector using a black thermal detector as reference.

Jones *et al.*, 1960). The mirror can be moved so that the essentially mono-chromatic beam from the monochromator falls alternately on the test de-tector and the reference thermal detector. The switch is synchronized with the mirror position so that the output of the appropriate detector is amplified and recorded. As shown in the figure the input beam to the monochromator is chopped. The lock-in amplifier provides synchronous amplification of the chopping frequency. To stay within the frequency response of the thermal detector the chopping frequency would ordinarily be approximately 10 Hz. However, since the $1/f$ noise in the test detector will decrease as the modula-tion frequency f is increased, it is sometimes necessary to increase the chopping frequency to 100–1000 Hz and provide amplification of the lower level signal. The principal requirement for the broadband source is that it provide an essentially continuous spectrum of radiant energy so that after transmission through the monochromator the detected energy is well above the detector noise level.

   If the test detector produces a voltage v_λ at each wavelength and the reference detector produces the voltage v_λ' at the same wavelength, then the relative spectral power responsivity of the test detector is $r_\lambda = (v_\lambda/v_\lambda')r_\lambda'$, where r_λ' is the relative spectral power responsivity of the reference detector. When the reference detector is spectrally flat, then $r_\lambda = v_\lambda/v_\lambda'$. As mentioned previously the absolute spectral power responsivity can be determined by using a filter rather than the monochromator shown in Fig. 8.12. At the wavelength λ the corresponding absolute spectral power responsivity is $R_\lambda = (v_\lambda/v_\lambda')R_\lambda'$, where R_λ' is the reference detector absolute spectral power responsivity.

   The total responsivity of a detector will depend on the relative spectral power distribution of the incident beam. This dependence was shown in Section 6.3 in connection with the definition of the spectral matching factor (SMF). A common spectral power distribution is that of a blackbody and the corresponding total responsivity of a detector is known as the *blackbody responsivity*. Following the definition of responsivity in Section 6.3, the total voltage responsivity with a blackbody spectrum is

$$R_{bb} = v/\Phi = \int_0^\infty R(\lambda)\Phi_{\lambda,bb}\,d\lambda \bigg/ \int_0^\infty \Phi_{\lambda,bb}\,d\lambda \quad (\text{V W}^{-1}), \qquad (8.39)$$

where R_{bb} is the blackbody (voltage) responsivity (V W^{-1}), v the measured detector output voltage (V), Φ the total power incident on the detector (W), $R(\lambda)$ the spectral responsivity (VW^{-1}), and $\Phi_{\lambda,bb}$ the spectral power distribution of blackbody radiation (W μm^{-1}). A similar definition applies to an instrument rather than a detector if the spectral responsivity and spectral power are defined in terms of the instrument entrance aperture. If a blackbody simulator is used, the blackbody responsivity can be measured.

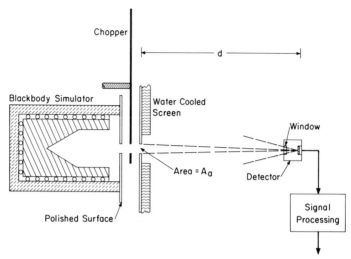

Fig. 8.13   Arrangement of a blackbody simulator and detector for measurement of black-body responsivity of detector.

An arrangement for performing this measurement is shown in Fig. 8.13. A blackbody simulator operating at temperature T emits radiant energy with a well-characterized spectral distribution as described in Chapter 4. If the detector spectral responsivity extends to wavelengths in the infrared beyond a few micrometers, it may be necessary to use a water-cooled screen to reduce background energy within the detector field of view. As shown in the figure the window determines the detector field of view. A chopper located as shown in the figure will produce a chopped beam in which the detector alternately sees the radiant energy from the simulator and the chopper blades. If the aperture in the screen acts as the limiting aperture subtended at the detector surface and when the distance $d \gtrsim 10\sqrt{A_a^{1/2}}$, so that the inverse square law applies, then the irradiance at the detector surface is

$$E_d = \tau_w k_c (\sigma T^4 A_a / \pi d^2) \quad (\text{W cm}^{-2}), \tag{8.40}$$

where τ_w is the window transmittance, k_c the constant associated with chopper waveform, σ the Stefan–Boltzmann constant, T the temperature of the blackbody simulator (K), A_a the area of the limiting aperture in the screen, and d the distance from the screen to the detector. For a square wave chopper the constant is $k_c = 0.45$. From (8.40) and the detector area the radiant power Φ at the detector can be found. From (8.39) the blackbody responsivity is then $R_{bb} = v/\Phi$.

A second important property of a detector is linearity. By definition, a detector is linear under the following conditions. Assume that a detector produces an output voltage $v_1 = R\Phi_1$ in response to incident power Φ_1, and an output voltage $v_2 = R\Phi_2$ in response to incident power Φ_2. When the response to an input $\Phi_1 + \Phi_2$ is $v = v_1 + v_2 = R(\Phi_1 + \Phi_2)$, then the detector is said to be linear. Most detectors will be linear over a limited range of power input Φ and the purpose of the linearity measurement is to determine this range. Techniques for determining linearity (Sanders, 1972) include (a) double aperture methods, (b) inverse square law methods, and (c) polarization attenuator methods. The emphasis in each technique is in precisely controlling the amount of radiant energy incident on the detector.

The first technique, the double aperture method (Sanders, 1972), uses two small adjacent apertures in the beam path. By opening either or both apertures the power quantities Φ_1, Φ_2, and $\Phi_1 + \Phi_2$ can be produced. The inverse square law method employs the dependence of power on distance from the source as described in Chapter 3. This method is highly reliable when large distances are available and when the distance from source to detector is at all times made large enough to justify the inverse square assumption (see Chapter 3). The polarization attenuator is a convenient alternative technique which does not require large distances and is continuously adjustable. Conditions under which it is accurate have been discussed in the literature (Mielenz and Eckerle, 1972).

The other detector properties of interest are the noise equivalent power (NEP) and normalized detectivity (D^*). As defined in Chapter 6 they are related by $D^* = (AB)^{1/2}/\text{NEP}$, where A is the detector active area and B the noise bandwidth. Since the NEP is defined by $\text{NEP} = v_{\text{rms}}/R$, it is necessary to measure both the rms noise voltage v_{rms} and the responsivity R to determine NEP and D^*. If the responsivity is that for a blackbody as described earlier in this section, then corresponding quantities are $\text{NEP}_{\text{bb}} = v_{\text{rms}}/R_{\text{bb}}$ and $D^*_{\text{bb}} = (AB)^{1/2}/\text{NEP}_{\text{bb}}$. The noise voltage v_{rms} is found by shielding the detector from incident radiant signal energy and measuring the root mean square output voltage from the detector. This noise voltage may be due to any of a number of noise mechanisms including (a) background shot noise, (b) dark current shot noise, (c) Johnson (thermal) noise, and (d) amplifier noise. As a result the conditions under which the measurement is made must be clearly defined.

REFERENCES

Bedford, R. E. (1972). Effective emissivities of blackbody cavities—A review, *In* "Temperature: Its Measurement and Control in Science and Industry," Vol. 4, p. 425. Instrument Society of America.
Bell, R. J. (1973). "Introduction to Fourier Transform Spectroscopy." Academic Press, New York.
Edwards, D. K. *et al.* (1961). *J. Opt. Soc. Am.* **51**, 1279.

Fellgett, P. (1951). Thesis, Cambridge Univ.

Fellgett, P. (1967). *J. Phys. C2* **28**, 165.

Franzen, D. L., and Schmidt, L. B. (1976). *Appl. Opt.* **15**, 3115.

Gunn, S. R. (1974). *Rev. Sci. Instrum.* **45**, 936.

Gunn, S. R. (1973). *J. Phys. E* **6**, 105.

Haycock, R. H., and Baker, D. J. (1974). Infrared physics, **14**, 259.

Heinisch, R. P. (1972). The emittance of blackbody cavities, *In* "Temperature: Its Measurement and Control in Science and Industry," Vol. 4, p. 435. Instrument Society of America.

Jones, R. C. *et al.* (1960). Standard procedure for testing infrared detectors and for describing their performance. Office of the Director of Defense Research and Engineering, Washington, D.C.

Mielenz, K. D., and Eckerle, K. L. (1972). *Appl. Opt.* **11**, 594.

Plumb, H. H. (1972). "Temperature: Its Measurement and Control in Science and Industry," Vol. 4. Instrument Society of America.

Potter, R. F. *et al.* (1959). *Proc. IRE* **47**, 1503.

Rabiner, L. R., and Gold, B. (1975). "Theory and Application of Digital Signal Processing." Prentice Hall, Englewood Cliffs, New Jersey.

Sakurai, K. *et al.* (1967). *IEEE Trans. Instrum. Measurement* **IM-16**, 212.

Sanders, C. L. (1972). *J. Res. Nat. Bur. Std.* **76A**, 437.

Sumpner, W. E. (1892). *Proc. Phys. Soc.* **12**, 10.

Steel, W. H. (1967). "Interferometry." Cambridge Univ. Press, London and New York.

Thacher, P. D. (1976). *Appl. Opt.* **15**, 1815.

Tietz, G. E. (1977). *Appl. Opt.* **16**, 1136.

Vanasse, G. A. (ed.) (1977). "Spectrometric Techniques," Vol. 1. Academic Press, New York.

Vanasse, G. A., and Sakai, H. (1967). Fourier spectroscopy, *Progr. Opt.* **6**.

Walsh, J. W. T. (1958). "Photometry." Dover, New York.

Watt, B. E. (1973). *Appl. Opt.* **12**, 2373.

West, E. D., and Churney, K. L. (1970). *J. Appl. Phys.* **41**, 2705.

West, E. D. *et al.* (1972). *J. Res. Nat. Bur. Std.* **76A**, 13.

9

Measurements of Reflectance, Transmittance, and Absorptance

9.1 INTRODUCTION

In this chapter we shall survey the processes that occur when materials are irradiated and the properties of materials that are important in determining optical behavior. The symbols and terminology used have been presented in Chapter 2.

9.1.1 Background

When a material is irradiated by optical radiant energy, a portion of the incident energy is reflected, another portion is absorbed, and a third portion is transmitted. This action can be described by various properties of the material (such as spectral, photometric, general, and optical). The properties of fluorescent materials are generally described in the same manner as those of nonfluorescent materials, although the former behave as primary sources for that portion of the radiant energy which is generated by fluorescence. The photometric properties are not specific properties of a material since they also depend on other parameters.

In general the properties depend on the following factors: the spectral composition of the radiant energy, the state of polarization (the reflected or transmitted energy is usually partially polarized even if the incident energy is unpolarized), the incident angle and angle of viewing, the angular extent of the incident radiation and the viewing beam, the thickness of the sample, the temperature, the state of the surface, and a weighting function.

9.1.2   Definition of Spectrophotometric Measurements

Spectrophotometry is the term commonly used to refer to the ensemble of techniques for measuring electromagnetic spectra generated by the interaction of a source of irradiance with a test sample of matter under laboratory conditions. For simplicity we sometimes use the term "light" in place of electromagnetic radiant energy. Light is radiant energy evaluated according to its capacity to produce human visual sensation, and this term is in common usage in spectrophotometry.

A spectrophotometric analysis always encompasses the measurement of an amplitude (intensity) and the measurement of energy (frequency), both in a quantitative way.

The interaction of light and matter produces two types of spectra in the electromagnetic spectral range: emission spectra and absorption spectra.

9.2   ORIGIN OF SPECTRA AND SPECTRAL RANGES

The electromagnetic spectra result from the transitions of matter from either ground-state or excited-state energy conditions into another state with concomitant absorption or emission of electromagnetic radiant energy to maintain the energy balance in compliance with the principle of energy conservation. An example of such a process is shown schematically in Fig. 9.1.

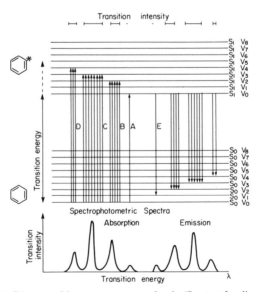

Fig. 9.1 State transitions among energy levels. (See text for discussion.)

The processes shown are explained below. The transition from S_0 to S_1 electronic level is shown by the vertical lines in the figure (this is the absorption process). The energy of the transition is represented by the length of the arrows. The broadening of the transition lines into a band is illustrated by the vibrational levels V_0, $V_1 \rightarrow V_n$ that are present both in the ground and the excited states of the molecule. The intensity of the transition (proportional to the square of the probability of the transition) is represented by the relative number of arrows. The reverse transitions (from $S_1 \rightarrow S_0$) represent the emission process.

The resulting spectra (absorption and emission) are given in the lower part of the figure. The coordinates are the two important spectrophotometric quantities: transition energy (expressed in frequency or wavelength) and transition intensity (expressed in units of absorptance). The conventional spectrophotometric domain is confined to ultraviolet, visible, and infrared spectral ranges. The spectra in this domain are known as atomic, electronic, and molecular spectra.

9.2.1   Atomic Spectra

Atomic spectra are typically intense lines produced by electronic transitions in unassociated atoms. They fall mostly into the ultraviolet spectral range; however, at times they can reach even the near-infrared region of the spectrum. These spectra are generated by electronic transitions between two or more energy levels available in the electronic structure of the atoms (atomic orbitals and atomic shells). These types of spectra can normally be observed both in absorption and in emission forms without any spectral shift. They are studied by atomic absorption and atomic fluorescence methods (Robinson, 1966; Grotrian, 1928; Dean and Rains, 1971), apart from traditional spectrophotometry.

9.2.2   Electronic Spectra (Herzberg, 1945a, 1966)

The electronic spectra are typically band spectra produced by electronic transitions in associated atoms. When these associated atoms are organic molecules, the spectra are known as molecular electronic spectra. Molecular electronic spectra are observed in the ultraviolet and visible spectral regions and only exceptionally in the near infrared. Their general features are broad unresolved bands.

Molecular electronic spectra rank from the very intense (extinction above 300,000) (Murrel, 1963) to extremely weak (forbidden transitions with extinction smaller than 100). These spectra are generated by electronic transitions between various energetic levels available in the electronic structure of the molecules (molecular orbitals).

The broadening of the electronic transition lines into bands is mostly due to the superposition of the vibrational transitions (infrared) and to a lesser extent of rotational and translational transitions (microwaves).

In their emission forms (fluorescence, phosphorescence, delayed fluorescence) they are generally shifted toward longer wavelengths (they are bathochromically or "red" shifted) relative to their absorption counterpart. This phenomenon is the well-known "Stokes shift" of the electronic spectra (Stokes, 1862). Exceptionally this type of spectra may exhibit a shift toward shorter wavelengths (hypsochromic or blue shift). This not very common phenomenon is known as the "anti-Stokes" shift of the electronic spectra.

The molecular electronic spectra are the most studied and most used of all the electronic spectra.

9.2.3   Semiconductor Electronic Spectra

The electronic spectra can be generated also among bands in solids. This phenomenon is common in semiconductive, photoconductive, and even ionic conductive crystalline lattices (Seraphin, 1976; DiBartolo, 1968). They may be generated by "direct" transitions between the valence band (ground state) and the conduction band of the solid. In this case the absorption spectrum assumes the shape of a continuous absorption for all energies higher than a specific one (bandgap energy). Direct transitions are intense and are commonly expressed in terms of absorption coefficient (κ), typically of the order of 10^5 cm^{-1} (10^4–10^6 for silicon) (Dash and Newman, 1954).

Alternatively, transitions in solids may involve the conduction band and a localized center of higher energy lying above the valence band. Such centers can be an impurity trapped in the lattice or a defect in the lattice (Bassani and Parravicini, 1975). These centers, known as *impurity centers* and/or *color centers*, may supply very active electrons for absorption transitions, or they may be very efficient electron acceptors in emission transitions (recombination centers and luminescence centers) (Goldberg, 1966).

Other transitions occurring in semiconductive solids are the "indirect transitions" or "thermally assisted" transitions to the conduction band. In the transitions of this type the thermal energy, in the form of lattice vibrations or "phonons," is utilized for promoting a valence-band electron to the conduction band. They produce relatively broad bands of subbandgap energy, the intensity of which is dependent on the temperature.

9.2.4   Electronic Spectra of Ionic Solutions

Electronic spectra are also commonly observed for ionic solutions (Petrucci, 1971). This is the case for many metal sulfates (Ni, Ca, Co). These spectra are observable in the visible spectral region and are produced by charge-transfer transitions among anions and cations. The spectra are very broad structureless bands, easily distinguishable from the molecular electronic spectra. Electronic spectra are also generated by charge-transfer transitions occurring among organic molecules. They exhibit the same general features as the charge-transfer spectra among inorganic ions; however, they are generally very weak and not easily observable. The observation of this type of spectra is hindered by the interference of the much stronger (overlapping) molecular spectra (Foster, 1969).

Yet another type of electronic spectra is ligand field spectra. These are generated by ligand field transitions that occur among metals and organic molecules (ligands) (Jorgensen, 1963, 1971).

9.2.5   Molecular Spectra

The molecular spectra are typically narrow-band spectra that are extremely well resolved. They are produced by vibrational transitions in the chemical bonds of molecules (transitions among discrete energy levels into which the vibrational energy of the bond is distributed) (Herzberg, 1945b,c). This type of spectra occurs in the infrared spectral region. The line-broadening into the narrow bands is produced by the superposition of the vibrational and translational transitions. Molecular spectra are weaker than molecular electronic spectra ($\varepsilon \approx 1000$). Vibrational transitions may take place in chemical bonds typical of a molecular functional group (i.e., —O—H, —N—H, —C=O). In this case they generate a few characteristic and easily interpretable bands generally called "functional bands." They may also take place in "connecting" bonds such as those in

$$-\overset{|}{\underset{|}{C}}-\overset{|}{\underset{|}{C}}-, \qquad -\overset{|}{\underset{|}{C}}-N-, \qquad -\overset{|}{\underset{|}{C}}-O-,$$

and in this case they produce very complex and characteristic bands called "skeletal" bands. These bands are very useful in molecular identification (fingerprinting).

Infrared spectrophotometry is, due to its high resolution capability, a very useful tool for studying and distinguishing the differences among molecules. Infrared spectra are, to date, used only in the absorption mode. However, vibrational spectra can be measured by a Raman technique that is similar to emission spectroscopy. Raman spectra carry spectroscopic information very similar to that of infrared absorption spectra, but since the intensity of Raman spectra depends on different "selection rules," they are often used as complementary to infrared spectra. It should be pointed out that Raman

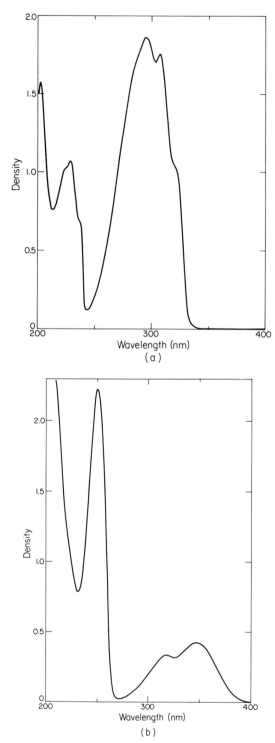

Fig. 9.2 Molecular electronic spectrum: (a) trans-stilbene in cyclohexane; (b) quinine fate.

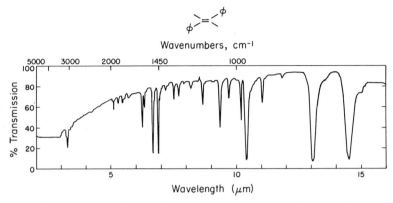

Fig. 9.3 Molecular vibration spectrum (trans-stilbene, KBr pressing).

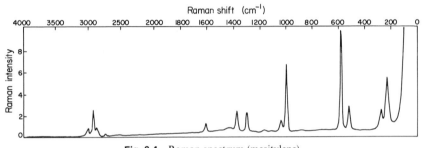

Fig. 9.4 Raman spectrum (mesitylene).

spectroscopy is essentially vibrational broadening of the light-scattering process, produced generally in the visible and the near-infrared regions. Figures 9.2–9.4 represent the various types of spectra just discussed.

9.3 PROCESSES INVOLVED IN SPECTROPHOTOMETRY

9.3.1 General Review

Spectrophotometric analysis has two distinct goals:

(1) to determine the amount of electromagnetic radiant energy absorbed by the sample;

(2) to determine the modifications imposed by the sample on a given flux of radiant energy.

In practical applications, analytical, photochemical, and photophysical, the determination of absorption is normally sought. However, when the interest is in spectral selectivity, attenuation and evaluation of color, and appearance of materials, the second goal becomes important.

Since conventional spectrophotometry does not measure the absorption directly, it must be determined from quantities that are experimentally obtainable.

The most recent techniques, based on the calorimetric effect of absorption, do indeed offer a direct measurement of the radiant energy absorbed by a sample. These new techniques are optoacoustic and thermooptical blooming spectroscopic techniques (*Adams et al.*, 1976a,b). These techniques, however valid in principle, have not been sufficiently developed and are not at present of practical use in spectrophotometry. Further details on these two techniques are available in the literature (Long *et al.*, 1976; Swofford *et al.*, 1976).

9.3.2 Distribution of Fluxes

A beam of collimated light passing through a turbid material is dispersed into many components as shown schematically in Fig. 9.5. As evident from the figure, one deals with six different fluxes. Flux $\Phi_0(\lambda_1)$ is the incident flux as it falls on a sample; $\Phi_{\tau_r}(\lambda_1)$ represents the flux transmitted through the sample without modification of the direction (regularly transmitted flux); $\Phi_{\tau_d}(\lambda_1)$ represents the transmitted flux scattered in all directions (diffusely transmitted flux). Similarly, there are regularly reflected $\Phi_{\rho_r}(\lambda)$ and diffusely reflected $\Phi_{\rho_d}(\lambda_1)$ fluxes. Finally, there is the emitted flux $\Phi_L(\lambda_2)$ which is the flux emitted by the sample at wavelength λ_2 when irradiated with $\Phi_0(\lambda_1)$. All of the quantities discussed here have been defined in Chapter 2.

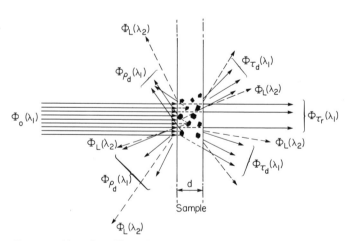

Fig. 9.5 Decomposition of a collimated beam of light through a turbid, fluorescent sample.

9.3.3 Beer–Lambert Law

A very important case in spectrophotometric measurements is the case when all of the fluxes emerging through a sample are zero with the exception of regularly transmitted flux Φ_{τ_r} and the absorber is an unassociated atom and/or molecule. In such a case there exists a very important relationship among the transmitted flux and certain fundamental physical parameters of the sample. This relationship is the well-known Beer–Lambert law. The parameters that must be known in this application are: optical path length (or sample thickness of a solid sample), the concentration of the absorbing species, and the molar specific absorption. The Beer–Lambert law is one of the most important laws in (absorption) spectrophotometry. In its original form it is written as

$$\tau = \Phi_{\tau_r}/\Phi_0 = e^{-kc'l} \tag{9.1}$$

where k is the molecular cross section in square centimeters, c' the number of molecules per cubic centimeter, and l the optical path length in centimeters. In one of its more modern versions this law can be written as

$$A = \log(1/\tau) = a_c'cl, \tag{9.2}$$

where a_c' is the molar absorption coefficient, c the molar concentration, and A the transmitted density (absorbance). The main application of the Beer–Lambert law is in the determination of concentration of the absorbing species from the transmission density data when the molar absorption coefficient a and the optical path length are known, or if the concentration is given, then a can be determined.

9.3.4 Validity of the Beer–Lambert Law

The Beer–Lambert law is valid when the absorbing species are noninteracting atoms and/or molecules. This condition is fully satisfied only with ideal gases at a reduced pressure (Chandrasekhar, 1950). In practice, however, the Beer–Lambert law is successfully applied to dilute solutions for a relatively broad range of concentrations. At high concentrations the absorbers start to form aggregates thus causing a change in the molar absorption coefficient. The other condition for validity of the Beer–Lambert law is that all the fluxes, with the exception of (Φ_{τ_r}), are zero. This condition is easily satisfied with quasi-ideal solutions. The problem arises if the sample exhibits luminescence and/or optical turbidity. There can be any number of apparent exceptions from the Beer–Lambert law, but they are due only to poor experimental procedures or to uncontrolled sample conditions (chemical instability, photochemical reactivity, solvent impurity, stray light, dirty cells, misaligned optics, poor cells, etc.)

9.4 METHODS OF MEASUREMENT

There are several methods for measuring spectrophotometric characteristics of materials. In this chapter we shall deal with those that are the most commonly used in practical applications.

The spectrophotometers can be either single- or double-beam instruments. In single-beam instruments the measurements are based on the substitution method, in which the sample and the standard are interchanged in the same position. On the other hand, the measurements with double-beam spectrophotometers are based on a continuous comparison between sample and standard, which are positioned into the sample and reference beams, respectively, and evaluated either simultaneously or sequentially.

The double-beam method conveniently uses a single light source, and the beam splitting for sample and reference channels is achieved either optically, by means of a mirror or a prism, or mechanically. One of the many advantages of the double-beam system is that one can interchange the sample and the standard to test the equality of the two beams. This interchange is particularly useful for comparing two standards of reflectance.

In spectrophotometric measurements attention should be given to the spectral distribution of the light source, the spectral characteristics of optical elements (including the spectral reflectance factor of the integrating-sphere wall), and the spectral (relative) responsivity of the detector.

In critical measurements and in standardization work it is also essential to know to what degree the measurements are affected by polarization. For techniques to determine the effect of polarization, the reader should consult the literature in this area (Buc and Stearns, 1945; Kay and Holland, 1971; Robertson, 1972; Grum and Spooner, 1973, Grum and Costa, 1974). For a more complete description of the distribution of optical radiant energy with respect to polarization, the reader is referred to the work of Shurcliff (1962) and Shumaker (1977).

The instruments available for photometric measurements are by no means perfect. All of them are subject to systematic errors, such as those caused by finite apertures in integrating spheres, selectivity and nonuniform reflectance of sphere walls, finite convergence or divergence of light beams, and the like. Measurements with most instruments, therefore, yield reflectometer values rather than reflectances (or transmittances) or reflectance factors.

The techniques used for the determination of various spectral quantities correspond to those used in the determination of photometric quantities, when a monochromator is used for illumination. For nonfluorescent samples, there is no difference between the method in which the sample is irradiated monochromatically and the measurements are made with a detector of any spectral response and that in which the sample is irradiated heterochromatically with monochromatic detection. Fluorescent materials, however, must be irradiated with heterochromatic light and the measurements made monochromatically.

9.5 REFLECTION (CIE, 1977)

9.5.1 Modes of Reflection Measurement

In principle, there are nine different ways of measuring reflected radiant energy. These are described in Table 9.1 (Judd, 1967) and consist of all possible

TABLE 9.1 Values of Nine Kinds of Reflectance for the Perfect Mirror and for the Perfect Diffuser

Kind of reflectance	Symbol[a]	Evaluation of quantity	
		Perfect mirror	Perfect diffuser
Bihemispherical	$\rho(2\pi; 2\pi)$	1	1
Hemispherical–conical	$\rho(2\pi; g')$	$(1/\pi)\int_{\omega'}\cos\theta_r\, d\omega'$	$(1/\pi)\int_{\omega'}\cos\theta_r\, d\omega'$
Hemispherical–directional	$\rho(2\pi; \theta_r, \phi_r)$	$(1/\pi)\cos\theta_r\, d\omega'$	$(1/\pi)\cos\theta_r\, d\omega'$
Conical–hemispherical	$\rho(g; 2\pi)$	1	1
Biconical	$\rho(g; g')$	$\dfrac{\int_{\text{overlap}}\cos\theta_r\, d\omega'}{\int_{\omega}\cos\theta_0\, d\omega}$	$(1/\pi)\int_{\omega'}\cos\theta_r\, d\omega'$
Conical–directional	$\rho(g; \theta_r, \phi_r)$	$(\cos\theta_r\, d\omega')/\int_{\omega}\cos\theta_0\, d\omega$ for θ_r, ϕ_r within mirror image of g; 0, for θ_r, ϕ_r not within	$(1/\pi)\cos\theta_r\, d\omega'$
Directional–hemispherical	$\rho(\theta_0, \phi_0; 2\pi)$	1	1
Directional–conical	$\rho(\theta_0, \phi_0; g')$	1, for $\theta_0 =$ some θ_r and $\phi_0 =$ some $\phi_r + \pi$; 0, for $\theta_0 \neq$ any θ_r or $\phi_0 \neq$ any $\phi_r + \pi$	$(1/\pi)\int_{\omega'}\cos\theta_r\, d\omega'$
Bidirectional	$\rho(\theta_0, \phi_0; \theta_r, \phi_r)$	1, for $\theta_0 = \theta_r$ and for $\phi_0 = \phi_r + \pi$ 0, for $\theta_0 \neq \theta_r$, or for $\phi_0 \neq \phi_r + \pi$	$(1/\pi)\cos\theta_r\, d\omega'$

[a] The parameters in parentheses are the incident angular geometry and the viewing angular geometry, respectively. The symbol g denotes conical geometry.

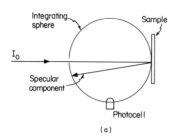

(a)

Fig. 9.6   Schematic for measuring (a) total and (b) diffuse reflectance.

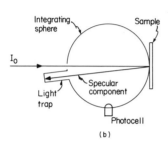

(b)

combinations in which the incident and reflected energy are unidirectional, are contained within a small solid angle, or encompass the entire hemisphere. Practically, only three of these are widely used: hemispherical–conical, conical–hemispherical, and biconical.

The hemispherical methods are carried out with the use of an integrating sphere as described in Chapter 3. All such measurements are based on the theory of the integrating sphere (Ulbricht, 1900; Taylor, 1920; Brudole, 1958; Wendlandt and Hecht, 1968), which assumes uniform and diffuse reflection over the entire wall of the sphere. To meet these conditions requires special care in the selection and application of the coating to the sphere (Sanders and Middleton, 1953; Brudole, 1960; Grum and Luckey, 1968).

Two different modes of reflection can be measured with integrating-sphere instruments: the specular component may be included or excluded, as indicated in Fig. 9.6 (Grum, 1972a). Practically, many specimens have broadened specular peaks which cannot be completely excluded with light traps of normal size. Not only is it necessary to give details of the size, shape, and location of the light trap, but measurements made under these conditions should be interpreted with caution.

9.5.2   Reflectance

The method of measuring reflectance depends on whether the sample exhibits diffuse, mixed, or specular reflectance.

9.5.2.1   DIFFUSE REFLECTANCE

For materials exhibiting scattering, the CIE has recommended (CIE, 1971) that the reflectance ρ be measured with conical–hemispherical geometry (normal/diffuse geometry, abbreviated 0/d). The specimen is illuminated by a beam whose axis is at an angle not exceeding 10° from the normal to the specimen. The reflected flux is collected by means of an integrating sphere. The angle between the axis and any ray of the illuminating beam should not exceed 5°. The integrating sphere may be of any diameter provided the total area of the ports does not exceed 10% of the internal reflecting sphere area. The reference for reflectance measurement is the incident energy. To comply with the definition of reflectance, the specular component must be included in this measurement.

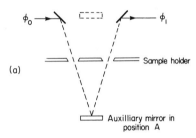

Fig. 9.7 Schematic for measuring specular reflectance: (a) sample removed and mirror in position A; (b) sample in place and mirror in position B.

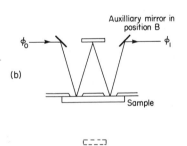

9.5.2.2   REGULAR REFLECTANCE

Regular reflectance can be measured in several ways. In one method (Fig. 9.7) (CIE, 1977) the incident energy is compared to the energy after two reflections from the specimen, with the aid of an auxiliary mirror. In Fig. 9.7a, with the sample removed and the auxiliary mirror in position A, the reflected flux is

$$\phi_1 = K\phi_0\rho_m,$$ (9.3)

where K is a geometrical factor, ϕ_0 the incident flux, and ρ_m the reflectance of the mirror. In Fig. 9.7b with the sample in place and the auxiliary mirror in position B,

$$\phi_2 = K\phi_0\rho_r\rho_m\rho_r,$$ (9.4)

where ρ_r is the regular reflectance of the sample. The ratio $\phi_2/\phi_1 = \rho_r{}^2$ (Ulbricht, 1900).

In another method, use is made of an instrument with a receiver rotating around an axis through the front surface of the sample. The incident flux ϕ_0 is measured with the sample removed, and the reflected flux ϕ_1 is measured with the sample in place and the receiver rotated to the specular reflection angle. The value of ρ_r is given by the ratio ϕ_1/ϕ_0.

A third method requires the use of a standard whose specular reflectance is calculated from its refractive index by means of Fresnel's equations.

9.5.3   Reflectance Factor

The CIE has recommended (CIE, 1971) that the reflectance factor be measured in one of the three following geometries[†]:

(a)   45°/normal (*abbreviation* 45/0).   The specimen is illuminated by one or more beams whose axes are at an angle of $45° \pm 5°$ from the normal to the specimen surface. The angle between the direction of viewing and the normal to the specimen should not exceed 10°. The angle between the beam axis and any ray of an illuminating beam should not exceed 5°. The same restriction should be observed in the viewing beam.[‡]

(b)   Normal/45° (*abbreviation* 0/45).   The specimen is illuminated by a beam whose axis is at an angle not exceeding 10° from the normal to the specimen. The specimen is viewed at an angle of $45° \pm 5°$ from the normal. The angle between the axis and any ray of the illuminating beam should not exceed 5°. The same restriction should be observed in the viewing beam.

[†] The reflectance factor approaches the radiance factor if the cone of the viewing beam becomes infinitely small. The 45°/normal condition gives the radiance factor $\beta_{45/0}$. The normal/45° condition gives the radiance factor $\beta_{0/45}$. The diffuse/normal condition gives the radiance factor $\beta_{d/0}$. The normal/diffuse condition gives the reflectance ρ.

[‡] In the normal/45° and normal/diffuse conditions, specimens with mixed reflection should not be measured with strictly normal illumination (CIE, 1971).

(c)   *Diffuse/normal* (*Abbreviation d/0*).   The specimen is illuminated diffusely by an integrating sphere. The angle between the normal to the specimen and the axis of the viewing beam should not exceed 10°. The integrating sphere may be of any diameter provided the total area of the ports does not exceed 10% of the internal reflecting sphere area. The angle between the axis and any ray of the viewing beam should not exceed 5°.

The difference between reflectance factor and reflectance lies in the selection of the reference: for the reflectance factor, the perfect reflecting diffuser is used as the reference, whereas the reference for the reflectance is the incident flux.

In measurements with any biconical geometry, the reflectance factor can exceed unity, whereas the reflectance cannot. If the calorimetric properties of the material are to be determined, the use of values exceeding unity may be inconvenient, so measurement of the reflectance is recommended. In this geometry, determination of the reflectance can be approximated by measuring the incident radiant energy with the use of a white standard of known reflectance.

9.5.4   Total Radiance Factor

Fluorescent specimens must be measured by a spectrophotometer or by a colorimeter, in which the sample is irradiated with undispersed white light of the controlled spectral power distribution. The energy reflected and emitted by the sample must then be detected monochromatically (Berger, 1958; Grum and Wightman, 1960; Fukuda and Sugiyama, 1961; Cappod, 1966; Grum, 1972b). A typical instrument setup for measuring total radiance factor is shown in Fig. 9.8, and an example of the measured results is given in

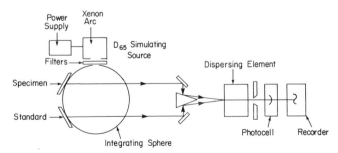

Fig. 9.8   Schematic diagram of heterochromatic sample illumination.

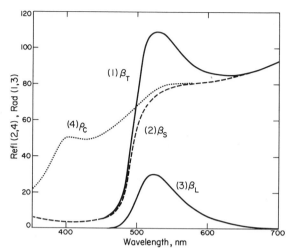

Fig. 9.9   Total radiance factor (β_T), reflected (β_S) and fluorescent (β_L) factors, and conventional reflectance of a green fluorescent sample [curve (4)].

Fig. 9.9. In Fig. 9.9, β_T is the total radiance factor, curve (2) the reflected radiance factor β_S, curve (3) the fluorescent radiance factor β_L, and curve (4) the reflectance ρ_C as measured on a conventional spectrophotometer with monochromatic irradiation and total detection. Since the sample (shown in the example) exhibits fluorescence, the ρ_c is grossly in error in the spectral region that is responsible for fluorescence excitation. In measurement of ρ_c a portion of the incident energy is absorbed and then emitted at longer wavelengths. The emitted energy is detected as though it were reflected energy at the same wavelength as the incident energy, and an erroneous curve results.

As stated in Chapter 2, the total radiance factor is defined as

$$\beta_T(\lambda) = \beta_S(\lambda) + \beta_L(\lambda), \tag{9.5}$$

where $\beta_T(\lambda)$ is the total radiance factor at wavelength λ and $\beta_S(\lambda)$ and $\beta_L(\lambda)$ the reflected and luminescent radiance factors, respectively.

There are several ways to measure the $\beta_S(\lambda)$ component; perhaps the best way is to use a dual monochromator system in which the fluorescent sample is irradiated monochromatically and the energy reflected from the sample is also detected monochromatically (Donaldson, 1954; Hisdal, 1956; Grum, 1971). The only requirement in such a system is that the bandwidth of the irradiating monochromator be kept fairly narrow in order to avoid excitation of wavelengths within the illuminating bandwidth.

Several approximations are available for measuring the separated components, two of which are well documented in the literature. One such method uses sharp cutoff filters in front of the source, so that no excitation wavelength can reach the sample (Eitle and Ganz, 1968). By careful filter selection, β_S can be obtained for most of the spectral region of interest.

Another way to measure β_S is to make two types of measurements of the sample; one measurement is conventional reflectance, giving $\rho_c(\lambda)$, and the other is $\beta_T(\lambda)$ described earlier. The $\rho_c(\lambda)$ represents $\beta_S(\lambda)$ at wavelengths longer than the excitation wavelengths, whereas $\beta_T(\lambda)$ represents the $\beta_S(\lambda)$ for wavelengths shorter than the emission region. With these two measurements there will be only the narrow region of overlap between excitation and emission spectra that must be interpolated (Simon, 1972).

Once $\beta_S(\lambda)$ is determined, $\beta_L(\lambda)$ can be deduced from Eq. (9.5), i.e.

$$\beta_L(\lambda) = \beta_T(\lambda) - \beta_S(\lambda). \tag{9.6}$$

9.5.5   The Remission Function

A sample that is not homogeneous but contains small particles of a size comparable to the wavelength of light may be called a *turbid medium*. When light passes through such a medium, the energy is decreased not only as a result of absorption but also due to scatter. For such a case the previously discussed Beer–Lambert law does not apply. Many opaque (reflecting) materials (papers, plastics, textiles, etc.) can be considered turbid media since most of the energy of the reflecting fluxes is diffused, due to multiple scatter.

By far the most important mechanism of the interaction of light with matter is scattering (Mie, 1908). This process is responsible for a number of phenomena such as the blueness of the sky, the colors of the rainbow, the color of clouds, and color of many nonmetallic materials.

It is not our intention to discuss here various complex turbid media theories, for these are well documented in the literature. Several representative references are given (Kubelka, 1948; Judd, 1938; Richards, 1970; Billmeyer and Abrams, 1973; Billmeyer and Richards, 1972). Here we mention only the Kubelka–Munk relationship (Kubelka, 1948), which is used in many practical applications and particularly in spectrophotometric evaluations of such materials as plastics, textiles, and papers.

The remission function k/s, the ratio of the Kubelka–Munk absorption coefficient to the Kubelka–Munk spectral scattering coefficient, is related to the reflectance ρ (or reflectometer value R) of an opaque layer of a

TABLE 9.2 Values of the Remission Function

%ρ	k/s	%ρ	k/s	%ρ	k/s	%ρ	k/s	%ρ	k/s
0.0		5.0	9.02	10.0	4.05	15.0	2.41	20.0	1.600
.5	99.0	.5	8.12	.5	3.81	.5	2.30	.5	1.542
1.0	49.0	6.0	7.36	11.0	3.60	16.0	2.21	21.0	1.486
.5	32.3	.5	6.73	.5	3.41	.5	2.11	.5	1.433
2.0	24.0	7.0	6.18	12.0	3.23	17.0	2.03	22.0	1.383
.5	19.01	.5	5.70	.5	3.06	.5	1.945	.5	1.335
3.0	15.68	8.0	5.29	13.0	2.91	18.0	1.868	23.0	1.289
.5	13.30	.5	4.93	.5	2.77	.5	1.795	.5	1.245
4.0	11.52	9.0	4.60	14.0	2.64	19.0	1.727	24.0	1.203
:5	10.13	.5	4.31	.5	2.52	.5	1.662	.5	1.163
25.0	1.125	30.0	.817	35.0	.604	40.0	.450	45.0	.336
.5	1.088	.5	.792	.5	.586	.5	.437	.5	.326
26.0	1.053	31.0	.768	36.0	.569	41.0	.425	46.0	.317
.5	1.019	.5	.745	.5	.552	.5	.412	.5	.308
27.0	.987	32.0	.723	37.0	.536	42.0	.401	47.0	.299
.5	.956	.5	.701	.5	.521	.5	.389	.5	.290
28.0	.926	33.0	.680	38.0	.506	43.0	.378	48.0	.282
.5	.897	.5	.660	.5	.491	.5	.367	.5	.273
29.0	.869	34.0	.641	39.0	.477	44.0	.356	49.0	.265
.5	.842	.5	.622	.5	.463	.5	.346	.5	.258
50.0	0.250	55.0	0.1841	60.0	0.1333	65.0	0.0942	70.0	0.0643
.5	.243	.5	.1784	.5	.1290	.5	.0909	.5	.0617
51.0	0.235	56.0	0.1729	61.0	0.1247	66.0	0.0876	71.0	0.0592
.5	.228	.5	.1675	.5	.1205	.5	.0844	.5	.0568
52.0	0.222	57.0	0.1622	62.0	0.1165	67.0	0.0813	72.0	0.0544
.5	.215	.5	.1571	.5	.1125	.5	.0782	.5	.0522
53.0	0.208	58.0	0.1521	63.0	0.1087	68.0	0.0753	73.0	0.0499
.5	.202	.5	.1472	.5	.1049	.5	.0724	.5	.0478
54.0	0.1959	59.0	0.1425	64.0	0.1013	69.0	0.0696	74.0	0.0457
.5	.1899	.5	.1378	.5	.0977	.5	.0669	.5	.0436
75.0	0.0417	80.0	0.0250	85.0	0.0132	90.0	0.00556	95.0	0.00132
.5	.0398	.5	.0236	.5	.0123	.5	.00499	.5	.00106
76.0	0.0379	81.0	0.0223	86.0	0.0114	91.0	0.00445	96.0	0.00083
.5	.0361	.5	.0210	.5	.0105	.5	.00395	.5	.00064
77.0	0.0344	82.0	0.0198	87.0	.00971	92.0	0.00348	97.0	0.00046
.5	.0327	.5	.0186	.5	.00893	.5	.00304	.5	.00032
78.0	0.0310	83.0	0.0174	88.0	0.00818	93.0	0.00263	98.0	0.00020
.5	.0294	.5	.0163	.5	.00747	.5	.00226	.5	.00011
79.0	0.0279	84.0	0.0152	89.0	0.00680	94.0	0.00192	99.0	0.00005
.5	.0264	.5	.0142	.5	.00616	.5	.00160	.5	.00001

scattering and absorbing material by the relation

$$k/s = (1 - \rho_\infty)^2/2\rho_\infty, \tag{9.7}$$

where k and s are the absorption and scattering coefficients, respectively, and ρ_∞ the reflectance of an infinitely thick sample layer. Values of the remission function at intervals of 0.1% in ρ_∞ are given in Table 9.2.

9.5.6 Total Internal Reflectance and Attenuated Total Reflectance

So far we have discussed and dealt with external reflectance, i.e., the irradiation of the sample from the optically less-dense medium (air). The reverse can also occur, and in such a case we are dealing with internal reflectance. At the critical angle the refracted ray leaves at the grazing angle, and the internal reflectance becomes 100%. The total internal reflectance technique has been used extensively in infrared spectroscopic work. In this technique an interface between a nonabsorbing dielectric substance of high refractive index and the sample is used as the reflecting surface. The incidence angle in this arrangement is normally greater than the critical angle, hence the reflection is complete for all wavelengths for which the sample is non-absorbing. This technique is very useful in evaluating spectral properties of weak absorbing materials since the multiple-reflection (attenuated) technique can be employed to amplify the energy of the signal measured. The reflectance spectra obtained in this manner resemble the transmission spectra.

The reader is referred to an excellent treatment of this subject by Kortüm (1969).

9.6 TRANSMISSION

9.6.1 Modes of Transmission Measurement

In principle, there are nine different ways of measuring transmitted radiant energy, analogous to the ways of measuring reflected radiant energy described earlier, but in practice only two of these are widely used. They are measurement with conical–hemispherical geometry, using an integrating sphere, and with biconical geometry.

With integrating-sphere instruments, two different modes of transmission can be measured: the regular (specular) component may be included or excluded, as indicated in Figs. 9.10a and 9.10c, respectively (CIE, 1977). The precaution of Section 9.5.1 regarding measurement with the regular component excluded should be observed.

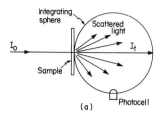

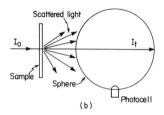

Fig. 9.10   Schematic diagrams for transmittance measurements: (a) total transmittance; (b) regular transmittance; (c) scattered light.

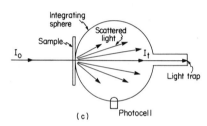

9.6.2 Transmittance

The method of measuring transmittance depends on whether or not the sample is turbid (exhibits light scattering).

9.6.2.1 DIFFUSE TRANSMITTANCE

For turbid materials, the specimen is illuminated, in conical–hemispherical (alternatively, hemispherical–conical) geometry, by a beam whose axis is at an angle not exceeding 10° from the normal to the specimen. The transmitted flux is collected by means of an integrating sphere. The angle between the axis and any ray of the illuminating beam should not exceed 5°. The integrating sphere may be of any diameter provided the total area of the ports does not exceed 10% of the internal reflecting sphere area. In this measurement it is imperative that the portion of the sphere irradiated by the incident beam has reflection characteristics identical to those of the remainder of the sphere coating.

Note.   The use of biconical geometry does not yield the same quantity.

9.6.2.2   REGULAR TRANSMITTANCE

Regular transmittance can be measured either with or without the use of an integrating sphere. Several methods are known for measuring regular transmission (or reflection) components. However, most of them are not practical for conventional use. The simplest way to obtain the regular components is from the difference between total transmittance (reflectance) and diffuse transmittance (reflectance).

9.6.3   Internal Transmittance

The internal transmittance of a nonscattering sample can be obtained in two ways: If a nonabsorbing specimen of the same material as the sample is available, it can be used as the reference, and the internal transmittance is measured directly. If such a specimen is not available, the Fresnel reflections can be calculated from the refractive index of the material, and the internal transmittance computed.

9.7   DETERMINATION OF THE SPECTRAL TRANSMITTANCE OF TRANSPARENT MATERIALS

The spectral characteristics of transparent, i.e., optically clear, materials are calculated from the spectral internal transmittance $\tau_i(\lambda)$ and the thickness d. The spectral internal transmittance $\tau_i(\lambda)$ is obtained from the spectral transmittance $\tau(\lambda)$ and the spectral refractive index $n(\lambda)$

$$\tau_i(\lambda) = \{[n^2(\lambda) + 1]/2n(\lambda)\}\tau(\lambda). \tag{9.8}$$

The spectral refractive index $n(\lambda)$ is measured by means of a refractometer. The spectral transmittance $\tau(\lambda)$ of the material having a thickness d is measured with collimated, essentially monochromatic radiant energy under normal incidence, as the ratio of the irradiances on the receiver with the sample $E(\lambda)_x$ and without the sample $E(\lambda)_0$.

$$\tau(\lambda) = E(\lambda)_x/E(\lambda)_0. \tag{9.9}$$

If an image of the light source is located near the reciver and the distance between this image and the receiver itself is small and d is large, then the reduction of the optical path length due to the sample has to be considered.

Since the measurement of transmittance when $\tau < 0.01$ is often less accurate than when $\tau > 0.01$, it is suggested that the sample thicknesses be selected for various wavelength regions so that the measured values for $\tau(\lambda)$ are near the middle of the range. The calculated values of the spectral internal transmittance $\tau_i(\lambda)$ for thickness d may then be recalculated by the Lambert–Bouguer law to obtain the transmissivity (internal transmittance for unit thickness d_0):

$$\tau_i(\lambda)_0 = \tau_i(\lambda)^{d_0/d}. \qquad (9.10)$$

In chemical absorption spectrophotometry, absorbing liquids are measured in glass or quartz cells, the sample and reference solutions being measured in identical cells. For the incident energy, the reflections at the sample and reference cells are identical and therefore cancel out. At the inner walls of both cells (reflectance $\bar{\rho}_2$) multiple reflections occur, and the radiant energy is attenuated differently in the sample and reference solutions. This is described by the sum of a power series:

$$\tau(\lambda)/\tau_0(\lambda) = [\tau_i(\lambda)/\tau_0(\lambda)][(1 - 10^{-2da'(\lambda)}\bar{\rho}_2^2)/(1 - \bar{\rho}_2^2)] \qquad (9.11)$$

Similarly, the spectral internal absorbance $A_i(\lambda)$ may be calculated from the measured value $A_s(\lambda)$ by the relation

$$A_i(\lambda) = A_s(\lambda) - \log[(1 - 10^{-2da'(\lambda)}\bar{\rho}_2^2)/(1 - \bar{\rho}_2^2)] \qquad (9.12)$$

The correction factor is a function of the absorption coefficient $a'(\lambda)$, and therefore the correction increases with increasing absorption; that is, the measured curves of the spectral transmittance are distorted by the absorption of the sample solution.

9.8   MEASUREMENT OF DIRECT ABSORPTANCE

Spectral absorptance is normally computed from two spectral quantities, the percent transmittance and the percent reflectance, and is expressed by

$$\alpha(\lambda) = 1.0 - [\tau(\lambda) + \rho(\lambda)], \qquad (9.13)$$

where $\tau(\lambda)$ is the total transmittance and $\rho(\lambda)$ the total reflectance, both measured with an integrating sphere. The law of conservation of energy applies to the degree of accuracy with which the two entities τ and ρ are measured. In any anisotropic scattering medium, it is necessary for the same area of the sample to be subjected to both measurements in order to obtain accurate results. This is not an easy condition to attain with conventional spectrophotometers.

As is evident from (9.13), the spectral absorptance is determined from separate $\tau(\lambda)$ and $\rho(\lambda)$ measurements. Hence, the sample is first positioned on the transmission port and then on the reflection port of the instrument or vice versa. The cross-sectional area of the incident beam may be different in the two positions; hence, a different area is examined in the two measurements, and this can cause a large error in the results, especially if the samples measured are turbid. This error, as well as geometrical error, can be eliminated if the 4π geometry is used in these measurements.

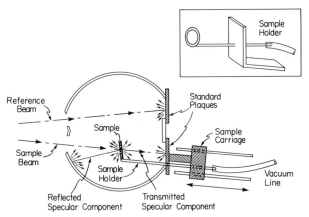

Fig. 9.11 A 4π geometry arrangement.

The 4π geometry is achieved by placing the turbid sample into the integrating sphere as shown schematically in Fig. 9.11. Also shown in the figure is a sample holder consisting of a ring made of 1.5-mm-diameter copper tubing. The tubing is soldered to a flat plate which forms the sample reflectance platen. This assembly can be coupled to a vacuum, which can hold the sample in place on the ring. The assembly (with the sample) is placed via a small optical bench into the proper position in the sphere. The sample holder must be coated with the same material as the walls of the integrating sphere.

With this arrangement only one measurement is made, representing $\tau(\lambda) + \rho(\lambda)$. The absorptance $\alpha(\lambda)$ can then be obtained directly by a simple electrical modification of the circuit controlling the photometric scale; hence the term "direct absorptance."

The advantages of this method of determination of absorptance are obvious. The measurements made in 4π geometry are absolute and very accurate. Optical integration is achieved identically for all components by making only a single measurement on the sample. Best of all, with this technique errors due to the sample irregularity, losses due to high refractive index, and interference among coatings are nearly eliminated. This technique with examples is described in detail elsewhere (Costa *et al.*, 1976).

9.9 SOURCES OF MEASUREMENT ERRORS

Sources of systematic error are invariably present to some degree in the measurements described above. The most important of these are described in the following paragraphs.

9.9.1   Photometric Scale Errors

Significant errors can arise from the setting of the upper and lower end points of the photometric scale and its deviations from linearity.

The upper end point of the photometric scale is normally set, for reflectance measurement, by the selection of an appropriate standard. The lower end point of the scale, for reflectance measurement, should be set not by blocking the light beam, but rather by the use of a highly efficient light trap at the sample port.

Secondary standards, such as filters with known transmittance or reflectance standards with known reflectometer values, are used widely to check the linearity of photometric scales (Menis and Shultz, 1970). For more precise work, however, use of the light-addition method is recommended (Reule, 1968).

9.9.2   Stray Light

Stray light, i.e., unwanted radiant energy measured along with the desired energy, most often arises from imperfect or dirty optical surfaces. Each instrument must be evaluated individually for stray light, using methods described in the literature (Pineo, 1940; Poulson, 1964).

9.9.3   Errors Arising in Measurement of Translucent Materials

Measurement of the transmittance or reflectometer values of translucent specimens is fraught with error under the best of circumstances. These errors arise mainly from loss of light toward or out of the sides of the specimen. The measured values are very sensitive to such details as sizes of apertures and their ratios to the sizes of the illuminated or viewed areas (Atkins and Billmeyer, 1966). Even relatively opaque materials such as the opal glasses used widely as secondary standards are subject to these errors.

9.9.4   Wavelength and Slit-Width Errors

When wavelength-dependent measurements are to be made, care must be taken that the wavelength scale is correct and that slit-width errors are recognized and corrected for (Keegan *et al.*, 1962; Buc and Stearns, 1945).

9.10   SPECIAL SPECTROPHOTOMETRIC TECHNIQUES

9.10.1   Introduction

The characterization of some of the optical properties of materials may require specific (nonconventional) measurement techniques. The complexity of the techniques and of the apparatus in highly specialized measurements depends on the nature of the problem, on the sample type, and on the information sought by the analysis. Among the specialized spectrophotometric techniques are vacuum ultraviolet techniques, microspectrophotometric techniques, cryogenic techniques, and various types of digital techniques.

Vacuum ultraviolet spectrophotometry, at wavelengths below 200 nm, is very cumbersome and requires not only special apparatus but also very careful measuring techniques and sophisticated methods of analysis. The major problems that must be considered in spectrophotometric investigation in the vacuum ultraviolet are: availability of proper optical components, selection of an appropriate detector, selection of suitable solvents, and selection of proper experimental techniques. One of the major problems connected with vacuum UV investigations is the spectral absorption of oxygen in that spectral region.

Regarding microspectrophotometric techniques, until recently there was no apparatus commerically available for this purpose. Zeiss, Leitz, and Nanometrics[†] now have microspectrophotometers for commerical use.

It is not our objective to describe here the above-mentioned and other special spectrophotometric techniques, for there are many and they are normally applicable only to specific research projects and/or problems. However, we shall mention briefly a few measurement techniques which may have somewhat broader applications.

9.10.2   Differential Absorption

A relatively new spectrophotometric method of measuring absorption spectra of dye coatings is a computer-assisted differential method, called DIFA for short. The method consists of separate spectral diffuse transmittance measurements of the reference and of the dyed sample and a subsequent computer manipulation of the two spectra. Appropriate computer software allows automatic scale expansion of the ordinate so that very small dye absorptions are amplified and readily detectable. The computer program also determines accurately all the peaks and valleys in the differential spectrum.

This method is particularly useful where the levels of dye are low and where the dye absorption overlaps the absorption of the dye matrix. Figure 9.12 shows an example of DIFA. In this figure, curve 1 is the absorptance of the matrix and curve 2 is the differentially obtained dye absorptance. Note that the absorption characteristics of the dye (curve 2) are not evident when measured by conventional means (curve 1).

Another type of differential spectrum is that obtained by the attenuation method. Measurement of spectral transmission characteristics of materials

† Nanospec/10 by Nanometrics, Sunnyvale, California.

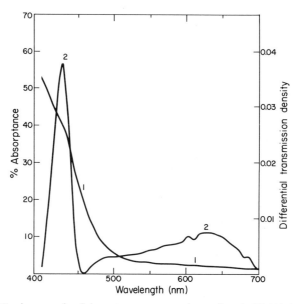

Fig. 9.12   An example of absorptance spectrum (curve 1) and of DIFA (curve 2).

with transmittance values of less than 0.0001% (6.00 density) cannot be made on conventional spectrophotometers unless special differential techniques are employed. The photometric scale range of most spectrophotometers is from 0.00 to 3.00 density (many do not go over a density of 2.00). To make optical density measurements over 3.00, a means of attenuating the reference beam is needed. This can be done by use of meshed screens of varying densities placed in the reference beam next to the transmission port. To calibrate the instrument for measurement of high density, two sets of glass (absorption) filters of the same melt and of varying thickness are used. The lowest density filter can be accurately calibrated by conventional means. The density per unit thickness is next computed at selected wavelengths, taking into account the necessary correction for Fresnel reflectance at the surface of the filter. Based on these computations, a theoretical density curve can be plotted as a function of filter thickness (the thickness must be measured by a micrometer to the nearest 0.01 mm).

Once this density per unit thickness calibration is completed, a value can be assigned to any thickness by multiplying this value by the thickness to determine the theoretical density, assuming the sample obeys the Beer–Lambert law.

Using this differential (attenuation) technique, one can measure the spectral transmission density of a sample in the high density range with an accuracy of 1% at the very high density level.

9.10.3   Derivative Absorption Technique

An instrumental method that provides resolution of overlapping bands (in absorption, transmission, or emission) is called *derivative spectrophotometry*. There are instrumental methods for directly recording a curve of the first- or second-order derivatives of absorptance and transmittance with respect to wavelength; i.e., $dA/d\lambda$, $d^2A/d\lambda^2$ or $d\tau/d\lambda$, and $d\tau^2/d\lambda^2$. (Saidel, 1955; Giese and French, 1955). Another approach to the same problem is to use digital spectrophotometric techniques and subsequent differentiation of the data (Grum, 1972). The advantage of a digital technique is that one can compute higher than second-order derivatives obtainable instrumentally. The best digital manipulation is based on the cubic spline technique. The original absorption (transmittance) data are first fitted by cubic spline, from which the first derivative is computed. The first-derivative spectrum is again fitted with the spline function after which the second derivative is computed. This procedure of higher-derivative computation is termed "cubic spline-on-spline." An example of such data is given in Fig. 9.13 where the absorption

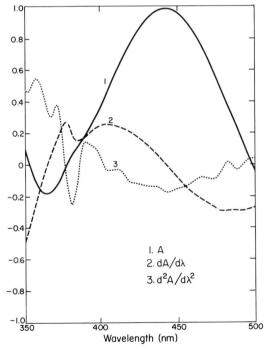

Fig. 9.13   Absorptance spectrum of a dye (curve 1: A); its first derivative (curve 2: $dA/d\lambda$); second derivative (curve 3: $d^2A/d\lambda^2$).

spectra of a dye and its first and second derivatives (spline-on-spline) are shown.

There are many merits in obtaining derivative spectra on absorbing species, some of which are: the overlapping absorption bands can be detected; the hidden absorption bands can be amplified (resolved); the complex spectra can be resolved into their components with proper calibration techniques.

Another type of quasi-derivative technique is the "dual wavelength" spectrophotometric technique (Shibath, 1976). This technique provides high sensitivity and selectivity in analysis of organic and inorganic materials. In this technique the light beams of different wavelengths (from two mono-chromators) are time-shared through a single cell by means of a rotating sector, and the difference of absorbance ΔA between the absorbances at λ_1 and λ_2 is measured. Very small absorbance changes can thus be detected.

REFERENCES

Adams, M. J., King, A. A., and Kirkbright, G. F. (1976a). *Analyst* **101**, 73.
Adams, M. J., Beadle, B. C., King, A. A., and Kirkbright, G. F. (1976b). *The Analyst* **101**, 553.
Atkins, J. T., and Billmeyer, F. W. Jr. (1966). *Mater. Res. Std.* **6**, 564.
Bassani, F., and Parravicini, G. P. (1975). "Electronic States and Optical Transitions in Solids." Pergamon, Oxford.
Berger, A. (1958). *Die Farbe* **8**, 189.
Billmeyer, F. W., Jr., and Richards, L. W. (1972). Scattering and absorption of radiation by lighting materials, CIE, TC-2.3. Subcommittee report on Turbid Media.
Billmeyer, F. W., Jr., and Abrams, R. L. (1973). *J. Paint Technol.* **45**(579), 31.
Brudole, W. (1958). *Die Farbe* **7**, 295.
Brudole, W. (1960). *J. Opt. Soc. Am.* **50**, 217.
Buc, G. L., and Stearns, E. I. (1945). *J. Opt. Soc. Am.* **35**, 658.
Buc, G. L., and Stearns, E. I. (1945). *J. Opt. Soc. Am.* **35**, 465.
Cappod, W. A. (1966). *Pulp and Paper Mag.* T495.
Chandrasekhar, S. (1950). "Radiative Transfer." Oxford Univ. Press (Clarendon), London and New York, Reprinted by Dover, New York, 1969.
CIE (1971). Colorimetry, Publ. No. 15 (E-1.3.1).
CIE (1977). TC-2.3 Tech. Rep.
Costa, L. F., Grum, F., and Wightman, T. E. (1976). *Color Res. Appl.* **1**, 193.
Dash, W. C., and Newman, R. (1954). *Phys. Rev.* **99**, 1151.
Dean, J. A., and Rains, T. C. (1971). "Flame Emission and Absorption Spectroscopy." Dekker, New York.
Di Bartolo, B. (1968). "Optical Interactions in Solids." Wiley, New York.
Donaldson, R. (1954). *Brit. J. Appl. Phys.* **5**, 210.
Eitle, D., and Ganz, E. (1968). *Textil Verdlung* **3**, 389.
Foster, R. (1969). "Organic Charge-Transfer Complexes." Academic Press, New York.
Fukuda, T., and Sugiyama, Y. (1961). *Farbe* **10**, 73.
Giese, A. T., and French, C. S. (1955). *Appl. Spectrsc.* **9**, 78.
Goldberg, P. (1966). "Luminescence of Inorganic Solids." Academic Press, New York.
Grotrian, W. (1928). Graphische Darstellung Der Spectren Von Atomen Und Ionen Mit Ein, Zwei, Und Drei Valenzetronen." Springer, Berlin.

Grum, F. (1971). CIE-XVII Session, Barcelona, Document P-71.22.
Grum, F. (1972a). "Physical Methods of Chemistry" (A. Weissberger and B. Rossiter, eds.), Chapter III, pp. 207–427; "Techniques of Chemistry" (A. Weissberger, ed.), Vol. 1, Part IIIB, Figure 3.57, p. 338. Wiley, New York.
Grum, F. (1972b). *J. Color Appearance* **1**, 18.
Grum, F., and Costa, L. F. (1974). *Appl. Opt.* **13**, 2228.
Grum, F., and Luckey, G. W. (1968). *Appl. Opt.* **7**, 2289.
Grum, F., and Spooner, D. L. (1973). *J. Color Appearance* **2**, 6.
Grum, F., and Wightman, T. (1960). *TAPPI* **43**, 400.
Grum, F., Paine, D., and Zoeller, L. (1972). *Appl. Opt.* **11**, 93.
Herzberg, G. (1944). "Atomic Spectra and Atomic Structure." Dover, New York.
Herzberg, G. (1945a). "Molecular Spectra and Molecular Structure," Vol. 1, Spectra of Diatomic Molecules. Van Nostrand–Reinhold, Princeton, New Jersey.
Herzberg, G. (1945b). "Molecular Spectra and Molecular Structure," Vol. 2. Van Nostrand–Reinhold, Princeton, New Jersey.
Herzberg, G. (1945c). "Infrared and Raman Spectra of Polyatomic Molecules." Van Nostrand–Reinhold, Princeton, New Jersey.
Herzberg, G. (1966). "Molecular Spectra and Molecular Structure," Vol. 3, Electronic Spectra and Electronic Structure of Polyatomic Molecules. Van Nostrand–Reinhold, Princeton, New Jersey.
Hisdal, B. (1956). *Opt. Acta* **3**, 139.
Jorgensen, C. K. (1963). "Inorganic Complexes." Academic Press, New York.
Jorgensen, C. K. (1971). "Modern Aspects of Ligand Field Theory." American Elsevier, New York.
Judd, D. B. (1938). *Paper Trade J.* **6**, 39.
Judd, D. B. (1967). *J. Opt. Soc. Am.* **57**, 445.
Kartachevskaya, V. E., Korte, H., and Robertson, A. R. (1975). *Appl. Opt.* **14**, 2694.
Kay, R. B., and Holland, R. J. (1971). *Appl. Opt.* **10**, 1587.
Keegan, H. J., Schleter, J. C., and Judd, D. B. (1962). *J. Res. Nat. Bur. Std.* **66A**, 203.
Kortüm, G. (1969). "Reflexionspektroskopie." Springer–Verlag, Berlin and New York.
Kubelka, P. (1948). *J. Opt. Soc. Am.* **38**, 448.
Long, M. E., Swofford, R. L., and Albrecht, A. C. (1976). *Science* **19**, 183.
Menis, O., and Shultz, J. I. (1970). Nat. Bur. Std. Tech. Note 544, p. 6.
Mie, G. (1908). *Ann. Phys.* **25**, 377.
Murrel, J. N. (1963). "Theory of Electronic Spectra of Organic Molecules." Wiley, New York.
Petrucci, S. (1971). "Ionic Interactions from Diluted Solutions to Fused Salts," Vol. 1–2, Physical Chemistry, Vol. 2–2. Academic Press, New York.
Pineo, O. W. (1940). *J. Opt. Soc. Am.* **30**, 276.
Poulson, R. E. (1964). *Appl. Opt.* **3**, 99.
Reule, A. (1968). *Appl. Opt.* **7**, 1023.
Richards, L. W. (1970). *J. Paint Technol.* **42**, 276.
Robinson, J. M. (1966). "Atomic Absorption Spectroscopy." Dekker, New York.
Robertson, H. R. (1972). *Appl. Opt.* **11**, 1436.
Saidel, L. J. (1955). *Arch. Biochem. Biophys.* **54**, 185.
Sanders, C. L., and Middleton, E. E. K. (1953). *J. Opt. Soc. Am.* **43**, 58.
Seraphin, B. O. (1976). "Optical Properties of Solids, New Developments." American Elsevier, New York.
Shibath, S. (1976). *Angew. Chem.* **88**, 750.
Shumaker, J. B. (1977). "Self-Study Manual on Optical Radiation Measurements: Part I," Chapter 6. The Nat. Bur. Std. Note 910.

Shurcliff, W. A. (1962). "Polarized Light." Harvard Univ. Press, Cambridge, Massachusetts.

Simon, F. T. (1972). *J. Color Appearance* **1**, 5.

Stokes, G. G. (1862). *Proc. Roy. Soc. London* **11**, 545.

Swofford, R. L., Long, M. E., and Albrecht, A. C. (1976). *J. Chem. Phys.* **65**, 179.

Taylor, A. H. (1920). *Sci. Papers NBS* **17**, 1–6.

Ulbricht, T. (1900). *Electrotech. Z.* **21**, 595.

Wendlandt, W. W., and Hecht, H. G. (1968). "Reflectance Spectroscopy." Wiley, New York.

10

Standards and Calibration

10.1 INTRODUCTION

The precise and accurate measurement of optical radiant energy requires measuring equipment which is properly standardized and measurement techniques based on sound procedures. These objectives can be achieved only through the use of calibrated reference standards and by following well-established measuring techniques.

There are two types of standards, documentary and physical. The documentary standards (normally voluntary standards) are documents produced by various national and/or international standards organizations such as ANSI, ASTM, and ISO. These documentary standards describe recommended procedures for specific measurements and applications. Such recommendations are based on the consensus of ideas brought forth by representatives from various industries, academic institutions, and governmental agencies.

Physical standards are actual material objects such as sources, detectors, or materials that have been calibrated, either by national standardizing laboratories or by commercial standardizing agencies, for a specific type of measurement. These materials are to be used for calibrating and/or checking the performance of the optical radiation measuring instrumentation.

The type of physical standard that needs to be used in standardizing the instrumentation varies from application to application and depends greatly on the accuracy of the measurement that needs to be achieved. The selection of the physical standard and the source of the standard also depends on the traceability requirements from the national standardizing laboratory.

In this chapter we shall limit our discussion to three major areas of optical radiation measurements: photometry, radiometry, and spectrophotometry, for it is nearly impossible to discuss in a single chapter the whole gamut of standards and existing techniques which are required in optical radiation measurements.

Why are standards and calibration techniques necessary? In research areas the comparability of measurements is a fundamental requirement. In industry inadequate standardization of measuring equipment often means inferior quality control of goods produced. Good standardization, on the other hand, increases the efficiency of production and distribution in many industries; it also tends to reduce any possible legal disputes, improve product quality, and reduce the waste of resources. Proper standardization techniques not only solve many measurement problems but also ensure adherence to the requirements of various governmental regulatory agencies. With the use of standard reference materials and measurement techniques all measurements can be simpler, less expensive, and more accurate.

The accuracy and efficiency with which photometry, radiometry, and spectrophotometry can be performed affect everyone. Such measurements are important in: the lighting industry, the treating of various diseases by phototherapy, specifying radiation hazards, pollution monitoring, quality control operations, weather forecasting, etc.

10.2   PHOTOMETRY AND RADIOMETRY

10.2.1   Primary Standards in Photometry

In photometry the physical quantities measured are weighted by a function that approximates the spectral response of the human eye. This spectral response, normally called the $V(\lambda)$ function, is given in Appendix 5.2 as $\bar{y}(\lambda)$. The radiant power detected by such a $V(\lambda)$ function detector is referred to as *luminous flux*. The unit of luminuous flux is the lumen. It is defined in terms of the radiant power emitted by a blackbody at the temperature of melting platinum. More specifically, a lumen is the luminous flux emitted per unit solid angle in a perpendicular direction by $1/600,000$ m^2 of a blackbody at the freezing point of platinum (approximately 2045 K).

The relationship between radiometric and photometric quantities is given by

$$\text{photometric quantity} = K_m \int_\lambda (\text{spectral radiometric quantities})V(\lambda)\,d\lambda,$$

$$(10.1)$$

where K_m is a constant relating lumens to watts ($K_m \cong 680$ lm/W); $V(\lambda)$ was defined in Chapter 2.

One of the most commonly used photometric quantities is the luminous intensity (of a source). This quantity is important in the manufacture and use of lamps. The unit of luminous intensity is the candela (lumens per steradian) and is defined in terms of the platinum point blackbody.

The reader may be interested in the evolution of this standard, hence a brief historical review is given here. In 1924 the CIE recommended international adoption of the luminance of a blackbody, under conditions to be determined later, as the primary standard of light. Later the CIE encouraged the Comité International des Poids et Mesures (CIPM) (CIE, 1935) to establish the full radiator at the freezing point of platinum as the primary standard of light (CIPM, 1946). Two years later the candela was defined as the unit of luminous intensity (Conférence Generale, 1948). The definition adopted was as follows: "The magnitude of the candela is such that the luminance of a full radiator at the temperature of solidification of platinum is 60 candelas per square centimeter." In 1967 this definition was further revised (Conference des Poids, 1967–1968). The magnitude of luminous flux was defined as: "One lumen is the flux emitted in a unit solid angle of one steradian by a point source having a uniform intensity of one candela."

It is desirable to be able to realize the primary standards with an accuracy of $\pm 0.1\%$ and to have secondary standards derived from this with an accuracy of $\pm 0.3\%$. Recent work with the primary standard of light has been reported from a number of material standardizing laboratories (Sanders and Jones, 1962; Fischer and Krönert, 1963).

Most recently, Blevin and Steiner (1975) proposed a redefinition of the candela and the lumen. They proposed (1) that the basic photometric unit be redefined so as to provide an exact numerical relationship between it and the SI† unit of power, the watt, for a specified monochromatic radiation, and (2) that the unit of luminous intensity, the candela, be replaced as a basic unit by the unit of luminous flux, the lumen. Their argument is that by this redefinition a desired closer link between photometry and spectroradiometry can be achieved. Thus the photometric values could be derived from spectroradiometric data by exact computation and the need for a primary standard of light would be removed.

The proposal of Blevin and Steiner aroused a lot of attention and the CCPR in 1975 (Comite Consultatif, 1975) submitted their proposal to the CIPM for possible adoption in 1977. The CCPR recommendation is that the photometric units be defined in terms of the watt and agreed-upon values for the $V(\lambda)$ function and K_m. In the context of this recommendation, the lumen is the luminous flux that produces a human visual response equivalent to that produced by $1/680$ W of monochromatic radiation of frequency 540.0154×10^{12} Hz.

---

† SI is the Système International D'Unités.

10.2.2   Secondary Standards

The accuracy of realizing a platinum point blackbody standard of light has been about 1%. In addition one should remember that such a realization is very tedious. Therefore, the national standardizing laboratories maintain the candela on a group of lamps. The lamps from different national laboratories have been periodically compared at the International Bureau of Weights and Measures (BIPM). By doing so, international agreement of about 2% has been maintained with very infrequent use of platinum point blackbodies.

Because of the difficulty in practice of making a receiver with the exact spectral response of the standard observer, errors occur in comparing the luminous flux or intensity of a lamp having one spectral energy distribution with that of a standard having a different spectral distribution. It is therefore customary in routine work to compare lamps having similar spectral distributions. For this purpose incandescent standard lamps are maintained for use at a few agreed distribution temperatures.

10.2.3   Standards of Luminous Intensity

Gas-filled, inside-frosted lamps are normally used as standards of luminous intensity. These have either T-20 or T-24 bulbs, medium-bipost bases, and C-13B filaments. The life of these standards depends on the lamp wattage and on the operating voltage. The 500-W lamps are designed to have about 500 hr of life at 120 V. Before calibration the lamps must be seasoned by operating them at 120 V ac for approximately 5% of their valid life. The luminous intensity should not vary more than 0.1%.

The photometer is calibrated in terms of the illuminance produced at the detector. The detector can be either the eye (visual photometry) or a physical detector such as a selenium barrier–layer cell or a silicon photodiode equipped with a filter which modifies the spectral response to match the CIE luminous efficiency function (CIE, 1970a). Measurements are usually made at a detector illumination level of approximately 80 lm/m², and the detector surface is fully illuminated by either the standard or the test lamp. The detector's output is accurately measured with an operational amplifier, and the output of the amplifier is measured with an accurate digital voltmeter.

When visual photometric methods are used in the calibration, the Lummer–Brodhun photometer head, usually equipped with a contrast field, is used in calibration. Two juxtaposed photometric fields are viewed through an eyepiece. The photometer head is moved along the photometer bench until equal brightness of the two fields is achieved.

All measurements are made on a optical (photometer) bench equipped with an appropriate number of limiting baffles.

The test lamp intensities are compared by multiplying the measured illuminance by the square of the distance between the lamp and the detector. This is valid where the maximum lamp emitting-surface dimension is less than one-tenth of the distance from lamp to detector.

The variation of luminous intensity of a gas-filled tungsten lamp is approximately related to the variation in its current by the formula (NBS, 1972)

$$c \, dI/I = 6.25 \, di/i, \qquad (10.2)$$

where I is the luminous intensity and i the lamp current (Kingslake, 1965). Hence, it is important to control and measure current very accurately in the calibration of lamps for luminous intensity.

10.2.4   Color Temperature Standards

Incandescent lamp standards of color temperature are calibrated on a photometric bench either by visual methods or by physical means. For visual photometric methods, again the Lummer–Brodhun photometer is used with an eyepiece. The colors of the standard and the test lamps are observed simultaneously in the split photometric fields of the photometer. The current of the test lamp is adjusted until the two lamps appear to have the same color as seen through the eyepiece of the photometer head.

In the physical method of color temperature calibration, a substitution method based on a red–blue ratio color temperature comparator is used.

Airway beacon lamps (500T20/13) are normally used as standards of color temperature. They are 500-W, 120-V lamps with clear T-20 bulbs, C-13B filaments, and medium-bipost bases. Again, as in the case of luminous intensity, the lamps are properly seasoned prior to calibration.[†]

All measurements are made with lamps operating on dc power. A suitable digital voltmeter is used to make electrical measurements.

At NBS (1972), all routine color temperature measurements are made on a device that compares the ratios of the red portion of the visible spectrum to the blue portion of the visible spectrum for test and standard sources. Equality of these two red-to-blue ratios is interpreted to mean equality in color temperature for thermally emitting light sources. In this procedure it is essential that the spectral power distributions of the two lamps are similar in shape throughout the visible region of the spectrum.

Other devices based on the red–blue ratio principle have been used for color temperature measurement and calibration (Brown, 1954; Grum, 1977). In nearly all devices the radiant energy from test and standard sources is collected by an integrating sphere that can be rotated about the optical axis of its

---

[†] NBS also issues inside-frosted lamp standards of color temperature (2856 K).

exit aperture. Energy leaves through the exit aperture and passes through red and blue filters alternately and is detected by a photomultiplier. After passing the amplifier–integrator chains, the two signals are compared on the null meter. Suitable adjustments are made to the lamp electrical parameters to obtain equality between the two signals. This then is a substitution method. Both test and standard lamps are placed in the same geometrical position with respect to the comparator, and the comparator indicates a spectral match among sources.

The present NBS scale of color temperature was established in 1934 (Wensel *et al.*, 1934) by visual comparison of monoplane, coiled tungsten filament, incandescent lamps with three melting-point blackbodies: platinum at 2045 K, rhodium at 2236 K, and iridium at 2720 K. An empirical approach was adopted for interpolating among the three fixed temperature points. Taking advantage of the cavity effect due to filament coiling, pyrometric determinations of the radiance temperature were made at a position on the inside of a single turn of the coiled filament. (Radiance temperature, also called luminance temperature and brightness temperature, is defined as the temperature of a blackbody for which the spectral radiance at the specified wavelength is the same as that of the radiator considered.) It was observed that the difference between the radiance temperature so determined and the color temperature of the lamp as a whole was a smoothly varying function of the voltage applied to the lamp. Pyrometric determinations of the radiance temperature of the three different coil turns of the test lamp yielded consistent color temperature values for the interpolated region. From these data NBS established an empirical equation relating the color temperature T_c and the applied voltage V(NBS, 1972):

$$T_c = A + B(V)^{1/2}. \tag{10.3}$$

The NBS scale was adjusted in 1949 (Judd, 1950) and in 1970 (NBS, 1970) and the extrapolations were made up to 3000 K and down to 1500 K. In addition to the calibrations at fixed color temperature points, associated empirical equations of the same form as Eq. (10.3) are used to establish the scale in interpolated and extrapolated regions (NBS, 1972):

$$I = A + BT_c + CT_c^2. \tag{10.4}$$

Differentiating,

$$dI = (B + 2CT_c)\,dT_c, \tag{10.5}$$

and noting that typical values of the constants A, B, and C for the airway beacon lamps are $A = -0.4T$, $B = 1.0 \times 10^{-3}$, and $C = 1.4 \times 10^{-7}$, and that these lamps draw about 4 A at 2856 K, we see that a current setting accuracy of at least 0.1% is required if the color temperature of the lamp is not to be affected by more than 2 K.

10.2.5   Distribution in Temperature, Color Temperature, and Correlated Color Temperature

In the interests of uniformity in specifying distribution temperature, color temperature, and correlated color temperature, these terms should be used in accordance with definitions of the CIE Lighting Vocabulary (CIE, 1970b).

The term "distribution temperature" is preferred in all cases where the spectral distribution curve is important. Although in general the values of color temperature or correlated color temperature cannot be used to predict the relative spectral distribution curve of the test source, if the source is an incandescent tungsten filament in a nonselective envelope, the correlated color temperature and the distribution temperature will not differ significantly.

Distribution temperature is the temperature of the full radiator for which the ordinates of the spectral distribution curve of its radiance are proportional, in the visible region, to those of the distribution curve of the radiator considered.

A mathematical expression for the determination of distribution temperature is

$$\int_{\lambda_1}^{\lambda_2} [1 - [M_\lambda'(\lambda)/aM_\lambda(\lambda, T)]]^2 \, d\lambda, \qquad (10.6)$$

where M_λ' is the exitance of the test source, $M_\lambda(\lambda, T)$ is given by the Planckian equation, and a is an arbitrary constant. The values of a and T, the Planckian temperature, are adjusted simultaneously until the integral is minimized. When this has been done, the final value of T is the distribution temperature. It is customary to restrict the use of the term "distribution temperature" to cases where the ratio $M_\lambda'/M_\lambda(\lambda, T)$ does not change by more than $\pm 5\%$.

For most photometric applications λ_1 and λ_2 may be 400 and 760 nm, respectively. If the source is a tungsten lamp, the distribution temperature may be calculated within ± 4 K by reducing the integral to a summation extending from 400 to 760 nm with $\Delta\lambda$ equal to 50 nm. If the distribution temperature does not closely describe the spectral energy distribution, it is desirable to specify the maximum deviation between the spectral distributions of the full radiator and of the test source.

Color temperature is the temperature of the full radiator that emits radiant energy of the same chromaticity as the radiator considered. (The corresponding relative spectral distribution curves may not be similar, in which event there will be a metameric match.)

Since the chromaticity of most practical sources differs significantly from that of any Planckian radiator, it should be emphasized that correlated color temperature should be used in most cases. Correlated color temperature is that temperature corresponding to the point on the Planckian locus that is nearest to the point representing the chromaticity of the illuminant when considered on an uniform chromaticity scale diagram, such as the CIE 1960 UCS diagram (CIE, 1959). The equations for u and v given below specify the 1960 UCS diagram in terms of the CIE x, y diagram:

$$u = 4X/(X + 15Y + 3Z) = 4x/(-2x + 12y + 3), \qquad (10.7)$$

$$v = 6Y/(X + 15Y + 3Z) = 6y/(-2x + 12y + 3). \qquad (10.8)$$

Mathematical procedures for determination of correlated color temperature are well documented in the literature (Kelly, 1963; Mori, *et al.*, 1964; Robertson, 1968).

10.2.6 Standards of Luminous Flux

Opal bulb lamps operating at a color temperature of approximately 2720 K are used at the NBS as standards of luminous flux. The lamps are calibrated by measuring their luminous intensity distributions relative to their luminous intensity in a single direction. The luminous intensity in the single direction is measured by comparison with the standard of luminous intensity. The total luminous flux is then determined by integrating this luminous intensity with respect to the angle of view.

All photometric measurements of luminous flux Φ_v should be made in conformity with the definition of luminous flux, i.e.,

$$\Phi_v = K_m \int_0^\infty \Phi_{e,\lambda} V(\lambda)\, d\lambda, \qquad (10.9)$$

where $\Phi_{e,\lambda} = d\Phi_e/d\lambda$ is the spectral concentration of radiant flux and $V(\lambda)$ the photopic luminous efficiency function. Similarly, luminance measurements should conform to the relationship

$$L_v = K_m \int_0^\infty L_{e,\lambda} V(\lambda)\, d\lambda, \qquad (10.10)$$

where $L_{e,\lambda} = dL_e/d\lambda$ is the spectral radiance for which the luminance is to be determined. In practice, a summation is made at convenient intervals of $\Delta\lambda$ (usually 10 nm).

10.3 SPECTRAL IRRADIANCE AND RADIANCE STANDARDS

10.3.1 Primary Standards

In radiometry, the physical quantity measured is the radiant power incident on a portion of a surface. From this measurement and the geometry involved, various power densities are calculated. These include the radiant power incident per unit area, called irradiance, and the radiant power incident per unit projected area and unit solid angle, called radiance. The quantities usually of interest are: the intensity of the source, i.e., the radiant power emitted by a source per unit solid angle, and the total flux, which is the radiant power emitted by a source in all directions. Since the measurement of radiant power often requires knowledge of how the energy is distributed relative to wavelength, the spectral concentration of these quantities is also of great interest.

Spectral radiance standards are derived from blackbodies, and radiant quantities are derivable from spectral radiance. (See Chapter 2 for units, symbols, and definitions.) The spectral radiance of a blackbody can be calculated from the Planckian radiation equation provided that the temperature of the blackbody is known.

When these standards are based on a full radiator of known temperature, the absolute spectral concentration of radiant exitance of such full a radiatior should be determined from the Planckian equation, where T is the temperature in kelvins in the thermodynamic scale of temperature. The Planckian equation is

$$M_\lambda(\lambda, T) = \partial M(\lambda, T)/\partial\lambda = C_1 \lambda^{-5} (e^{C_2/\lambda T} - 1)^{-1}. \qquad (10.11)$$

The values of C_1 and C_2 are

$$C_1 = 2\pi h c^2 = (3.74150 \pm 0.00009) \times 10^{-16} \quad \mathrm{W\,m^2},$$
$$C_2 = hc/k = (1.43880 \pm 0.00006) \times 10^{-2}\,\mathrm{m\,K}.$$

For T one should accept the value as measured in accordance with the International Practical Temperature Scale (Conference Generale, 1968). [See also Terrien and Preston-Thomas for the small differences which may arise between the thermodynamic temperatures, and corresponding values of the IPTS 1968 (Terrien and Preston-Thomas, 1967).] The uncertainty of the best thermodynamic temperature measurements in the region between 1500 and 3000 K produces an uncertainty in spectral radiance in the visible spectral range of about 1%, and the total uncertainty is about 0.75%.

The radiometric quantities can also be derived from electrical measurements by using an electrically calibrated radiometer (Geist and Blevin, 1973; Geist et al., 1976). Such instruments are designed to respond in the same way to both radiant heating and electrical heating. Then, if an electrical signal is adjusted until the output of the radiometer is the same as that when the radiant signal is incident, the radiant energy absorbed is equal to the power dissipated in the radiometer by the electrical signal. Since electrical power measurements can be made much more accurately than temperature measurements, the inherent accuracy of the radiometer is increased. The recent development of pyroelectric electrically calibrated radiometers makes this approach much more practical than before, and it is being investigated extensively (NBS, 1976).

10.3.2   Secondary Standards

To meet the increasing demand for absolute measurements of spectral irradiance, national standardizing laboratories have developed and inter-compared secondary standards of spectral radiance and spectral irradiance for the spectral range of 250 to 2500 nm.

The National Bureau of Standards realized a scale of spectral irradiance in 1963 with a set of 200-W, quartz–iodine, tungsten coiled-filament lamps (Stair *et al.*, 1963). Later the scale of spectral irradiance was realized through a comparison of a group of four 1000-W DXW quartz–halogen lamps.

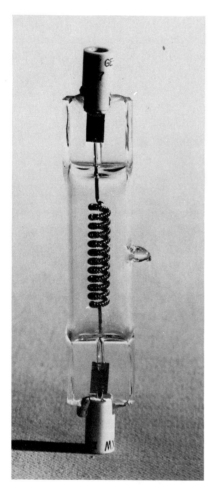

Fig. 10.1 A photograph of a typical spectral irradiance standard lamp. (Type: DXW, 1000-W tungsten–halogen.)

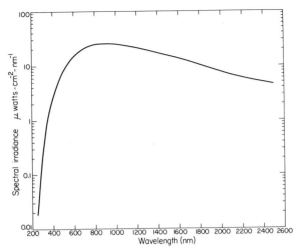

Fig. 10.2 Spectral irradiance curve for the DXW, 1000-W tungsten–halogen lamp in Fig. 10.1.

The DXW quartz–halogen, tungsten lamps are coiled-filament commercial lamps. They have recessed single-contact bases and a rated life of 150 hr at 120 V. A photograph of such a lamp is shown in Fig. 10.1, and its typical spectral irradiance is shown in Fig. 10.2.

Before calibration the lamps must be seasoned on dc for 30 hr at 120 V. For proper selection the lamp output is monitored at 654.6 nm for 24 hr to determine the drift rate. Lamps with a drift rate larger than 0.5% are considered unsatisfactory for standards. Many lamps of this type exhibit several emission lines and an irregular absorption band near 280 nm. The emission lines have been attributed to the presence of neutral atoms of sodium and aluminum, impurities introduced during the manufacturing process. These weak emission lines occur in the near-ultraviolet and green–yellow regions of the spectrum. The absorption band (280 nm) is approximately 4 nm wide and can be as large as 60% of the interpolated continuum output of the lamp. When the absorption band is so great, the lamp is not suitable for a standard.

As with photometric calibrations, the standard lamps must be properly oriented and aligned in the optical system.

The uncertainty of reported values varies from about 2% in the ultraviolet to less than 1% in the visible and near-infrared regions.

In 1975, 1000-W quartz–halogen, modified FEL lamps were made available as standards of spectral irradiance (NBS, 1974). A sketch of such a lamp is shown in Fig. 10.3. FEL lamps are 1000-W, clear-bulb, quartz–halogen,

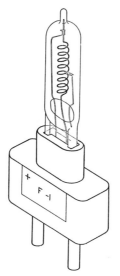

Fig. 10.3 A sketch of an FEL tungsten–halogen lamp in the mount.

tungsten, coiled-filament (CC-8) lamps. They have a rated life of 500 hr at 120 V. The lamps are manufactured with a two-pin base. Before calibration, the lamp base is converted to an iridium bipost base, and the base structure is encapsulated in an epoxy compound.

The advantages of the FEL lamps are ease of adjustment and, most important, freedom from emission lines and absorption bands.

The operating current for these lamps is determined by matching their radiant output to that of the working standard lamps at 654.6 nm. Typical current used with these lamps is 7.6–8.0 A.

10.4 ABSOLUTE RADIOMETRY

10.4.1 Photometry versus Radiometry

By using the recommended K_m value and an absolute radiometer, and with a $V(\lambda)$ filter of known transmittance, photometric evaluation can be made by a single measurement (Preston, 1969). Alternatively, a source whose relative spectral distribution is known may be measured in terms of power units by means of an absolute radiometer, then its spectral radiant flux can be calculated. Two methods can be used in this computation; in one it is assumed that the absolute spectral radiant flux produced by a standard source is known, and in the second it is assumed that both the relative spectral distribution of the standard source and the luminous flux produced by it are known.

First the following ratio is measured:

$$R(\lambda) = \frac{\Phi'_{e,\lambda}}{\Phi_{e,\lambda}}, \tag{10.12}$$

where $\Phi_{e,\lambda}$ and $\Phi'_{e,\lambda}$ are the spectral concentrations of radiant flux from the standard and test lamps, respectively. Then, from the known basic photometric equation

$$\Phi_v'/\Phi_v = \int \Phi'_{e,\lambda} V(\lambda)\, d\lambda \Big/ \int \Phi_{e,\lambda} V(\lambda)\, d\lambda. \tag{10.13}$$

Alternatively, using $R(\lambda)$,

$$\Phi_v' = \Phi_v \int R(\lambda)\Phi_{e,\lambda} V(\lambda)\, d\lambda \Big/ \int \Phi_{e,\lambda} V(\lambda)\, d\lambda, \tag{10.14}$$

which may be solved since $R(\lambda)$ is measured and $\Phi_{e,\lambda}$ and Φ_v are given, where Φ_v and Φ_v' are luminous fluxes of standard and test lamps, respectively. Note that (10.14) requires only relative values of $\Phi_{e,\lambda}$.

10.4.2 Absolute Detectors

In absolute radiometric measurements one has to have a standard (calibrated) source, as discussed earlier, or one can use a calibrated detector (thermopile or photon detector).

Electrically calibrated detectors have been recently developed for this purpose and can be used in absolute radiometric measurements (Geist and Blevin, 1973). Also available from NBS and other national standardizing laboratories are calibrated silicon photodiodes (NBS, 1975).

Since some of the silicon detectors show a degradation of the uniformity of response across the detector surface and may also vary in responsivity when subjected to UV radiation, care must be taken when measurements are made with these devices (NBS, 1976).

An excellent reference on precision measurement and calibration in photometry and radiometry is NBS Special Publication 300 (Hammond and Mason, 1971), which brings together many published papers, monographs, and bibliographies by NBS authors dealing with the precision measurement of specific physical quantities and the calibration of related metrology equipment.

10.5   SPECTROPHOTOMETRIC STANDARDS

10.5.1   Introduction

Standard reference materials for spectrophotometry have a dual purpose.

(1)   They are used for checking the performance of spectrophotometers, i.e., to check the instruments for various inherent errors and to ensure the optimum performance of such instrument.

(2)   They are used to calibrate the instruments and to perform the analyses relative to given standard reference materials.

Most standard reference materials are available through various standards organizations, instrument manufacturers, and/or from measurements and testing laboratories. The National Bureau of Standards and other national standards laboratories provide many different types of standard reference materials intended for specific purposes or measurement types.

For the purpose of this discussion we shall divide the standard reference materials into two categories:

(1)   materials that are used for checking and calibrating wavelength scales of spectrophotometers, and

(2)   standard reference materials designed for checking and calibrating photometric scales and for determining various instrumental errors discussed earlier in this chapter.

The materials in the second category are further divided into transmittance and reflectance standards.

A recommended practice for accurate measurements is to check the instrument performance daily, both when the instrument is cold and after it has been used (warmed up) for some time. Accurate results are impossible unless spectrophotometers are properly calibrated and kept in good calibration. This can be done only by the use of the appropriate standard reference materials.

10.5.2   Standards and Methods for Checking Wavelength Scales of Spectrophotometers

For measurements in which wavelength is critical (it usually is critical), the wavelength can be calibrated using stable line emission sources. The

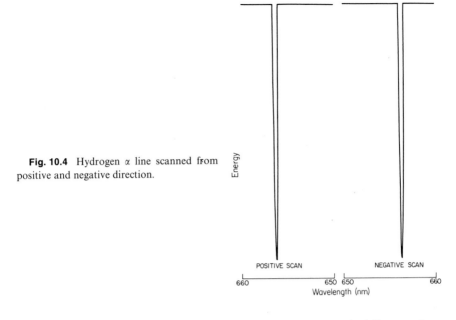

Fig. 10.4 Hydrogen α line scanned from positive and negative direction.

emission line transmitted by the monochromator is recorded while scanning the monochromator across the wavelength of the line source. The centroid wavelength scale setting corresponding to the line wavelength can then be determined mathematically. In some instruments, such as Cary 14 or 17 spectrophotometers, this type of check, at least for one wavelength, is easily done as follows. The instrument is set for operation on the energy mode and with the hydrogen or deuterium source in place the instrument is scanned across the wavelength of 650 nm. With proper expansion of the abscissa on the chart and with proper slit setting the hydrogen line (656.3 nm) can be readily recorded. An example of such a record is shown in Fig. 10.4.

In day-to-day operations, the standard reference materials are normally used for checking the wavelength scale of the spectrophotometers. Standard didymium glass (Keegan and Gibson, 1944) has been used for a long time for wavelength checks in the visible and near-infrared spectral regions. The wavelength accuracy attainable with such a glass is considered to be within ± 1.0 nm. More recently holmium glass has been used for wavelength checks in the visible and ultraviolet regions of the spectrum (Vandenbelt, 1961). Holmium has strong absorption lines at about 241, 279, 287, 361, 453, and 536 nm; didymium has much broader absorption bands at about 402, 530, 573, 586, 685, 741, and 803 nm. Benzene vapor is an excellent material for calibration of the wavelength scale in the ultraviolet region (225–275 nm) (ASTM, 1966). One should remember that wavelength calibration standards

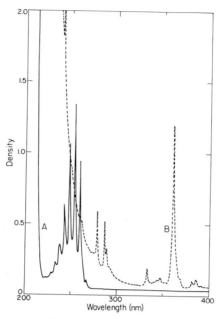

Fig. 10.5   Absorption spectra of benzene vapor (a) and holmium oxide (b) (200–400 nm).

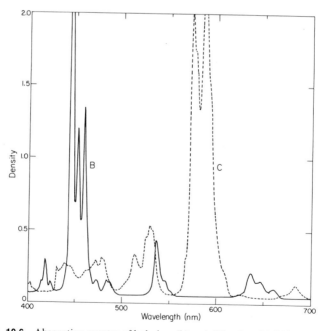

Fig. 10.6   Absorption spectra of holmium (b) and didymium (c) (400–700 nm).

should be measured at the desired instrument bandpass. Figures 10.5 and 10.6 present the spectra of some of these materials.

In the infrared spectral region (2–15 μm) the wavelength scale may be calibrated by using the absorption bands of either polystyrene or indene (Tables, 1961).

10.5.3 Standards for Checking Photometric Scale Transmittance

The accuracy of the photometric scale is the most important performance factor in the spectrophotometric measurements, yet this factor is very difficult to define since the measurement of transmittance and/or absorbance is a function of many instrumental parameters such as wavelength accuracy, slit width, and stray light.

Various national standardizing laboratories have long recognized the need for methods and materials to check the accuracy of the photometric scales of spectrophotometers. Various materials made available for this purpose were either inorganic glass or liquid samples (standard solutions). Standard solutions most commonly used for this purpose are cobalt ammonium sulfate, copper sulfate, and potassium dichromate solutions, normally supplied in sealed vials. Of these, potassium dichromate solutions have been used most extensively (Menis and Schultz, 1971). Potassium dichromate in moderate concentration in dilute perchloric acid gives absorption maxima at 257 and 350 nm and minima at 235 and 313 nm and can be used as an absorption standard in the spectral range from 230 to 390 nm. These solutions are stable (photochemically) and obey Beer's law over a considerable range. NBS also supplies liquid standards (SRM-931) consisting of high-purity cobalt and nickel in a mixture of nitric and perchloric acid (Burke et al., 1972).

Solution standards, however, are not the most suitable to use. They must be prepared very accurately and carefully. They are normally prepared under nitrogen and contained in sealed quartz cuvettes. If a slight loss of solvent occurs during sealing, an increase in concentration occurs which changes the calibration.

Glass filters are much easier to use, and hence are the preferred standard materials. Unfortunately, glass filter standards are presently available only for the visible spectral region. The glass standards (cobalt blue, canary yellow, and Jena B. G.) which were available for a long time from the National Bureau of Standards were colored glasses having quite selective absorbances (Keegan et al; 1962). The spectral transmittance curves of a set of such filters are shown in Fig. 10.7. These glasses were carefully selected; they are non-scattering and are designed to detect instrumental errors resulting from stray light, slit width, and back reflectance.

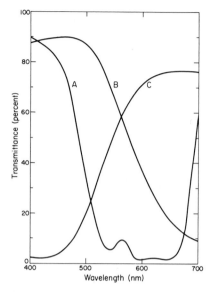

Fig. 10.7   Spectral transmittances of a set
of standard glasses: cobalt blue (a), Jena B.G.
(b), and canary yellow (c).

More recently another set of glass filter standards was made available.
These glasses are identified as Standard Reference Material 930, available
at NBS (Mavrodineanu, 1972, 1973; Clarke, 1968). They are a set of three
neutral-density filters, designed for accurate calibration of the photometric
scale of spectrophotometers. The transmittance of these filters (nominally
10, 20, and 30%) is certified with a relative uncertainty of $\pm 0.5\%$ at selected

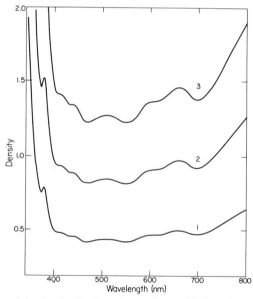

Fig. 10.8   Spectral density distribution of a set of neutral Schott glasses NG-4 (SRM-930).

wavelengths. These filters are supplied with filter holders to ensure best positioning in the instrument. Spectral transmission density curves for one set of such standard filters are shown in Fig. 10.8.

Attempts are also being made to produce filter standards for the ultraviolet. Inconel alloy on quartz may offer such a possibility (see the example in Fig. 10.9).

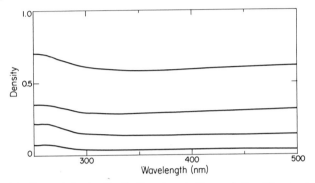

Fig. 10.9 Spectral density distribution of a set of Inconel alloy filters on quartz.

There are no readily available standard materials for checking the photometric scale of spectrophotometers for the infrared region. Some possibilities for such standards are shown in Figs. 10.10 and 10.11. These again are neutral-density filters especially selected for the purpose.

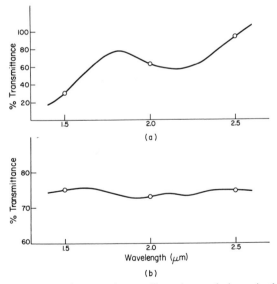

Fig. 10.10 Spectral transmittances of two calibrated neutral glasses in the near-infrared: (a) Sample 1, Pittsburgh 2043 glass, 1.94 mm thick; (b) Sample 2, Bausch and Lomb neutral glass, 2.36 mm thick.

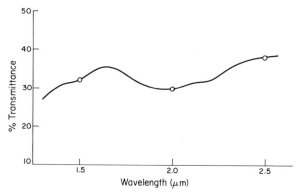

Fig. 10.11   Spectral transmittance of Bausch and Lomb neutral glass, 5.66 mm thick.

10.5.4   Reflectance Standards

Several white materials are now available for working standards of reflectance. The most prominent are magnesium oxide or barium sulfate pressings and/or coatings and various glasses.

Magnesium oxide, prepared by thermal deposition, has for a long time been used as a standard of reflectance (ASTM, 1965). Although it is not too difficult to prepare a fairly good coating of magneisum oxide, it is quite time consuming. It is still used, however, because of its unique optical properties, although it has been recognized for some time that good, reproducible, smoked magnesium oxide standards are very difficult to prepare and are not very stable (Hammond, 1955). It is for this reason that repeated attempts have been made to use other materials and other preparations as standards.

The reflectance values of most materials used as standards of reflectance depend heavily on the method of manufacture, the degree of impurity, environmental conditions, and many other factors. To maintain a reflection scale over a period of weeks or to transfer a scale from one instrument to another, the standard material used should fulfill the following important properties.

(1)   The material must be rigid and mechanically robust.
(2)   It must be stable under ambient conditions met in practice.
(3)   The material must have a uniform, flat surface.
(4)   It must be easy to clean.
(5)   It must be opaque and should not exhibit:
    (a)   interference with the sample backing,
    (b)   edge-losses, i.e., the lateral diffusion of radiation within the sample should be small.

(6)   The surface of the sample should have either a matte finish (should approximate the surface of a uniform diffuser) or a polished finish (should approximate the uniform diffuser except at angles near regular reflection).

(7)   The spectral selectivity of the sample should be very small especially in the visible and near-ultraviolet spectral regions. Its absolute reflectance should be high (0.9–1.0).

(8)   The sample should be free of luminescence.

(9)   The absolute spectral reflectance properties should be independent of incident radiation, temperature, and humidity.

All these criteria are important in one way or another. However, in each specific case one has to decide which criteria are the most important for a particular application.

10.5.4.1   WHITE STANDARD MATERIALS

In view of the problems associated with thermally deposited magnesium oxide that is also not very stable (Priest, 1930; Middleton and Sanders, 1951), it has been more and more frequently replaced with various preparations of barium sulfate. Considerable variations among various barium sulfate preparations, attributed to surface structure, methods of preparation, impurities, and particle size distribution, have been reported in the literature (Budde, 1959). New advances have been made in preparation of barium sulfate so that quite good and reproducible $BaSO_4$ pressings are now available (Grum and Luckey, 1968; Richter and Terstiege, 1970).

Investigations were also made to produce MgO pressings in place of thermally deposited samples (Hammond *et al.*, 1962). Magnesium oxide pressings are more easily reproduced than smoked standards and are also more stable.

Typical white materials that are used as standards of reflectance are: barium sulfate pressings (Grum and Luckey, 1968; Richter and Terstiege, 1970), barium sulfate coatings (one part of gelatin and 1000 parts of barium sulfate) (Mischer and Rometsch, 1950), carboxymethyl–cellulose (CMC) coatings of barium sulfate (Middleton and Sanders, 1953), potassium sulfate coatings (Shutt *et al.*, 1974), polyvinyl alcohol (PVA) coatings of barium sulfate (Grum and Luckey, 1968), opal glass MS-14 (Voishvillo, 1962) opal glass MS-20 (Voishvillo, 1972), and white structural glass (Vitrolite tile) (Gable and Stearns, 1949). The last three materials mentioned have been quite extensively used as working standards of reflectance (Kartachevskaya *et al.*, 1975). The reflectances of coating materials with various binders depends too critically on the coating thickness, the substrate, the absorption of the binder, and on the manner of preparation. However, the aforementioned white coatings have been used extensively as materials for coating the integrating spheres of spectrophotometers and photometers (Erb, 1975).

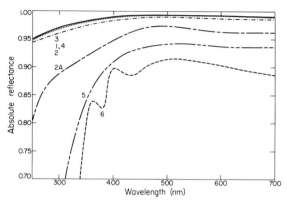

Fig. 10.12   Absolute reflectances of typical white materials (250–700 nm).

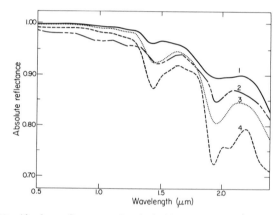

Fig. 10.13   Absolute reflectances of typical white materials in the near infrared.

The spectral reflectance data on some of the mentioned materials are given in Figs. 10.12 and 10.13 for UV–visible and near-infrared regions, respectively.

Recently another material has been introduced whose optical properties, together with physical and chemical characteristics, compare favorably with those of the established standard of reflectance. The material is an organic fluorinated polymer of microcrystalline structure formed under high pressure. (Grum and Saltzman, 1975). This material, Halon tetrafluoroethylene polymer, is highly reproducible, stable, easy to clean, and suitable for use in any geometrical arrangement. The pressed disks of this material are rugged and easy to use. The spectral absolute reflectance of Halon polymer is given in Figs. 10.12 and 10.13 along with those of other white materials, and its geometric properties are given in Fig. 10.14. (For the techniques of absolute reflectance measurement, the reader is referred to an excellent review on this subject by W. Budde, "Calibration of Reflectance Standards,") (Budde, 1976) [see also Grum and Wightman (1977)].

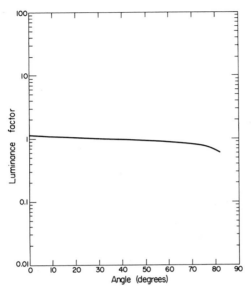

Fig. 10.14 Goniphotometric curve of Halon polymer pressing.

10.5.4.2 COLORED STANDARD MATERIALS

There are very few colored materials available for use as standards for checking photometric scales of reflectance spectrophotometers. The noted exception is a set of colored ceramic tiles (Clarke and Samways, 1968; Clarke, 1973). This set of tiles, available from the National Physical Laboratory, Teddington, England, consists of twelve colored and/or gray ceramic tiles suitable as reference surfaces for checking the performance of colorimeters and spectrophotometers in industrial use. These tiles can be used in diagnosing specific types of errors such as nonlinearity of photometric scale, spectral response errors, and geometrical errors.

Also mentioned in the literature are ceramic fibers (reflectance about 70% as measured versus MgO) (Trytten and Flowers, 1966; Watson, 1971) and tiles used for calibrating colorimeters and densitometers (McCamy *et al.*, 1976).

10.5.4.3 BLACK REFERENCE MATERIALS

The need for materials having a very low reflectance over a large spectral region is necessary for many scientific applications, most notably in aerospace, spectrophotometry, and spectroradiometry. The validity of spectral reflectance measurements is based on the assumption of infinite thickness of the sample being measured. This means that the sample is thick enough so that additional thickness would not affect the reflectance value of the sample. This condition rarely prevails in actual practice, especially with paper or fabric samples or with photographic emulsions. Hence, the sample has to be backed by some standard material. If the sample does not have an infinite optical thickness the measured spectral reflectance data will be to some degree influenced by the reflectance of the backing. This type of measuring error is particularly noticeable in the near-infrared spectral region. When reflectance measurements are made for computing absorptance of the sample, the backing materials must absorb all the energy that may be transmitted by the sample. Such a condition is achieved if a low-reflectance spectrally nonselective blackbody is used as a backing material.

Materials normally used for this purpose are black Norzon cloth (Behr-Manning), flat black paint (Krylon; 3M Company, or black glasses such as Carrara glass (Pittsburgh Plate Glass Co.). All of these materials exhibit a reflectance of about 0.5–1.0% in the visible and much higher reflectance values in the near-infrared spectral region (up to 30%). Typical values of reflectance of samples mentioned are given in Table 10.1.

TABLE 10.1   Reflectance Values of Black Materials
at Selected Wavelengths

	Percent reflectance at λ (μm)				
Material	0.5	1.0	1.5	2.0	2.5
Norzon cloth Ultra flat	0.4	0.9	11.0	14.0	8.0
Black Krylon paint on aluminum	0.3	0.9	1.2	1.4	1.4
KRL black	0.04	0.06	0.07	0.10	0.10
Carrara glass	0.16	17.00	50.00	47.00	24.00

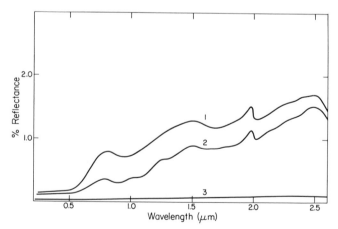

Fig. 10.15   Percent reflectance of black materials. Curve 1, ultra-flat black enamel on aluminum plate; curve 2, ultra-flat enamel sprayed on Eastman white reflectance paint on aluminum plate; curve 3, KRL blackbody.

The most efficient black material is KRL black (Grum, 1976) that consists of a cylindrical cavity coated with Eastman white reflectance coating and oversprayed with ultra-flat black Krylon paint. In this case the $BaSO_4$ subtrate provides a good diffusing property of the cavity. The reflectances of KRL black and of some of the other black materials mentioned are shown in Fig. 10.15.

10.6 CONCLUDING REMARKS

The measurement of reflectance is complex and many times such measurements are not made correctly. The complexity is further intensified when samples subjected to reflectance measurements also exhibit fluorescence. In such a case one should refer to the total radiance factor measurement described in an earlier chapter. The use of standard reference materials is essential in reflectance measurement for meaningful results. With the selection of proper procedure accuracy and precision of reflectance calibration of the order of $\pm 0.3\%$ can be achieved.

Since the reflectance value depends on the geometry of incidence and collection of the light, it is imperative that the reference materials used are calibrated for the desired geometry. Many times an assumption is made that the reflectance factor $\beta_{d/0}$ is equal to a reflectance $\rho_{d/0}$, which has the same value as the inverse reflectance $\rho_{o/d}$. This may not always be true; however, it is true for such materials as $BaSO_4$.

It should also be stressed that at present there is no internationally accepted (unique) standard of reflectance; hence one has to use one of the recommended materials that is most suitable for the particular application.

For general and theoretical discussions on reflectance spectroscopy the reader is referred to the work of Wendlandt and Hecht (1966) and the work of Kortüm (1969).

REFERENCES

ASTM (1965). Designation E259-65.
ASTM (1966). ASTM Manual on Recommended Practices in Spectrophotometry, ASTM Committee E-13.
Blevin, W. R., and Steiner, B. (1975). *Metrologia* **11**, 97.
Brown, W. J. (1954). *J. Sci. Instrum.* **31**, 469.
Budde, W. (1959). *J. Opt. Soc. Am.* **50**, 217.
Budde, W. (1976). *J. Res. Nat. Bur. Std.* **80A**, No. 4.
Burke, R. W., Deardorf, E. R., and Menis, O. (1972). *J. Res. Nat. Bur. Std.* **76A**.
CIE (1935). Compte Rendu, p. 2.
CIE (1959). Compte Rendu, p. 107.
CIE (1970a). Publ. No. 18 (E-1.2).
CIE (1970b). Publ. No. 17 (E-1.1).

CIPM (1946). *Proc. Verbaux, 2nd Ser.* **20**, 119.

Clarke, F. J. J. (1968). National Physical Laboratory, England, Rep. No. 3042, November 4.

Clarke, F. J. J. (1973). *Congr. Int. Colour Assoc., 2nd, York, England.* Survey Lectures, p. 346.

Clarke, F. J. J., and Samways, R. R. (1968). NPL Rep. No. MC2, August.

Cohen, E. R., and Dumond, J. W. M. (1965). *Rev. Mod. Phys.* **37**, 590.

Comité Consultatif de Photometrie et Radiometrie (1975). 8th Session, September.

Conf. Gén. Poids et Mesures, 9th (1948). Compte Rendu, p. 54.

Conf. Poids Mesures, 13th (1967–1968). Compte Rendu, p. 104.

Conf. Gén. Poids Mesures, 13th (1968). Compte Rendu, Annex II.

Erb, W. (1975). CIE TC-1.2 Subcommittee on Reflection and White Diffusers, Subcommittee Rep.

Fischer, B., and Krönert, R. (1963). *Feingeratetechnik* **12**, 405.

Gabel, J. W., and Stearns, E. I. (1949). *J. Opt. Soc. Am.* **30**, 481.

Geist, J., and Blevin, W. R. (1973). *Appl. Opt.* **12**, 2532.

Geist, J., Dewey, H. J., and Lind, M. (1976). *Appl. Phys. Lett.* **28**, 171.

Grum, F. (1976). Private communication.

Grum, F. (1977). Private communication.

Grum, F., and Luckey, G. W. (1968). *Appl. Opt.* **2**, 2289.

Grum, F., and Saltzman, M. (1975). 18th Session of CIE, London, September 10–18, Paper No. 75.77.

Grum, F., and Wightman, T. (1977). *Appl. Opt.* **16**, 2775.

Hammond, H. K., III (1955). *J. Opt. Soc. Am.* **45**, 904.

Hammond, H. K., III and Mason, H. L. (1971). Precision measurement and calibration, selected NBS papers on radiometry and photometry, NBS Spec. Publ. 300, Vol. 7.

Hammond, H. K., III, Caldwell, B. P., and Goebel, D. G. (1962). *J. Opt. Soc. Am.* **52**, 1321.

Judd, D. B. (1950). *J. Res. Nat. Bur. Std.* 44.

Kartachevskaya, V. E., Korte, H., and Robertson, A. R. (1975). *Appl. Opt.* **14**, 2694.

Keegan, H. J., and Gibson, K. S. (1944). *J. Opt. Soc. Am.* **34**, 770.

Keegan, H. J., Schleter, J. C., and Judd, D. B. (1962). *J. Res. Nat. Bur. Std.* **66A**(3), 203.

Kelly, K. L. (1963). *J. Opt. Soc. Am.* **53**, 999.

Kingslake, A. (1965). "Applied Optics and Optical Engineering," p. 61. Academic Press, New York.

Kortüm, G. (1969). "Reflexionsspektroskopie." Springer–Verlag, Berlin and New York.

Mavrodineanu, R. (1972). *J. Res. Nat. Bur. Std.* **76A**, 405.

Mavrodineanu, R. (1973). NBS Special Publ. 378, p. 31.

McCamy, C. S., Marcus, H., and Davidson, J. G. (1976). *J. Appl. Photo. Eng.* **2**, 95.

Menis, O., and Schultz, J. I. (1970). NBS Tech. Notes 554 and 584.

Middleton, W. E. K., and Sanders, C. L. (1951). *J. Opt. Soc. Am.* **41**, 419.

Middleton, W. E. K., and Sanders, C. L. (1953). *Ill. Eng.* **48**, 254.

Mischer, K., and Rometsch, R. (1950). *Experimentia* **6**, 302.

Mori, L., Sugiyama, H., and Kambe, N. (1964). *Acta Chromat.* **1**, 396.

NBS Tech. News Bull. (1970). **54**, 206.

NBS (1972). Tech. Notes 594-3.

NBS Opt. Radiation News (1974). Introduction of a New Lamp Type for Spectral Irradiance Calibrations, No. 5, September.

NBS (1975). Optical Radiation News, No. 7, p. 2.

NBS Optical Radiation News (1975). No. 10.

NBS Optical Radiation News (1976). No. 13, January.

NBS Optical Radiation News (1976). Photodetector News Notes, No. 17, September.

Preston, J. S. (1969). *Light. Res. Technol.* **1**, 95.

Priest, I. G. (1930). *J. Opt. Soc. Am.* **20**, 157.

Richter, M., and Terstiege, H. (1970). *Lichttechnik* **22**, 447.

Robertson, A. R. (1968). *J. Opt. Soc. Am.* **58**, 1528.

Sanders, C. L., and Jones, O. C. (1962). *J. Opt. Soc. Am.* **52**, 731.

Schutt, J. B., Arens, J. F., Shai, C. M., and Stromberg, E. (1974). *Appl. Opt.* **13**, 2218.

Stair, R., Schneider, W. E., and Jackson, J. K. (1963). *Appl. Opt.* **2**, 1151.

Tables of wave numbers for the calibration of infrared spectrophotometers (1961). *Pure Appl. Chem.* **1**, No. 4, 684.

Terrien, J., and Preston–Thomas, H. (1967). *Metrologia* **3**, 29.

Trytten, G., and Flowers, W. (1966). *Appl. Opt.* **5**, 1895.

Vandenbelt, J. M. (1961). *J. Opt. Soc. Am.* **51**, 802.

Voishvillo, N. A. (1962). *Opt. Spectrosc.* **12**, 244.

Voishvillo, N. A. (1972). *Sov. J. Opt. Technol.* **39**, 305.

Watson, R. D. (1971). *Appl. Opt.* **10**, 1685.

Wendlandt, W. W., and Hecht, H. G. (1966). "Reflectance Spectroscopy." Wiley (Interscience), New York.

Wensel, H. T., Judd, D. B., and Roeser, W. F. (1934). *J. Res. Nat. Bur. Std.* **12**, 527.

Index

329